Animals, Ethics, and Engineering

Animals, Ethics, and Engineering

edited by

Rosalyn W. Berne

JENNY STANFORD
PUBLISHING

Published by

Jenny Stanford Publishing Pte. Ltd.
101 Thomson Road
#06-01, United Square
Singapore 307591

Email: editorial@jennystanford.com
Web: www.jennystanford.com

British Library Cataloguing-in-Publication Data
A catalogue record for this book is available from the British Library.

Animals, Ethics, and Engineering: Intersections and Implications

Cover image: Courtesy of Nora Prindle.

ISBN 978-981-5129-56-4 (Hardcover)
ISBN 978-1-003-65231-1 (eBook)

Contents

Preface

In 1999, after finishing a PhD in Religious Studies (with a focus on bioethics), I joined the faculty of the University of Virginia's School of Engineering and Applied Sciences (SEAS). My department, Engineering and Society, is unusual in being located inside an engineering school, and rarer still that it teaches required engineering ethics courses. In learning how and preparing to teach those courses, it didn't take me long to identify the canon of engineering ethics literature, formed from the writings of leading scholars in the field. I quickly came to realize that conventional approaches to engineering ethics draw heavily on case studies and ethics theory. But in all these years of teaching undergraduate engineers, I have yet to see a corpus of cases or ethics theory in the engineering ethics textbooks that focus specifically on non-human animals. I've come to see that the field is definitively anthropocentric.

I care about animals in general but have a deep personal interest in horses. I see them as beautiful, majestic creatures, and being with them gives me a profound sense of joy. For centuries, horses have been put to work for human purposes, i.e., hunting, mining, warfare, transportation, exploration, entertainment, and sports such as rodeos and racing. It is primarily due to engineered systems, gadgets and devices (i.e., saddles, spurs, metal shoes, reins, bits, bridles, crops, racetracks, etc.) that humans have been able to control the movements and even the wills of the equine species. (There are exceptions, of course; some people ride horses without putting bits in their mouths, saddles on their backs, or bridles around their heads and necks.) I want horses to be appreciated and treated with honor and respect, cared for and supported by those who use and ride them. It breaks my heart when horses die or are seriously injured on the racetrack. Perhaps, if engineering ethics were taught with animals in mind, then racetracks (if humans continue such a practice of "sport") would be designed with the health and care of horse's hoofs and joints as a primary engineering criterion.

There are many examples of the intersection of other-than-human animals and engineering: factories that produce chicken eggs and chicken bodies for human consumption; engineered chemicals and machines for tanning animal hides; the fences, cages, watering troughs, etc., engineered for zoos; automatic feeders and watering systems for pets, just to name a few. Meanwhile, humans are learning more and more about and from other animals, leading to new ways of engaging them for human purposes. Indeed, the more we learn about their sentience, intelligence, and genetic relationships to humans, the more questions we need to answer and practices we need to develop to better align engineering pedagogy and practice with ethical principles.

There are good reasons to include consideration of animals in engineering ethics education. One reason is practical; human flourishing in our ecosystems depends on the well-being of non-human animals, and engineering practices are implicated in threats to animal well-being, including species extinction. Another reason why engineering education ought to bring attention to non-human animals is philosophical, having to do with the moral standing of non-human animals, and how they are treated by the effects of engineering practice. I want engineering students to become aware of these issues, and to grapple with what it may mean to apply engineering in ways that impact other animals. Engineering education ought to be able to provide future engineers with the skills and awareness to build ethical consideration of animals, their welfare, and their role in the world into their designs, and into the development and deployment of new engineered devices and systems.

There's plenty of literature available for teaching about animal ethics, animal welfare, and animal rights, but little that connects the dots of engineering and the lives of these creatures. What's needed are scholarly resources to build a practical ethics approach for bringing animals into engineering education. That's the why of *Animals, Ethics, and Engineering*: to provide material for engineering students, educators, professionals and others to consider, through a lens of practical ethics, ways in which non-human animals can and ought to be factored into engineering design and practice.

Acknowledgements

This project received significant support from Madeline Kibler, student research assistant, and was made possible by the Anne Shirley Carter Olsson Chair for Applied Ethics, an endowed professorship at the University of Virginia, currently held by Rosalyn W. Berne.

Chapter 1

Introduction: Considering Non-Human Animals in Engineering Ethics Education

Rosalyn W. Berne
Department of Engineering and Society, University of Virginia, Virginia, USA
rwb@virginia.edu

For centuries, non-human animals and engineering have intersected. Dogs, for example, were domesticated by our early hunter-gatherer ancestors, probably through breeding disparate populations of the ancestral wolf, for purposes of hunting big game [1]. Engineered through the centuries by selective breeding for desired characteristics and features pleasing to humans, we now use dogs for therapeutic purposes, companionship, hunting, sleigh pulling and other kinds of work, and as pets.[1] Another example of the intersection of engineering and non-human animals is the study of melting glaciers, in which the head of a southern elephant seal is tagged with an ocean sensor to indicate unusual movement into Antarctic waters that would normally be inaccessible due to glacial ice, providing evidence to researchers that melting of the Denman Glacier may be a major threat to global coastlines [3]. There are

[1]A process that might be better understood as co-evolution, "as a relationship between populations of two species, rather than as a state into which one puts the other [2]."

Animals, Ethics, and Engineering: Intersections and Implications
Edited by Rosalyn W. Berne

ISBN 978-981-5129-56-4 (Hardcover), 978-1-003-65231-1 (eBook)
www.jennystanford.com

many such examples, which is why there are many good reasons to include consideration of animals in engineering ethics education.

One reason is practical; human flourishing depends on the well-being of non-human animals, and engineering practices are implicated in threats to animal wellbeing, such as with species extinction. As Engineering Without Borders -UK notes, "industries that use engineered machinery, processing, and chemicals are wreaking havoc on bio-diversity [4]." Earth's animal populations have declined by an average of 69% since 1970 [5], including the *nearly 30%* (almost three billion) of North American birds that have vanished since 1970 [6]. In his documentary on extinction, David Attenborough declares that "The consequences of these losses for us as a species are far reaching and profound [7]." Honeybee colony loss would be a case in point, as most humans are fed by crops that are pollinated by bees. Another is the loss of salmon species due to engineered dams [8]. Engineering may have to provide solutions to these problems. Such as through animal tracking systems for identifying phenotypes, species, or ecosystems vulnerable to climate change. As Lennox et. al suggest [9], these systems can help manage interventions such as fisheries management, habitat restoration, invasive species control, "or enhancement measures that prevent extinction and strengthen the resilience of communities against the most damaging effects of climate change" (pp. 4).

Another reason why engineering education ought to bring attention to non-human animals is philosophical, having to do with the moral standing of non-human animals, and how they are treated by and/or are a result of engineering. A perspective that aims to avoid or minimize harm to animals begs the question, what would it mean for engineering practices if non-human-animals are in fact sentient, as many scientists are beginning to discover [10]? For the UK, it has meant a complete revision of its Animal Welfare Act [11].

Christine Korsgaard asks whether people are more important than other animals [12]? The answer she gives to that philosophical question is a complex one, and not directly taken up in this book. But the prevailing belief is yes; people *are* more important than other animals, and this belief is implicit in most engineering design and practices. And, by omission from the curriculum, is an implicit belief in engineering education.

The primary aim of *Animals, Ethics, and Engineering* is to provide material for engineering students and engineering professionals to consider, through a lens of practical ethics, ways in which non-human animals can, are, and ought to be factored into engineering design and practice. Practical Ethics, as explained by D. Thompson, is "a linking discipline" seeking to bridge theory and practice in drawing on other disciplines and other forms of knowledge in addition to philosophy. It's an approach to ethics that "tries to relate professional rules and clinical experience to the broader social context in which professionals practice, and to the deeper moral assumptions on which professions depend [13]."

With that in mind, this introductory chapter begins with a quick look at historic, ideological roots of contemporary engineering, then considers "the animal turn," and how increasing attention is being paid to animals in scholarly literatures of the humanities and social sciences, but not so in engineering education. It next suggests the need for a new ethics in engineering and technology, to include concerns of their intersections with lives of non-human animals. A few resources for engineering ethics that do consider animals are identified, then brief introductions are given for each of the 18 chapters of this volume.

1.1 The Cartesian Roots of Engineering

Lynn White Jr. wrote in his seminal paper over 50 years ago, "What we do about ecology depends on our ideas of the man-nature relationship. More science and more technology are not going to get us out of the present ecologic crisis until we find a new religion or rethink our old one" [14]. Agreed: technology alone is not the solution to our ongoing crises. But whether or not a new religion is needed, as White suggests, a new approach to engineering education is required; one that can dismantle the Cartesian ideology of humans as being separate and distinctive, thusly superior to and more important than other animals.

Descartes's thinking marked a major break in human cultural evolution. Most cultures to that point did not, and many indigenous cultures still don't, see humans and animals as unequal or fundamentally different. Yet the Cartesian separation between mind and body – famously summed up as *Cogito ergo sum* ("I think, therefore I exist"), led to the assertion that other animals don't exist in the way that humans do. This anthropocentric assumption implicitly placed a boundary between "animals" and "humans" by assuming that since animals lack souls, intelligence, feelings, and consciousness, they can ethically be treated as objects to be used solely as a means for human purposes. The Cartesian assertion further supported a human supremacy worldview, with humans deemed superior by divine decree and given dominion over the natural world, including all animals. As Franco notes:

> René Descartes's (1596–1650) description of animals as "machine-like" was heavily criticized by many of his contemporaries, but nevertheless provided scientists a way to justify what would now be considered extremely gruesome experiments in a time when anesthesia, for humans and animals alike, was not available. [15] (p. 241)

Franco [16] contrasts Descartes thinking to other early philosophers of influence in the ethical treatment of animals. There was Baruch Spinoza (1632–1677) who "did not deny animals' ability to feel, but considered we should nevertheless "use them as we please, treating them in a way which best suits us; for their nature is not like ours." (p. 241), and Immanuel Kant (1724–1804), who:

> . . . would reject Cartesian mechanistic views, thus acknowledging sentience to other animals. However, Kant would not extend his concept of human intrinsic and inalienable dignity to other species. In his of Duties to Animals and Spirits, and mirroring Thomas Aquinas's views on the subject, Kant observed that "all animals exist only as means, and not for their own sakes, in that they have no self-consciousness, whereas man is the end (. . .) it follows that we have no immediate duties to animals; our duties towards them are indirect duties to humanity." (p. 241).

Though these are musings of men long deceased, their influence persists. Acts of omission in the disregard of non-human animals fall across the spectrum of engineered systems and technological devices. These are often deployed at animal's expense for solely human purposes, e.g., deep-sea mining [17] and missile testing that disrupt habitat and migration patterns. Anthropogenic noise pollution (caused by airplanes, automobiles, trains, watercraft, construction, mining, etc.), affects acoustically oriented animals, and the noise may reverberate through biological communities, even in adjacent quiet environments" [18]. Primates are used in crash tests for auto safety engineering research. Oil spills are another example of passive disregard for animals subjected to engineered systems. After the Santa Barbara oil spill of 1969, the declaration of environmental rights stated that ethics must be extended beyond social relations to govern human contact with all forms of life and the environment [19]. Had those ethics been exercised, the *Amoco Cadiz* Oil Spill of 1978, the Exxon Valdeze spill in 1989, the Kolva River oil spill of 1994, BP's Deep Water Horizon spill of 2010, and others may have averted disaster and their devastating effects on marine life.

Milstein et al. suggest that:

> . . . the human/nature binary is a dead-end – both conceptually and materially – in that it leads to epistemological failure, culminates in a fractured and fractious sense of self divested from the more-than-human, and results in actual death of planetary systems and human and nonhuman lives that in fact are interrelated parts of, and depend upon, these systems [20].

And yet, the influence of Descartes and other early thinkers lingers in acts of commission: where engineering is intentionally used to employ non-human animals for human purposes, such as to be eaten, as materials in household goods and clothing, for entertainment and sport, transportation, labor, or in research. If animal ethics had a more prominent place in the education of engineers, perhaps the engineered devices and systems used for horse racing, a multi-billion-dollar sport that can cause serious damage to horses including their death, would have different designs. Without exposure to animal ethics, how are engineers to ethically design industrial, high-density poultry, pork, and milk production, systems which can cause sentient animals to suffer greatly [21]. These kinds of issues need to be put before future engineers to ponder, wrestle with, and discuss, so that they are at least aware of animals, perhaps even conscientious, when they become practicing engineers. Practicing engineers who are willing to do so, might ask themselves these same kinds of questions.

1.2 The "Animal Turn"

In 1964, attention turned to non-human animals with Ruth Harrison's *Animal Machines*, which had a profound effect on public awareness and the treatment of caged and farmed animals. Peter Singer's *Animal Liberation* (1975) was considered the founding philosophical statement of the animal liberation movement, but concern about animals remained largely nascent in the literature over the decades to follow. In her seminal piece of 2007, "On the Animal Turn," Ritvo [22] pointed to the emergence of animals as a more frequent focus of scholarship in the humanities and social sciences. Grusin [23] and others have since further documented the animal turn Ritvo identified. They point to how Animal Studies, including those found in STS programs, have proliferated in higher education. For example, Cornell's STS course on *The Entangled Lives of Humans and Animals* draws on readings from Anthropology, Science & Technology Studies, and animal trainers and behaviorists [24]. And Brown University's STS Animal Studies program includes a course in *Animals & Ethics* [25]. It Is apparent that STS and other academic fields have begun to take an interest in non-human animals. But few engineering programs include STS courses in their curriculum[2], and few STS courses are animal focused.

There are many relatively recent documents, reports, and studies that could be well adapted for engineering ethics instruction about animals. One key resource is *The Animal Turn: Digitizing Animal Protection and Human-Animal Studies Collections* from NC State, a project funded by a grant from the Council on Library and Information Resources [26]. Released in 2019, *The Routledge Handbook of Animal Ethics* (Routledge Handbooks in Applied Ethics) is a broad-reaching volume about human interactions with animals,

[2]The University of Virginia's School of Engineering and Applied Sciences is an exception, wherein STS courses are part of the required undergraduate curriculum.

including contributions from scholars in a variety of fields [27]. And there are many, many other books in print about animal ethics.

But there is no "animal turn" apparent in the engineering ethics literature, where animals are rarely considered. A quick scan of the table of contents of engineering ethics textbooks reveals a trend of sustainability and ecological matters as being increasing topics of concern, but few such documents refer specifically to non-human animals. Vesilind and Gunn's 1998 edited volume, *Engineering, Ethics, and the Environment*, includes a case called "Running Over Box Turtles," and a section on "Extending the Moral Community to Include All Life [28]." A chapter called "Cultivating the Virtue of Respect for Nature" is included in the 6th edition of Harris and Pritchard's 2019 *Engineering Ethics: Concepts and Cases* [29]. Chapter 8 in the 5th edition of Martin, Zhu & Schinzinger's *Ethics in Engineering* is titled "Engineering and Environmental Ethics in the Anthropocene" and includes a small section explaining the theory of sentient-centered ethics [30]. Searching Amazon.com books for 'Engineering Ethics, Animals,' brings up titles such as *Animals and the Fukushima Nuclear Disaster* (The Palgrave Macmillan Animal Ethics Series) [31]. A Google Scholar search under the same topic turns up multiple articles pertaining to genetic engineering of animals, and animal experimentation. Refining that search for publications since 2019, an article on "Welcoming Robots into the Moral Circle: A Defence of Ethical Behaviourism" [32] reflects the push by some to expand the established boundaries of life as defined, to include technological creations. Another topic of recent interest is laboratory grown meat [33]. While these are all potentially useful and important resources, the lack of scholarly, curricular material about ethics and animals in the development and deployment of new technologies and engineered devices, or of existing engineered systems, persists.

That said, ethical concerns about engineering, technology, and animals have been reflected in feature films, and science fiction writings, and these can be useful tools in engineering ethics instruction. But the pedagogical gap should not be left to popular culture to fill. Finding Nemo's scene of loud boat motors and dangerous propellors surely left an impression on many children who are now adults working as engineers. But as Tidwell argues:

> [The film's] "re-visioning of nature compromises the environmental message of the movie by not allowing it to stand on its own but instead leading us to feel good about our mere awareness of the problems at hand. It also allows us to avoid dealing with these problems because humans are seen as well-meaning, ignorant (and therefore dangerous), but also as separate from nature. Any damage to nature is seen only to affect the ocean world, not the human world; there are no human consequences to our actions" [34].

The story told in the sequel to the film *Avatar, Avatar: The Way of Water* [35], centers on a high-tech system designed to hunt "Tulkuns," whale-like animals whose brain enzymes are harvested for human anti-aging remedies called *amrita. (Amrita* is a Sanskrit word meaning "immortality.") [36]) Tapping deep mythological roots, both *Avatar* films raise critical questions about the sentience and moral agency of non-human animals and is a film that can be used in an engineering ethics course to provide an ecofeminist analysis of the constructs depicted in the film. But engineering students also need scholarly material that is meant for them.

1.3 The Need for a New Ethics for Engineering Education

Ecologists, biologists, anthropologists, and other scientists would likely agree that humans and non-human animals are integrally connected in a larger web of life. What must also be acknowledged is that the impacts of engineering on non-human animals can and do have profound implications for their lives, and for ours (i.e., plastics pollution [37] and light pollution [38]). So, it might well be argued that engineering education has significant ethical responsibility, and an important contribution to make, towards healing the ecosystem for the sake of our survival and theirs, with non-human animals being essential to a sustainable ecology. Another perspective is of finding creative ways to engage animals through engineering design, for their sake and ours.

Other questions are relevant to the enterprise of bringing animals into view in engineering ethics education. Such as, what would it mean if it turns out that Descartes was mistaken, and the entire premise of the mechanistic universe was a misunderstanding? If engineering writ large were able to move towards a holistic frame of understanding, wherein life is not composed of a set of discrete machines but functions as an interconnected whole, then what would be considered "good," or "harmful," "duty," or "virtuous," in the design of engineered systems and the practice of engineering in its effects on non-human animals? Is the engineering profession able and willing to embrace sentient-centric ethics? Which would mean giving credence to the moral significance of sentience in animals; the capacity to perceive and experience feelings such as pleasure, joy, pain, and distress. These (and other) questions are worth posing, in engaging conversations with engineering students.

Students enrolled in one of my undergraduate engineering ethics classes were assigned an "ethical autobiography" essay during the first week of the class. One of the essays returned included the following paragraph:

> *I feel extremely connected to living beings. In my lab experience, I've had to morally consider the use of animals for research. I will not be very graphic here, but some of the lab procedures have been very shocking to me. I found that my elephant automatically made the decision for me. I carried on with the actions according to the people in my environment and did not question the research methods. After the fact, my rider really had to break down why I so easily accepted the reasoning behind this ethical decision. These animals are being used for research purposes in shocking ways, but at my core I know that there is the potential for creating a therapeutic that could save many human lives from life-threatening diseases. I found it necessary to completely separate my two moral identities: my empathetic, creature-saving side and my clinical, scientific side.*

Another student, also a biomedical engineering major, during a class discussion said that she must disconnect herself from her feelings when injecting toxic nerve agents used for military research, into rats. The student went on to say that xenotransplantation, in the application of pig organs grown and used for transplantation into human bodies, has been a topic of much discussion among students majoring in biomedical engineering. How can engineering educators help these and students in other fields of engineering to navigate the rapidly changing terrain of biotechnology, AI, nanotechnology and other emerging and converging technologies that are bringing to the forefront new capabilities in engineering design, particularly when these intersect with the lives of non-human animals?

We can trust that engineering students and professionals pursue technological solutions and engineering challenges largely for the sake of "the good." But these endeavors are largely homocentric in orientation. It was the needs of people, not of migrating birds, that were the focus of engineering the "Tribute in Light" installation commemorating September 11, by illuminating lower Manhattan. But then:

> By twist of fate, the anniversary of the Sept. 11 attacks coincides with a much older yearly ritual: the migration of birds across New York City as fall approaches. The convergence creates a spectacle that is eerily beautiful, yet according to one study, endangers some 160,000 birds a year, starkly illustrating the perils of humans and animals sharing an urban ecosystem. . . The tiring detour through the beams of light can put birds at risk of starvation or injury to populations already threatened by light pollution, collisions with buildings, habitat destruction and climate change [39].

Without question, the natural environment is of increasing concern to engineering professional societies and that concern is reflected in recent revisions to their professional codes. But, again, explicit reference to non-human animals is largely omitted from those codes. Animal welfare policies and initiatives in the USA are minimal in their standards of care and limited in which animals are included. (By contrast, in the UK, standards are high and inclusive based on animal sentience.)

Making the world a better place and improving the quality of life are commonly stated intentions of engineering research and practice, as reflected in the varied professional codes of engineering. But those codes explicitly give preference to the welfare of humans, and are otherwise mute on professional responsibilities pertaining to non-human animals:

- The National Society of Professional Engineering decrees that professional engineers should hold paramount the safety, health, and welfare of *the public*" [40].
- The American Society of Civil Engineers puts forth as a fundamental principle that they "utilize their knowledge and skills to enhance the quality of life *for humanity*" [41].
- "In recognition of the importance of our technologies in affecting the quality of life throughout the world," the IEEE includes in its code "to improve the understanding by *individuals and society* of the capabilities and societal implications of conventional and emerging technologies, including intelligent systems" and "to treat *all persons* fairly and with respect. . ." [42].
- For the NSF, "the responsible and ethical conduct of research involves not only a responsibility to generate and disseminate knowledge with rigor and integrity, but also a responsibility to:
    - conduct peer review with the highest ethical standards,
    - diligently protect proprietary information and intellectual property from inappropriate disclosure,
    - and treat *students and colleagues* fairly and with respect." [43]

Engineering education ought to be able to provide future engineers with the skills and awareness to build consideration of animals, their welfare, and their role in the world into their designs, development and deployment of new engineered devices and systems, as an ethical responsibility. And it should be able to help students to address animal-related nodes of ethical concern in existing engineered systems. But we are not yet well-equipped to do so; a new "ethics" is needed.

In 1984, Hans Jonas wrote in *The Imperative of Responsibility: In Search of an Ethics for the Technological Age* that "new kinds and dimensions of action require a commensurate ethic of foresight and responsibility which is as novel as the eventualities that arise out of the works of Homo Faber in the era of technology" [44] (p. 18). This writing is still pertinent after all these years. Jonas feared "an apocalypse threatening from the nature of the unintended dynamics of technical civilization," (p. 202) leading to the desolation of the planet. Jonas included man as an object of technology, having turned on himself and imposed himself on nature in such a way that "calls upon the utter resources of ethical thought, which never before has been faced with elective alternatives to what we considered the definite terms of the human condition" [44] (p. 18). It seems that Homo Faber's technology includes non-human animals, too, who for Jonas, "are psychophysical individuals with common but varied forms of organic identity and living capacities" (p. 6) with important and valuable significance for themselves and perhaps for others.

English Literature and Cultural Studies scholar S. Bender sees Hans Jonas as one who focused on the subjective body, as consisting of matter and spirit and therefore connects all living beings [45] (pp. 40). In an evaluation of possible futures she draws from science fiction, Bender seeks to raise ethics to a position that incorporates humanities relations with its non-human others and technologies, (pp. 36).

As with Jonas, Bender and other contemporary thinkers have called for new approaches to ethics [45]. Such as Basart, et al. who suggest the inclusion of a 'feminine' perspective in ethics as being "... essential for the complete development of the moral life of the engineering profession." [46] (p. 412) Similarly, Riley [47] (p. 198) sought to provide avenues for how to approach integrating feminist ethics in engineering. Concerns over the need for new approaches to engineering ethics have also emerged over the insufficiency of traditional engineering research ethics to address developments in bioengineering R&D. Hyun calls for a new approach that utilizes contemporary engineering ethics, which accepts that "engineering itself is a value-laden activity and that the values that drive design decisions are often themselves ethical in nature," arising from his concern about advances in stem cell science and bioengineering [48]. Nienke de Graeff et al. [49] have indicated a need for more scholarly consideration on the ethics of genetic engineering for non-human animals saying:

> Our findings illuminate several key considerations about the academic debate, including a low disciplinary diversity in the contributing academics, a scarcity of systematic comparisons of potential consequences of using these technologies, an underrepresentation of animal interests, and a disjunction between the public and academic debate on this topic. (p. 10)

And they call for academics to get more involved in the discussion about genome editing, and to incorporate animal interests (p. 10).

But what scholarly resources might be used to begin to build a new practical ethics approach for bringing animals into engineering education?

1.4 Resources for Animal-Oriented Engineering Ethics

With the aim of introducing a practical ethics approach into engineering education, that pays heed to the lives of non-human animals, case studies of engineering projects that specifically attempt to address animal welfare can be useful. For example, painting windmill blades to increase their visibility at the Smøla wind-power plant in Norway, where the annual bird fatality rate was reduced at the turbines with a painted blade by over 70% [50]. Temple Grandin's design of a more humane cattle handling system [51] is another possible use case.

Reaching beyond the discipline of philosophy to include other fields, such as the fields of Animal Geographies and Multispecies Ethnography offers pertinent approaches and resources. As Hovorka explains:

> Animal geographies are at their core grounded in ethical commitments and emancipatory practices to improve the lives of animals. Since the late 20th century, animal geographers have addressed the silence around animals in social theory and have remapped the moral landscape and balance of power alongside efforts from environmentalists, those championing rights of marginalized human groups, and animal advocates. Core ethical concepts

> have emerged from animal geographies, emphasizing context-specific, place-based, and embodied encounters as part of moral assessments of human-animal relations. [52] (p. 12)

As such, Animal Geographies could be used to form a determination of the value conflicts and cultural factors which put animals at risk of harm under the impact of engineering and technological development. This seems to be an important place to start given that:

> Animal geographers 'investigate how humans think about, place, and engage with animals, how animals shape human identities and social dynamics, as well as how broader social cultural, political economic, and ecological processes influence animal distributions, circumstances, behaviours, experiences, and well-being. [53] (p. 127)

And that:

> Animal geographies at their core explore human-animal relations through attention to animality, animal spaces, and beastly places as grounded in eclectic and integrative methodological approaches and ethical commitments to improving more-than-human lives. [54] (p. 131)

Multispecies Ethnographers might be considered another helpful resource to draw on in the creation of a new ethics for engineering education. As Münster and Locke claim:

> Multispecies ethnography must thus be seen as a part of a larger quest in the social sciences and humanities to replace dualist ontologies by relational perspectives, to overcome anthropocentrism by pointing to the meaningful agency of nonhuman others, and to highlight the intersections between ecological relations, political economy and cultural representations. [54] (p. 1)

In the dynamic intersection of technology and ethics, one of the most pressing frontiers of inquiry is the ethical treatment and integration of animals in engineering practices. *Animals, Ethics, and Engineering* is a compendium of diverse perspectives, each shedding light on different facets of this complex relationship. From philosophical musings to practical applications, each chapter delves into the nuances of animal ethics within the realm of engineering, offering both theoretical frameworks and real-world insights as a project of practical ethics.

1.5 Contributions to This Volume

The rich tapestry of insights reflected in this edited volume challenge us to reevaluate our relationship with animals and the ethical responsibilities inherent in technological progress.

Chapters 2–7 of the collection provide a theoretical landscape that brings animal ethics into focus for engineering and technological design. Chapter 2, "The Ethical Quandary of Animal Well-Being and Its Relevance for Engineers," by Paul B. Thompson, examines the ethical dimensions of animal well-being and its implications for engineering practice, advocating for a more conscientious approach to technological innovation. In "Engineering in a Multispecies World: Animal Geographies, Ethics, and Place" (Chapter 3), Julie Urbanik, Jenny Issacs, David Tovar, and William Lynn offer a comprehensive exploration of animal geographies and geoethics, underscoring the interconnectedness of engineering, ethics, and the natural world. This is followed by "Missing Voices: Non-Human Animals as Stakeholders in Engineering Design" (Chapter 4), in which Gary Halada prompts us to consider the voices of non-human animals as stakeholders in engineering design, advocating for a more inclusive and ethically informed approach to technological innovation. In "Bridging the Divide: Toward a Holistic Approach to Animal Ethics" (Chapter 5), Philipp Aerni advocates for a holistic approach to animal ethics, drawing inspiration from Spinoza's monumental work, *Ethics: Proved in Geometrical Order*. He challenges the prevailing notion of speciesism and emphasizes the importance of context in understanding the ethical implications of technology on animal welfare. In Chapter 6, "The Value of Life: Resuscitation,

Resurrection, and the Ethics of De-Extinction," Kurt Reymers delves into the ethical complexities of de-extinction, probing the underlying philosophies and moral implications of reviving extinct species. This cluster of chapters ends with a question posed by Clara Mancini: how should our species address the responsibility that arises from our ability to develop technology that allows us to overcome biological and environmental limitations, and exert seemingly unlimited power over other beings? Mancini addresses that question in "Animal-Centered Technology and Sustainable Development: An Earth Odyssey" (Chapter 7) making the case that the field of Animal-Computer Interaction (ACI), which focuses on researching and designing technology *for* and *with* animals, could provide the animal-centered perspective needed to address this question.

Chapters 8–13 focus on applied ethics, highlighting intersections of non-human animals with computer systems, technological devices, AI, and biotechnological processes. It starts with Chapter 8, "Ethical and Societal Issues Occasioned by Xenotransplantation," in which Bernard E. Rollin tackles the ethical dilemmas surrounding xenotransplantation, arguing for a holistic consideration of the moral implications of such biotechnological advancements. Chapter 9, "Understanding Ethical Issues Raised by Veterinary Medical AI" by Simon Coghlan and Thomas Quinn, navigates the ethical landscape of AI in veterinary medicine, emphasizing the need for engineers to grapple with the ethical dimensions of their innovations. Samuel Camenzind and his colleague Matthias Eggel dissect the foundational principles of Replacement, Refinement, and Reduction (3Rs) within the context of modern biotechnology, arguing for an ethical reevaluation in light of contemporary scientific advancements in "The 3Rs Principles and Genetic Pain Disenhancement" (Chapter 10). In "Animals and AI: The Role of Animals in AI Research and Application" (Chapter 11), Leonie Bossert and Thilo Hagendorff delve into the ethical considerations surrounding the utilization of animals in AI research, advocating for a nuanced understanding of their role beyond mere resources for technological advancement. In "Touchscreen Uses for Animals: Teaching Discrimination Tasks" (Chapter 12), Jennifer M. Cunha, Corinne C. Renguette, and Sierra Ouimette propose innovative approaches to enhance animal welfare through the ethical use of technology, focusing on touchscreen devices as tools for research and enrichment. In "An Investigation into the Encounter Between Social Robots and Animals" (Chapter 13), Oliver Bendel delves into the evolving landscape of social robots designed to interact with animals, dissecting the intricate dynamics of these interactions and their ethical ramifications.

Chapters 14–17 offer a lens of specific species on non-human animals, either as inspiration or with direct concern. "Lead a Horse to Water: Beach Ponies in the American South and the Cultivation of Ferality" (Chapter 14), written by Abraham Gibson, presents a historical examination of beach ponies, challenging traditional notions of engineering and inviting reflection on ethical stewardship of animal populations. Chapter 15, "The Story of Robotics in Coral Restoration," by Nicole Carey, explores the potential of robotics in coral restoration efforts, raising pertinent questions about the ethical implications of integrating technology into conservation practices. In "To See the Bees: Reflections of a Beekeeper on Her Journey of Healing" (Chapter 16), Diane Hellwich-Wellons shares personal reflections on beekeeping and the profound connections between humans and animals, inviting readers to contemplate the ethical dimensions of our interactions with the natural world. This is followed by the writing of Fiona French, who in "Designing and Crafting Systems for Non-Human Animals: Inspirations from Elephants (Chapter 17), draws inspiration from elephants to explore the intersection of design and animal welfare, and offers insights into crafting animal-centered computing systems.

The final chapter of *Animals, Ethics, and Engineering* circles back to the first, with another call for animal ethics in the education aspiring engineers. In "Why Are Non-Human Animals Neglected in Engineering Ethics Education?" (Chapter 18), Justin Hess delves into the lacuna of animal-centricity in engineering ethics education, identifying root causes and advocating for a more inclusive pedagogical approach.

Dispersed throughout the volume are three vignettes of varied animal-engineering design solutions, written by Madeline Kibler.

As we navigate the complex terrain of animals, ethics, and engineering, may this collection serve as a beacon guiding us and our students toward a more animal-conscious, engineered world.

References

1. Morell, V. 2015. How Wolf Became Dog. *Scientific American*. July 1.
2. Russell, E. 2014. Coevolutionary History. *The American Historical Review* 119: 1514–1528. https://doi.org/10.1093/ahr/119.5.1514.
3. Mooney, C., and S. Ducroquet. 2023. When Scientists Tagged A Curious Seal, He Led Them to Signs of a Potential Climate Disaster. *UCI Department of Earth System Science*. January 20.
4. UK, Engineers without Borders. 2020. Extinction: The facts. *Engineers Without Borders UK*. September 18.
5. The 2022 Living Planet Report. 2023. https://livingplanet.panda.org/en-US/. Accessed March 14.
6. State of the Birds 2022. 2024. *State of the Birds 2022*. https://www.stateofthebirds.org/2022. Accessed May 10.
7. Attenborough D. 2024. Extinction: The Facts. *BBC*. https://www.bbc.co.uk/programmes/m000mn4n. Accessed May 10.
8. Dams: Impacts on Salmon and Steelhead. 2024. https://www.nwcouncil.org/reports/columbia-river-history/damsimpacts/. Accessed May 10.
9. Lennox, R. J., P. Afonso, K. Birnie-Gauvin, L. S. Dahlmo, C. I. Nilsen, R. Arlinghaus, S. J. Cooke, et al. 2024. Electronic Tagging and Tracking Aquatic Animals to Understand a World Increasingly Shaped by a Changing Climate and Extreme Weather Events. *Canadian Journal of Fisheries and Aquatic Sciences* 81: 326–339. https://doi.org/10.1139/cjfas-2023-0145.
10. Dastidar, P. G., A. Khan, and A. Sinha. 2022. Possible Self-Awareness in Wild Adélie Penguins Pygoscelis adeliae. bioRxiv. https://doi.org/10.1101/2022.11.04.515260.
11. Harvey, F., and F. Harvey Environment Correspondent. 2021. Animals to be Formally Recognised as Sentient Beings in UK Law. *The Guardian*, May 12, Sec. World News.
12. Korsgaard, C. M. 2018. *Fellow Creatures: Our Obligations to the Other Animals*. Oxford University Press. https://doi.org/10.1093/oso/9780198753858.001.0001.
13. What Is Practical Ethics? 2024. https://ethics.harvard.edu/what-practical-ethics. Accessed May 10.
14. White, L. 1967. The Historical Roots of Our Ecologic Crisis. *Science* 155. *American Association for the Advancement of Science*: 1203–1207. https://doi.org/10.1126/science.155.3767.1203.

15. The Origin of Descartes' Mechanical Philosophy - ProQuest. 2023. https://www.proquest.com/openview/64ae7c2d9b8430d62f8cbf57bebe3318/1?pq-origsite=gscholar&cbl=47203. Accessed March 25.
16. Franco, N. H. 2013. Animal Experiments in Biomedical Research: A Historical Perspective. *Animals* 3: 238–273. https://doi.org/10.3390/ani3010238.
17. Zimmer, K. 2022. The Oceanic Fallout of Deep-Sea Mining. *Nautilus*. December 14.
18. Senzaki, M., T. Kadoya, and C. D. Francis. 2020. Direct and Indirect Effects of Noise Pollution Alter Biological Communities in and Near Noise-Exposed Environments. *Proceedings of the Royal Society B: Biological Sciences* 287. https://doi.org/10.1098/rspb.2020.0176.
19. Merchant, C. 1990. Environmental Ethics and Political Conflict: A View from California. *Environmental Ethics* 12: 45–68. https://doi.org/10.5840/enviroethics19901217.
20. Milstein, T., M. O. Thomas, J. Hoffmann, and J. Carr. 2023. "Even I am a Part of Nature": Unraveling the Human/Nature Binary to Enable Systems Change. *Environmental Communication* 17. 421–436. https://doi.org/10.1080/17524032.2023.2199946.
21. Hampton, J. O., T. H. Hyndman, B. L. Allen, and B. Fischer. 2021. Animal Harms and Food Production: Informing Ethical Choices. *Animals: An Open Access Journal from MDPI* 11: 1225. https://doi.org/10.3390/ani11051225.
22. Ritvo, H. 2007. on the Animal Turn. *Daedalus* 136: 118–122. https://doi.org/10.1162/daed.2007.136.4.118.
23. Grusin, R. 2015. *The Nonhuman Turn*. U of Minnesota Press.
24. STS 4101 - The Entangled Lives of Humans and Animals - Acalog ACMS™. 2023. https://courses.cornell.edu/preview_course_nopop.php?catoid=31&coid=498934. Accessed March 25.
25. Animal Studies | Science and Technology Studies. 2023. https://www.brown.edu/academics/science-and-technology-studies/animal-studies. Accessed March 25.
26. NC State University Libraries' Rare and Unique Digital Collections. 2023. https://d.lib.ncsu.edu/collections/catalog?_=1543444883344&f%5Bispartof_facet%5D%5B%5D=Animal+Turn. Accessed March 25.
27. Fischer, B. 2019. *The Routledge Handbook of Animal Ethics*. Routledge.
28. Vesilind, P. A., and A. S. Gunn. 1998. *Engineering, Ethics, and the Environment*. Cambridge University Press.

29. Jr, C. E. H., M. S. Pritchard, M. J. Rabins, R. James, and E. Englehardt. 2013. *Engineering Ethics: Concepts and Cases*. Cengage Learning.
30. Martin, M., Q. Zhu, and R. Schinzinger. 2021. *Ethics in Engineering*.
31. Itoh, M. 2018. *Animals and the Fukushima Nuclear Disaster*. Springer.
32. Danaher, J. 2020. Welcoming Robots into the Moral Circle: A Defence of Ethical Behaviourism. *Science and Engineering Ethics* 26: 2023–2049. https://doi.org/10.1007/s11948-019-00119-x.
33. Mouat, M. J., R. Prince, and M. M. Roche. 2019. Making Value Out of Ethics: The Emerging Economic Geography of Lab-Grown Meat and Other Animal-Free Food Products. *Economic Geography* 95: 136–158. https://doi.org/10.1080/00130095.2018.1508994.
34. Tidwell, C. 2009. "Fish Are Just Like People, Only Flakier": Environmental Practice and Theory in Finding Nemo. *Americana: The Journal of American Popular Culture: 1900 to Present* 8.
35. *Avatar: The Way of Water*. 2023. https://movies.disney.com/avatar-the-way-of-water. Accessed March 25.
36. Erdmann, K. 2022. Real Science Behind Avatar: The Way Of Water's Anti-Aging Drug. *ScreenRant*. December 20.
37. Agathokleous, E., I. Iavicoli, D. Barceló, and E. J. Calabrese. 2021. Ecological Risks in a 'Plastic' World: A Threat to Biological Diversity? *Journal of Hazardous Materials* 417: 126035. https://doi.org/10.1016/j.jhazmat.2021.126035.
38. Hölker, F., T. Moss, B. Griefahn, W. Kloas, C. C. Voigt, D. Henckel, A. Hänel, et al. 2010. The Dark Side of Light: A Transdisciplinary Research Agenda for Light Pollution Policy. *Ecology and Society* 15: 13. https://doi.org/10.5751/ES-03685-150413.
39. Barnard, A. 2019. The 9/11 Tribute Lights Are Endangering 160,000 Birds a Year. *The New York Times*, September 9, Sec. New York.
40. Code of Ethics | National Society of Professional Engineers. 2023. https://www.nspe.org/resources/ethics/code-ethics. Accessed March 18.
41. Code of Ethics. 2023. https://www.asce.org/career-growth/ethics/code-of-ethics. Accessed March 18.
42. IEEE Code of Ethics. 2023. https://www.ieee.org/about/corporate/governance/p7-8.html. Accessed March 18.
43. Responsible and Ethical Conduct of Research (RECR) |NSF - National Science Foundation. 2023. https://www.nsf.gov/od/recr.jsp. Accessed March 18.

44. Jonas, H. 1984. *The Imperative of Responsibility: In Search of an Ethics for the Technological Age*. University of Chicago Press.
45. Bender, S. 2023. *Ethics for the Future: Perspectives from 21st Century Fiction*. Bielefeld: Transcript publishing.
46. Basart, J. M., M. Farrús, and M. Serra. 2015. New Ethical Challenges for Today Engineering and Technology. *Telematics and Informatics* 32: 409–415. https://doi.org/10.1016/j.tele.2014.05.008.
47. Riley, D. 2013. Hidden in Plain View: Feminists Doing Engineering Ethics, Engineers Doing Feminist Ethics. *Science and Engineering Ethics* 19: 189–206. https://doi.org/10.1007/s11948-011-9320-0.
48. Hyun, I. 2020. Towards a New Bioengineering Ethics. News. *Center For Bioethics. Harvard Medical School*. January 1.
49. Graeff, N., K. Jongsma, J. Johnston, S. Hartley, and A. Brendenoord. 2018. The Ethics of Genome Editing in Non-Human Animals: A Systematic Review of Reasons Reported in the Academic Literature. *Philisophical Transactions, The Royal Society*.
50. May, R., T. Nygård, U. Falkdalen, J. Åström, Ø. Hamre, and B. G. Stokke. 2020. Paint It Black: Efficacy of Increased Wind Turbine Rotor Blade Visibility to Reduce Avian Fatalities. *Ecology and Evolution* 10: 8927–8935. https://doi.org/10.1002/ece3.6592.
51. Grandin, T. Livestock Handling Systems, Cattle Corrals, Stockyards, and Races. https://www.grandin.com/design/design.html.
52. Hovorka, A., S. McCubbin, and L. V. Patter. 2021. *A Research Agenda for Animal Geographies*. Edward Elgar Publishing.
53. Barrett, M. J., V. Hinz, V. Wijngaarden, and M. Lovrod. 2021. Speaking with Other Animals through Intuitive Interspecies Communication: Towards Cognitive and Interspecies Justice. *A Research Agenda for Animal Geographies*. Edward Elgar Publishing: 149–165.
54. Münster, U., and P. Locke. 2015. Multispecies Ethnography. *Entry for Oxford University Bibliographies Online*. https://doi.org/10.1093/OBO/97801997665670130.

Chapter 2

The Ethical Quandary of Animal Well-Being and Its Relevance for Engineers

Paul B. Thompson
Department of Philosophy, Michigan State University, East Lansing, Michigan 48824, USA
thom649@msu.edu

2.1 Introduction

Like other contributions to this volume, the subject matter for this chapter sits at the intersection of major historical trends, the future trajectory of which is uncertain. There is a growing impact of humanity's tools and techniques on natural systems. Climate change is only the most obvious and comprehensive example. Humans have long had significant impacts on the composition of species existing in a given ecosystem, but the trend is for greater domestication of previously wild environments, heightened challenges to the survival of wildlife and an increasing rate of species extinction. At the same time, the perception of ethical duties or responsibilities to the non-human world of living things has also undergone dramatic change.

Animals, Ethics, and Engineering: Intersections and Implications
Edited by Rosalyn W. Berne

ISBN 978-981-5129-56-4 (Hardcover), 978-1-003-65231-1 (eBook)
www.jennystanford.com

A general ethic of humane or anti-cruel treatment of vertebrate species has arguably prevailed across all cultures, but the last half-century has witnessed a virtual explosion of interest in revisiting traditional views on the norms for evaluating the treatment of animals. The new philosophies of animal ethics have sharpened our collective ability to understand the impact of human activity on animal others, but they have also created dilemmas.

Although biological classification systems unambiguously locate *homo sapiens* with the animal kingdom, a categorical dichotomy between humans and all other animals has long animated thought on the nature and extent of morality. The epic poet Hesiod (8th century BCE) held that humans and only humans are endowed by Zeus with a sense of justice. At the dawn of the 21st century, Jacques Derrida undertook a lengthy reflection on the way that our idea of humanity has been shaped by the unquestioned assumption that humans and animals occupy discrete moral categories defined in opposition to one another. Most philosophers thought of humans and other animals as occupying radically distinct categories. Derrida wanted us to imagine how our understanding of what it is to be human would be different today if challenges to this counterfactual assumption had been more prevalent [1]. Between these extremes, people have imagined the similarity and difference between humans and other members of the animal kingdom in an astonishing variety of ways.

Most notoriously, Rene Descartes described the animal body as a machine. Although Descartes' precise intent remains uncertain, his writings became the basis for denying any sense of subjectivity or feeling to animals, altogether. Vivisectionists were able to rationalize their practice by claiming that, appearances to the contrary, experimental animals did not suffer pain or emotional trauma [2, 3]. I recall my stepfather telling me that he was taught this as part of his medical education in the 1950s. Whatever the case may have been at that time, this extreme picture of animals and their utter lack of moral significance was under assault by the 1970s. The work of Konrad Lorenz and Nicholass Tinbergen gave rise to the science of cognitive ethology. In 1976, zoologist Donald Griffin broached the possibility of consciousness worthy of moral respect in his book *The Question of Animal Awareness* [4]. At the same time, philosophers at Oxford University were reacting to a work of popular science by Ruth Harrison. *Animal Machines: The New Factory Farming Industry* challenged a number of abusive practices, while (as the title suggests) singling out the industrialization of livestock farming [5].

The chapter first offers a context-setting overview of animal ethics, followed by a discussion of its relevance for engineering practice. When structures and processes for housing, producing and utilizing animals in laboratories, zoos, the pet industry, food, wildlife protection and entertainment are considered, the potential scope of both topics is enormous. The analysis here simplifies the task by emphasizing the confinement and husbandry of livestock species. Engineers design facilities for housing and for delivery of food and water with an eye toward the profitability of animal agriculture, but it is only recently that the economic interests of producers have diverged dramatically from those of animal health and well-being. Although other forms of animal use might raise different challenges, the engineering of production, transport and slaughter in the food industry provides a useful focus for discussing the engineering ethics of animal use.

Despite the enormous complexity in diverse philosophical approaches to animal ethics, contemporary ethologists have developed useful indicators of animal welfare in livestock production facilities. The chapter continues with a discussion of how these indicators might be used in ethically responsible engineering practice. However, as it also shows, the science of animal welfare has introduced new sources of philosophical tension, rather than resolving them. The chapter ends with a discussion of these tensions and suggestions on how they might be addressed within the nexus comprising food animal production, ethics and the engineering of animal production systems.

2.2 The Conceptual Landscape of Animal Ethics: 1975 to 2000

The Australian philosopher Peter Singer brought some of the new philosophical thinking to widespread public attention with an article in the *New York Review of Books* in 1972, as well as his own book *Animal Liberation*, first published in 1975 with many subsequent editions. Singer emphasized the moral significance of animal suffering, producing arguments intended to show that the benefits humans derive from most uses of animals do not offset the pain endured by the animals used for food, clothing and medical experiments [6]. Soon afterward, Tom Regan (1938–2017) developed an importantly different approach to animal ethics with an emphasis on animals as the subject of moral rights. Sociological studies of the growing public interest in animal abuse credit Singer and Regan with providing an important source of validity and gravitas for a social movement to reform various human practices that use or affect non-human animals [7].

At some risk of over-simplification, Singer takes a utilitarian perspective in which the ethically justified course of action is the one that produces the greatest potential quantity of welfare or well-being, once the impact of all potentially affected parties has been taken into consideration. Singer estimates the benefits and harms caused by any action, process or policy affecting animals. He evaluates the ethics of that action, process or policy as a function of happiness or satisfaction. On this view, sentience (e.g., the cognitive ability to experience pain or satisfaction) is what qualifies any being for moral significance, that is, for having its welfare included in the totalization of benefits and harms that determine the morality of an act. Singer presumed that the key philosophical task was to convince readers that animal interests should be included in cost/benefit style of ethical evaluation. He aimed to establish that many species (and certainly the vertebrate species raised in food production) experience forms of pain and dissatisfaction akin to that of human beings. Hence it is, on Singer's view, maliciously inconsistent to exclude impact on animals when estimating the benefit and harm of any practice affecting animals. Singer called this "speciesism" [6, 8].

After making the philosophical case for including animal interests in an ethical evaluation, *Animal Liberation* considers several domains of animal use, including research, food production and the fur industry. In each of these applications, Singer's argument aimed to establish the conclusion that benefits derived by humans were not sufficient to offset the pain endured by animals [6]. He did not specifically consider how alternative practices in these domains might reduce the suffering of animals. Nevertheless, including animal interests in a standard engineering risk assessment of design alternatives would be broadly consistent with Singer's understanding of animal welfare, and his utilitarian approach to evaluating the ethics of animal use.

In contrast, the notion of moral rights establishes rules either permitting or delimiting an agent's range of action. As in the civil and criminal law, a person acting "within his or her rights" cannot be sanctioned simply because others suffer. Yet a violation of another's right is morally wrong, even when the experienced harm is comparatively trivial. The philosophical approaches to deriving a comprehensive list of rights are too diverse to summarize. Regan addressed this problem by postulating that any "subject-of-a-life" possesses moral rights. The word *subject* designates the psychological unity of experiential awareness. Subjects are not only sentient; their perceptions are also internally connected with one another in a way that produces individualized identity over time. Survival behavior is evidence of caring or interest preserving this subjective thread, and Regan believes that for many animals, it is reinforced by emotional attachment. Human culture has many names for this subjective unity: self; ego; soul; atman; πνεῦμα; spirit. Regan's philosophical achievement lay in extending this notion to non-humans. Although he was aware that some members of the animal kingdom might lack the capacity for self-awareness, Regan was willing to argue that we have moral grounds for presuming that all vertebrates are subjects-of-a-life [9].

Regan's view has three key concrete implications. First, if the subject-of-a-life criterion has any moral persuasiveness at all, it supports a presumptive right not to be killed. Thus, Regan's version of animal rights poses a very high burden of proof regarding human uses for animals that result in death of the animal. The rights approach can sanction the taking of life, but only when a conflict of rights is at stake. Here, Regan might accept the loss of animal life to preserve a human being's right not to be killed. This is the second key point: rights function as a threshold, a bright line that cost/benefit comparisons are simply not allowed to cross.

Sacrificing one party's rights merely to secure benefits to another party will be prohibited. Only in the difficult cases where one party's rights cannot be preserved without violating the rights of another party does a sacrifice of rights become justifiable, and even here there is the need for moral principles to adjudicate which interests have priority. Finally, then, Regan's view holds that the type of cost/benefit calculation carried on by utilitarians is itself morally bankrupt. In a move that mimics the ethics of Immanuel Kant (1724–1804), Regan holds that regarding one's conduct toward any subject-of-a-life as a means to achieving benefits is itself wrong [9].

The views of Singer and Regan exemplify a longstanding point of contrast in philosophical ethics. Utilitarians have viewed pain, disease, death and discomfort as evils that can be offset by beneficial outcomes. Rights theorists find some types of action as completely off-limits and incapable of being rendered morally acceptable as the price that must be paid to achieve a greater good. Many teachers of ethics courses would regard this as one of the most fundamental points to master in an introductory course on ethics. However, Singer and Regan are similar in a different respect: both presume that animals deserve ethical consideration in virtue of the kind of beings that they are. For Singer, any being that experiences pain or suffering is sentient. Experiences of satisfaction and dissatisfaction are what qualify a being for inclusion in an ethical benefit-cost assessment. For Regan, the psychological unity implied by the subject-of-a-life criterion qualifies a being as the bearer of rights. In this respect, both approaches can be understood as promoting an ethics that turns upon metaphysical claims.

2.3 The Landscape of Animal Ethics: 2000–2025

Singer-style utilitarianism and Regan-style rights theory dominated animal ethics for at least two decades. However, by the beginning of the 21st century alternative approaches began to become persuasive. At the quarter-century mark, styles of animal ethics are so diverse that the field defies concise exposition. However, several important themes can be identified. The first is a shift from metaphysics to politics as the basis for animal rights. The views of Singer and

Regan view are metaphysical in the sense that both posit subjective, non-physical characteristics as the substrate that generates moral duties. In Singer's case, it is sentience; in Regan's, it is the subject-of-a-life. A political view of rights understands rights as cultural constructs that assist in the negotiation and performance of social life. A political subject is simply any agent whose interests are regulated or protected by law and/or social conventions. Bernard Rollin proposed such an approach to animal rights quite early, but the notion did not gain traction until it was advocated explicitly within the context of political theory [10, 11]. From a political view, legal and institutional rights for animals emerge from a polity's recognition that their interests are worthy of respect and protection.

Relational ethics is the second theme to challenge the dominance of the Singer/Regan dichotomy. Here the central idea is that human obligations and preferred forms of comportment toward animals arise in virtue of specific relationships, rather than following from an animal's possession (or non-possession) of features like sentience or subjectivity. Relational views are more attentive to cultural practice, including the long history of understanding human and animal as dichotomous, non-overlapping conceptual categories. Yet relational theorists also call for reflective change in categories. Some feminist ethical theorists argue for an attitude of care and solidarity with animals based on the shared history of oppression [12, 13]. Cora Diamond (b. 1937) suggests that regarding animals as "fellow creatures" is a more promising relational basis for animal ethics than the sentience or subjectivity criteria promoted by Singer and Regan [14]. Other theorists outline specific circumstances that characterize distinct human-animal relationships such as the relation to wild vs. domesticated animals, or to pets vs. food animals, arguing that each of these circumstances entails a different ethical framework [15, 16].

A third development arose from the work of ethologists and research on animal cognition. Successors to Lorenz, Tinbergen and Griffin accumulated evidence for more sophisticated cognitive abilities than implied simply by sentience, including reciprocal altruism [17] and a sense of fair play [18]. While Singer and Regan had stressed how similarity between humans and other animals serves as the basis for moral obligations, this stream of research also revealed ways in which animals might have experiences, interests and other forms of mentality quite *unlike* those of humans which were nonetheless material to their welfare. Again, this notion that species-specific traits might have moral significance was anticipated by Rollin, who referred to such traits as an animal's *telos* [19]. Recognition of a bat's ability to echolocate is a prominent example. By the quarter-century mark of the 21st century, popular science writer Ed Yong had brought the cognitive significance of electromagnetic sensory abilities, sensitivity to the polarity of solar radiation and highly developed olfactory and auditory abilities to widespread attention [20].

The upshot is that animal ethics is a complex domain where alternative ways to understand human beings' moral responsibility to animals vie with one another. Many of the aforementioned approaches still attempt to follow Singer and Regan in articulating a comprehensive or general ethic classifying use of or relations with animals as morally justified and legitimate, on one hand, or illicit and prohibited, on the other. In contrast, some (but not all) political and relational ethics are ameliorative: they provide guidance for improving animals' quality of life or a person's relationship with them, recognizing that improvement falls far short of what would be required to render human use of animals morally justifiable in an unqualified sense. As will be discussed presently, these so-called "non-ideal" approaches to ethics have had more engagement with the ethologists and cognitive scientists who are providing new insight into the perceptual world of non-humans. This approach may be especially useful to practicing engineers.

2.4 Animal Ethics and Engineering Practice

Every aspect of the built environment probably has some impact on animals inhabiting an environment along with human beings. For example, wind turbines have been found to cause habitat loss for migrating species of birds [21]. Mitigation of these impacts on birds has become a design parameter for energy systems. Responses include design changes intended to reduce collision risk [22, 23] as well as careful siting and regulation of wind farm operations [24]. As suggested earlier, Singer-style utilitarian animal ethics bears on engineering responses to the impact of energy systems on wildlife by providing the ethical rationale for including impact on sentient non-humans in an engineering risk assessment. The benefits of turbines might offset the impact on bird welfare, but a full ethical evaluation would require an estimate of the welfare impact from collisions and habitat loss. It would also consider the potential benefits to animals, if this potentially renewable energy source is able to displace coal and gas-fired electrical plants. This form of ethical reasoning also supports economizing on costs by increasing the efficiency of power generation per unit of impact on animal life.

However, as noted above, an animal rights perspective asks a different question. Trade-off reasoning cannot override a rights claim. The loss of life from collisions with wind turbines could only be justified if the energy generated by a turbine is essential to securing a comparable right to some other party, presumably a human being. Yet this is a high barrier to meet. Plausible arguments support the claim that in today's world, humans need energy in much the same manner that they need food. If there is a right to food, there would be a right to energy. But just as there are many kinds of food and different ways to satisfy a right to food (buying it, growing it, visiting a food bank) there are many ways to power a household. It would be difficult to argue that a wind turbine is a unique requirement for securing any given human being's right to life. Furthermore, Regan's version of the animal rights ethic suggests that the economizing logic of the utilitarian treats animal life as a tool for securing benefits to humans. This kind of thinking is itself a moral affront. As a matter of historical fact, animal rights have indeed been cited in opposition to wind power [25].

The wind turbine case serves as an object lesson illustrating how engineering projects can become inadvertently embroiled in the utilitarian vs. rights debate in animal ethics. Engineers may want to develop some facility in these basic philosophical approaches if only to achieve a better understanding of how their work will be perceived by members of the general public. While utilitarian welfare optimization is conceptually more congenial to risk management practices currently used by engineers, those who are impressed by a Regan-style rights view will not only reject the notion that benefits to humans can justify harm to animals, but they may also regard the use of trade-off reasoning as itself morally problematic. The shift to a more politically based understanding of animal rights could re-orient the focus of ethical analysis toward legal statutes or norms that a community has implicitly agreed to endorse, but this still might not make cost-benefit trade-off reasoning acceptable to those who argue for animal rights.

Some engineering projects affect animals' quality of life directly, rather than as unintended environmental consequences. They may even be intended to do so, and in these circumstances, animal welfare or well-being should be treated as a criterion for engineering design. Housing and transport facilities have obvious impacts on animal health and welfare. The engineering design for such facilities generally includes some parameters making direct reference to animal health. There is extensive engineering literature on airflow and manure management in animal housing, for instance. In the following section, housing and husbandry for food animals is discussed as an example in which animal ethics informs engineering practice at the level of technical design.

2.5 Food Animal Welfare

Recent developments in the engineering of concentrated animal feeding operations (CAFOs) demonstrate how concern for the welfare of livestock and poultry within these facilities can interact with other design elements [26]. Indoor housing for poultry and livestock arose in response to a diverse set of production problems. In some environments, animals needed to be sheltered from weather, while in other cases producers found it easier to control pathogens in confined settings. Ease of husbandry and processing also provided an incentive for expanding the use of built structures. In any case, the number of animals housed in these systems began to increase significantly over the last half of the 20th century. With larger herds, the amount of space per animal decreased. Crowding was among the first animal welfare issues noted in industrial production systems. By the century's end, welfare-oriented standards were being developed for almost every major animal commodity including pork, milk, eggs and poultry meat [27].

The Five Freedoms was promoted by the Farm Animal Welfare Council (FAWC) in the United Kingdom as an initial rubric for addressing animal welfare issues in CAFOs. *Freedom from thirst, hunger and malnutrition* stipulates water and feed access. *Freedom from discomfort* and *Freedom to express normal behavior* both stipulate that facilities must have adequate space to rest and perform any instinctual activity. *Freedom from pain, injury and disease* stipulates husbandry and veterinary care, as opposed to a design criterion, while *Freedom from fear and distress* references interaction with humans or other animals in the confined environment of the CAFO [28]. The Five Freedoms function much like rights in that they stipulate "musts" or minimal requirements for morally acceptable livestock protection. One could not offset the harm of thirst and hunger by delivering benefits to consumers, for example, though freedom from pain, injury and disease will prove difficult to assure for every animal. Each of the five freedoms bears on engineering design for animal production facilities, though in distinct and sometimes surprising ways.

To state some obvious examples, engineering standards specify water and feed access, as well as temperature control, airflow and manure management. They have been the focus of refinements in the design and construction of animal production facilities for many years. Freedom from discomfort is similar, though here it may be important to coordinate engineering design with decisions about breeds and husbandry practice, as a facility appropriate for smaller pigs or chickens may cause discomfort for larger breeds. Engineers may not think of these standards as ethical in nature, but they are. The economic interests of producers and the welfare of their animals were somewhat aligned in most livestock production systems in use prior to World War II. Systems contrary to the Five Freedoms were less profitable because the rate at which an animal converts feed into meat, milk and eggs depends on animal health. The economics of animal production changed when expensive CAFO technology came into use.

The profitability of any enterprise depends on many factors, but in every case, the most profitable firms are the ones that optimize return on their total investment. Prior to the development of CAFOs, the animals themselves, along with feed, water and veterinary services represented the lion's share of a livestock farmer's total investment. As buildings with automated temperature control and feed delivery came into use, optimizing return on capital costs amortized over the useful lifespan of these technical systems became more and more significant. The egg industry is a particularly dramatic example because capital costs include elaborate automation of egg collection and manure control, in addition to the feed and water access or climate control needed in all CAFOs. A 2010 multiple methods analysis of studies on the egg industry concluded that battery-cages maximizing egg production on a per hen basis were not competitive those economizing on capital investment in technology. Sacrificing hen health (and thus per hen productivity) was more profitable [29].

In the age of CAFOs and engineered animal production, the economic incentives of producers become imperfectly aligned with standards intended to promote animal health and well-being. Animal welfare must therefore be treated as a design criterion operating independently from criteria determined through economic productivity analysis. In some respects, this simply reflects a common feature of engineering ethics where a client's interests may not align with the engineer's responsibility to meet recognized safety and environmental standards. In a similar manner, human factors engineering becomes a part of the engineering problem. While engineers may have primary ethical responsibility for a poorly designed system of feed delivery or climate control, responsibility for assuring any of the Five Freedoms depends heavily on the operator. As such, addressing animal welfare needs also involves engineering judgements regarding workforce capability and willingness to comply with welfare-related procedures [30].

2.6 Farm Animal Welfare Beyond the Five Freedoms

Although the Five Freedoms continue to be influential, they have been criticized as being excessively focused on negative welfare states such as pain and distress. They do little to identify features that can substantially improve the quality of life for animals living under confinement. David Mellor argues that it is more helpful to think in terms of five domains, each of which can contribute beneficially or detrimentally to animal welfare. Mellor's five domains are: nutrition, environment, health, behavior and mental states [31]. Mike Appleby also offered a more simplified model with *animal bodies* encompassing all the standard measures of veterinary health, *animal minds* including affective or psychological welfare states such as pain or an experience of well-being, and *animal natures* consisting of behaviors typically performed by members of the species [32]. Appleby's framework is particularly useful for illustrating ethical tensions that arise in engineering for animal welfare.

The key tension arises in the interpretation of animal natures, or species typical behavior. Welfare indicators in this category emphasize behaviors thought to be important for welfare. For example, laying hens exhibit a strong inclination to rest on a perch and to retreat to a designated area when laying eggs. But there are two ways to understand why these behaviors are important to welfare. On one view, stifling an important behavior compromises welfare because it causes pain, discomfort or some other detriment to animal health. Here, the animal's welfare consists in veterinary health measures (Appleby's animal bodies) and cognitive affect, especially pain or discomfort as well as positive, satisfying experiences (Appleby's animal minds). Observing whether hens are able to perform perching and nesting behavior (animal natures) is a method for inferring positive or negative aspects of hen welfare, but the welfare itself resides in the cognitive distress of hens denied the opportunity for perching or nesting. It is an attractive indicator of welfare in part because of its reliance on observational data.

The alternative view regards these species typical behaviors as components of welfare, rather than as indirect indicators of cognitive affect or ill-health. This view aligns with Rollin's earlier identification of *telos* as a morally significant basis for deriving a conception of animal rights. At one point, Rollin characterized *telos* in terms of characteristics that make an animal what it is, using the phrase "the pigness of pig" [19]. At other times, he emphasized genetically based behavioral drives, arguing that while such drives must be respected, there is no moral wrong involved in changing an animal's *telos* [33]. Contrarily, many ethicists have held that at least some applications of bioengineering are immoral when applied to animals because they fail to respect the way that a specific genetic make-up and its species typical behaviors *constitute* animal welfare, irrespective of how the animal feels. Put differently, performing these behaviors is simply what it means to say that animals of a given species are living a good life [34].

The tension between these two interpretations of animal natures is most profoundly felt with respect to genetic modifications that might have profound effects on behavior or even the manner in which a given animal perceives and relates to its environment. A Danish group initiated ethical debate on this possibility with their study of congenitally blind chickens [35]. Does the ethical acceptability of this genetic change depend solely on how it affects veterinary and cognitive aspects of welfare? Put succinctly, do we think blind hens have good welfare simply because they are not experiencing pain and may even be leading a happy life, or is there something about this kind of change that vitiates the very possibility of having good welfare? Opinion among both ethicists and animal welfare scientists remains divided on this question [36, 37].

Aside from bioengineering, the ethical interpretation of animal natures has implications for the housing and husbandry aspects of food animal production. Sticking with laying hens, providing outdoor access is currently included in the design standards for organic egg production. The first approach evaluates outdoor access under the assumption that welfare is to be evaluated by its relationship to measures such as morbidity and mortality, on the one hand, or indicators of cognitive stress, on the other. On this view, current data suggests that outdoor access actually causes a decrease in welfare [38]. The contrary view, which may be reflective of the opinion held by organic egg producers and many members of the public, is hens denied access to the outdoors are simply not living the kind of life that chickens deserve to have. Engineers would then have a responsibility to redesign outdoor access facilities with an eye to reducing mortality and morbidity. Interestingly, an empirical opinion study of factors contributing to the welfare of farmed animals indicates that the U.S. public is evenly divided between these viewpoints [39].

2.7 Conclusion: Ethical Responsibilities for Engineers

Animal ethics has undergone a remarkable evolution over the last half-century, moving beyond the philosophical contest between Peter Singer's utilitarian ethic and Tom Regan's approach to animal rights. Singer and Regan introduced the philosophical rationale for taking animal interests seriously when making an ethical evaluation, but throughout the 20th century this largely meant making sweeping judgments about the moral acceptability of animal use. For example, their philosophical differences notwithstanding, both Singer and Regan advocated an ethically based vegetarianism, campaigned against the fur industry and in favor of curtailing the use of many animals in scientific research. Their positions did not provide a clear basis for integrating ethics into those engineering practices having the most direct effect on animal lives.

This changed over the first quarter of the 21st century, as more nuanced philosophical perspectives were articulated, and more importantly, through empirical research on indicators of animal welfare. As illustrated through discussions of agricultural and bioengineering in the food system, increasingly robust criteria allow engineers to adjust design and process standards in ways that have ethically significant (both positively and negatively) on the quality of animal lives. Engineered animal production systems have introduced efficiencies that reduce the consumer's cost for animal products, but they have also sundered the connection between animal welfare and the economic interests of producers, in many instances. As such, we have reached a juncture where attention to animal welfare has become ethically imperative for engineers developing systems of animal housing and provision of food, water and other critical resources.

Although animal welfare is an inherently ethical topic, many of its dimensions are clear and standards can be supported through close attention to veterinary and ethological research. At the same time, as the public's sensitivity to animal interests has grown, other aspects of animal care and use raise controversy and divide opinion on the kind of life animals under human care deserve to have. Future studies in engineering ethics will not only need to recognize animal interests as worthy of inclusion in exercises such as engineering risk assessment, but they will also need interaction among engineers, animal welfare scientists and bioethicists. Most significantly, engineers may be able to develop technical approaches that obviate what are now taken to be irresolvable conflicts of interest and philosophical quandaries in humanity's relationship with the non-human world. However, this type of innovation will require an appreciation of both empirical and philosophical aspects of animal well-being, as well as recognition of the challenges arising in all the arenas where humans and animals interact.

References

1. Derrida, J. 2008. *The Animal that Therefore I Am*. D. Wills, Tr. Fordham University Press.
2. Harrison, P. 1992. Descartes On Animals. *The Philosophical Quarterly* 42: 219–227. https://doi.org/10.2307/2220217.
3. Rollin, B. E. 1998. *The Unheeded Cry: Animal Consciousness, Animal Pain, and Science*. Iowa State University Press.
4. Griffin, D. R. 1976. *The Question of Animal Awareness: Evolutionary Continuity of Mental Experience*. Rockefeller U Press.
5. Ruth H. 1964. *Animal Machines: The New Factory Farming Industry*. V. Stuart.
6. Singer, P. 1975. *Animal Liberation*. Harper Collins.
7. Jasper, J. M., and D. Nelkin. 1992. *The Animal Rights Crusade: The Growth of a Moral Protest*. Free Press.
8. Singer, P. 2011. *Practical Ethics*. 3rd edition. Cambridge University Press.
9. Regan, T. 1983. *The Case for Animal Rights*. University of California Press.
10. Donaldson, S., and W. Kymlicka. 2011. *Zoopolis: A Political Theory of Animal Rights*: Oxford University Press.

11. Garner, R. 2006. Animal Welfare: A Political Defense. *Journal of Animal Law and Ethics* 1: 161.
12. Donovan, J. 1990. Animal Rights and Feminist Theory. *Signs* 15: 350–375.
13. Adams, C. J. 1990. *The Sexual Politics of Meat: A Feminist-Vegetarian Critical Theory*. Continuum.
14. Diamond, C. 2008. The Difficulty of Reality and the Difficulty of Philosophy. In *Philosophy & Animal Life*, ed. S. Cavell, C. Diamond, J. McDowell, I. Hacking, and C. Wolfe. Columbia University Press.
15. Palmer, C. 2010. *Animal Ethics in Context*. Columbia University Press.
16. McKenna, E. 2021. *Living with Animals: Rights, Responsibilities and to Respect*. Rowman and Littlefield.
17. de Waal, F., R. Wright, C. M. Korsgaard, P. Kitcher, and P. Singer. 2006. *Primates and Philosophers: How Morality Evolved*. Princeton University Press.
18. Bekoff, M., and J. Pierce. 2010. *Wild Justice: The Moral Lives of Animals*. Illustrated edition. University of Chicago Press.
19. Rollin, B. 1981. *Animal Rights and Human Morality*. Amherst, NY: Prometheus Books.
20. Yong, E. 2022. *An Immense World: How Animal Senses Reveal the Hidden Realms Around Us*. Random House.
21. Marques, A. T., C. D. Santos, F. Hanssen, A.-R. Muñoz, A. Onrubia, M. Wikelski, F. Moreira, J. M. Palmeirim, and J. P. Silva. 2020. Wind Turbines Cause Functional Habitat Loss for Migratory Soaring Birds. *The Journal of Animal Ecology* 89: 93–103. https://doi.org/10.1111/1365-2656.12961.
22. Eriksson, S., H. Bernhoff, and M. Leijon. 2008. Evaluation of Different Turbine Concepts for Wind Power. *Renewable and Sustainable Energy Reviews* 12: 1419–1434. https://doi.org/10.1016/j.rser.2006.05.017.
23. Shimada, Y. 2021. Reducing Bird Collision Risk Per Megawatt by Introducing Longer Wind Turbine Blades. *Ornithological Science* 20: 253–261. https://doi.org/ 10.2326/osj.20.253.
24. Singh, K., E. D. Baker, and M. A. Lackner. 2015. Curtailing Wind Turbine Operations to Reduce Avian Mortality. *Renewable Energy* 78: 351–356. https://doi.org/10.1016/j.renene.2014.12.064.
25. Martin, R. J., and R. Ballard. 2013. Reconciling the Migratory Bird Treaty Act with Expanding Wind Energy to Keep Big Wheels Turning and Endangered Birds Flying. *Animal Law* 20: 145–166.

26. Mench, J. A., E. A. Pajor, H. James, and P. B. Thompson. 2008. *The Welfare of Animals in Concentrated Animal Feeding Operations*. Pew Commission on Industrial Animal Production.
27. Fraser, D. 2017. Animal Welfare: Translating Science into Practice. In *Advances in Agricultural Animal Welfare: Science and Practice*, ed. J. Mench, 129–144. Elsevier Science & Technology.
28. Webster, A. J., 2001. Farm Animal Welfare: The Five Freedoms and the Free Market. *The Veterinary Journal* 161(3): 229–237. https://doi.org/10.1053/tvjl.2000.0563.
29. Millman, S. T., J. A. Mench, and A. E. Malleau. 2010. The Future of Poultry Welfare. In *The Welfare of Domestic Fowl and Other Captive Birds*, ed. Ian J. H. Duncan and Penny Hawkins, 279–302. Springer. https://doi.org/10.1007/978-90-481-3650-6_12.
30. Mellor, D. J., N. J. Beausoleil, K. E. Littlewood, A. N. McLean, P. D. McGreevy, B. Jones, and C. Wilkins. 2020. The 2020 Five Domains Model: Including Human-Animal Interactions in Assessments of Animal Welfare. *Animals* 10: 1870. https://doi.org/10.3390/ani10101870.
31. Mellor, D. J. 2016. Updating Animal Welfare Thinking: Moving Beyond the "Five Freedoms" towards "A Life Worth Living." *Animals* 6: 21. https://doi.org/10.3390/ani6030021.
32. Appleby, M. C. 1999. *What should We Do about Animal Welfare?* Blackwell Science.
33. Rollin, B. E., 1998. On Telos and Genetic Engineering. In *Animal Biotechnology and Ethics* (pp. 156–171). Springer.
34. Holland, A. 1995. Artificial Lives: Philosophical Dimensions of Farm Animal Biotechnology. In *Issues in Agricultural Bioethics*, ed. T. B. Mepham, G. A. Tucker, and J. Wiseman (pp. 293–206). University of Nottingham Press.
35. Sandøe, P., B. L. Nielsen, L. G. Christensen, and P. Sorensen. 1999. Staying Good While Playing God: The Ethics of Breeding Farm Animals. *Animal Welfare (South Mimms, England)* 8: 313–328.
36. Thompson, P. B. 2007. Ethics on the Frontiers of Livestock Science. In *Redesigning Animal Agriculture: the Challenge of the 21st Century*, ed. D. L. Swain, E. Charmley, J. W. Steel, and S. G. Coffey (pp. 30–45). CABI Books. https://doi.org/10.1079/9781845932237.0030.
37. Swanson, J. C. 2010. How Welfare Is Measured and Why Scientists Do It Differently. *Journal of Veterinary Medical Education* 37: 89–93. https://doi.org/10.3138/jvme.37.1.89.

38. Holt, P. S. 2021. Centennial Review: A Revisiting of Hen Welfare and Egg Safety Consequences of Mandatory Outdoor Access for Organic Egg Production. *Poultry Science* 100: 101436. https://doi.org/10.1016/j.psj.2021.101436.
39. Norwood, B, and J Lusk. 2012. *Compassion by the Pound: The Economics of Farm Animal Welfare*. Oxford University Press.

Chapter 3

Engineering in a Multispecies World: Animal Geographies, Ethics, and Place

Julie Urbanik,[a] Jenny Isaacs,[b] David Tovar,[c] and William S. Lynn[a,d]

[a] *PAN Works, Wilbraham, Massachusetts 01095, USA*
[b] *Department of Geography, Rutgers University Piscataway, New Jersey 08854, USA*
[c] *Department of Mechanical and Aerospace Engineering, Rutgers University Piscataway, New Jersey 08854, USA*
[d] *Marsh Institute, Clark University Worcester, Massachusetts 01610, USA*
julie.urbanik@gmail.com, jenny.isaacs@rutgers.edu, dat169@scarletmail.rutgers.edu, wlynn@clarku.edu

The things human societies build, make, and use have a tremendous impact on non-human animals (hereafter animals). There is increasing consensus that the future of life on Earth depends upon whether we can design systems and products that help all species flourish in healthy environments.

Engineering and Geography are two disciplines shaping our future. Engineering is about the use of math and science to design and develop products for human society. Engineers are the ones who design and build bridges, buildings, and materials. Geography is the study of where, how, and why things are the way they are around the world. Geographers are the ones asking, "why

Animals, Ethics, and Engineering: Intersections and Implications
Edited by Rosalyn W. Berne

ISBN 978-981-5129-56-4 (Hardcover), 978-1-003-65231-1 (eBook)
www.jennystanford.com

this design, this building, or this material in this particular place at this particular time?" In addition, geographers study the relationship between regulations, policies, and the built environment; how engineering products impact different human groups; and how these products positively or negatively impact the natural environment. Both disciplines have an interest in understanding spaces and places in terms of their physical attributes as well as their psychological and cultural attributes (e.g., sense of place, cultural/spiritual/commercial needs).

In this chapter, we will first introduce animal geography, a subfield of geography. We will share how those working from this perspective examine a key aspect of the natural world—namely animal species besides the human species. Second, we introduce the concept of geoethics, an ethical framework used by animal geographers to examine how we ought to live with other animals. Third, we survey the connections between the branches of engineering and animals. Finally, we discuss how geoethics can be applied by engineers to incorporate both geographical and ethical insights into their work.

3.1 Animal Geography

When some people think of geography, they think of maps or the National Geographic Society. Though maps have always been central to geography, and the Society is one of the public faces of geography through its magazine and TV channel, the overall discipline is much more complex. The word geography comes from Greek and translates as "earth writing" or "earth description" and it is, arguably, the oldest interdisciplinary tradition of modern academic knowledge. In this chapter, we emphasize geography's long history as the elder tradition of what we today call environmental studies, combining studies of how nature intersects with society to learn how humans do and ought to live on earth [1, 2]. Geography is loosely divided into four major, yet overlapping, areas: (1) physical geography studies the process of the planet itself (e.g., volcanoes, climate change), (2) human geography studies the relationships within and between human groups (e.g., religious affiliation, political boundary making), (3) human-environment geography focuses on the interaction of human societies with the natural world (e.g., pollution, the built environment), and (4) the mapping sciences focus on using technology to advance our ability to map our world (e.g., Geographic Information Systems (GIS), remote sensing). The sub-field of animal geography works at the intersection of human geography and human-environment geography while incorporating mapping technologies.

Animal geography is most succinctly defined as the study of where, when, how, and why animals intersect with humans and how humans can come to better know animals and their needs in their own right [3]. Because there are so many different animals and different ways that humans interact with them, animal geographers have found it helpful to use a five-part categorization system that describes the animals and/or human–animal interactions under investigation. The first four categories are: wild animals, farmed animals (for food or parts (e.g., fur)), companion animals (i.e., pets), and working animals (e.g., animals in science and medicine, donkeys/horses, police/military dogs). The expansive fifth category, cultural animals, includes the study of animals used in human entertainment (e.g., fighting, circuses, zoos, movies, sports mascots), in religious or cultural identity contexts (e.g., religiously-specific slaughter methods or cultural groups that prohibit the consumption of all/certain species), or in the digital world (e.g., avatars, memes).

Within these categories, animal geographers use four main approaches: a historical approach that explores human–animal relations over time, an economic approach that examines how animals, in whole or in part, circulate through monetary and barter systems, an ethical/political approach that looks at what is right/wrong or legal/illegal as it pertains to human treatment of other animals, and the cultural landscape approach which explores the visibility of animals in our daily lives. Methodologically, animal geographers, like all geographers, draw from a variety of data sources. In some cases, collecting quantitative data (large-scale surveys) is needed, in other cases qualitative data (interviews, focus groups, textual analysis) works best, and sometimes field-based methods are the right choice (observation, counts, multispecies ethnography, etc.). Materials that may be produced include maps, historical timelines, statistical analyses, and comparative economic or policy analyses.

Animal geographers use four key geographic concepts in their research. The first is place; place refers to a specific location (e.g., the Paris Zoo or Washington, D.C.) but place could also refer to an abstract location/in terms of relationality (e.g., the place of rats in human society—are they cuddly and smart pets in our homes or are they vermin that should be eradicated?). By focusing on the physical and conceptual place of animals for specific human–animal interactions, animal geographers examine the variety of cultural animal practices around the world. They ask questions about what might help or hinder better understandings of animals and human relations with them. The second concept is scale; scale can be understood first as a nested set of places, with smaller places located within larger ones (e.g., the home, neighborhood, community, town, city, county, state, nation, continent, globe). Secondly, scale is understood as the amount of something happening in a place or across space. The third concept is boundaries; boundaries are formal demarcation zones between territories (e.g., the boundary between Missouri and Kansas or between the United States and Canada), or they are informal/cultural markers that define insiders and outsiders (e.g., being "southern" or being vegetarian). Finally, whether looking at place, scale, or boundaries, animal geographers are also interested in power relations, examining who (which individuals, groups, or species) has control and authority in a place and why. It is essential to understand power relations when it comes to nonhumans because this helps researchers understand how decisions are made that impact animals, who gets to make them, and what these decisions reveal about social values.

Animal geographers are different from those who study the animals themselves (e.g., biologists), or those who study human–animal relations in other disciplines (e.g., psychology), because we approach the subject in a holistic way. For animal geographers, context is key. All geographers recognize that events, practices, and belief systems do not happen in vacuums, but unfold as processes embedded in relations of time and place, within nature and society.

3.2 Ethics and Animal Geography

Ethics, according to Socrates in Plato's *Republic*, is an exploration into "how we ought to live." It is an attempt to help us understand what is right or wrong in relation to our individual and collective lives. For animal geographers, how we ought to live in relation to the other animals we share our planet with is a central concern. Animal geographers begin with the premise that all animals have intrinsic value in and of themselves–their value lies in the fact that they are alive, autonomous beings with their own wants, needs, and capabilities as befits their particular species. As such, they should not be seen or treated as outside human systems of ethics and morality, or beings subject to the whims of people. Put simply, as living beings *in relation to* other beings, animals deserve respect and care for their well-being [4]. This fact of relationality means it is important to examine the geography (i.e., the where, the why, the when, and the how) of how ethical determinations are made regarding human actions that directly or indirectly impact other animal species.

In the 'analytical' tradition of ethics and philosophy, there is often an emphasis on universal rules of right and wrong that could be applied in all circumstances and in all locations. These rules, such as "you should not kill," are idealized notions of right and wrong that, in the abstract, seem fair and just in terms of how individuals and communities should be treated. The main problem with this tradition is that in the real world, these abstract maxims about what is right and wrong might not apply—or at a minimum become murky. Does "you should not kill" apply when another person (or nonhuman animal) is threatening your own life or the life of a loved one? Does this mean self-defense is unethical? It seems clear that there is a spectrum of ethical actions depending on practical circumstances. In essence, context matters.

Animal geographer and ethicist, William Lynn, has developed the concept of geoethics as a way to approach ethical issues from a context- and place-based position, rather than a position of difficult-to-follow, universalizing rules [5]. Combining geography's emphasis on the importance of context to human and environmental events, with ethics' desire to provide guidelines for how we ought to live to facilitate flourishing, geoethics begins with the recognition that "as geographic beings we are necessarily embodied as individuals, and ennatured, that is situated in nature's rich web of life-forms and life-forming processes" [5]. In other words, our existence is dependent upon and interrelated with the many beings and processes of the natural world. Geoethics recognizes that we, as humans, live in a multispecies world so we need a multispecies ethics to help guide our actions given the complexity of our interconnected existence.

Geoethics offers four principles that can help guide our decision-making and conduct as humans with respect to other animals. These principles are not absolute laws, but they provide a framework for analysis depending on the specific context. The principle of *geocentrism* means to recognize the moral value of animals, humans, and the rest of nature. With this principle, humans are neither placed outside of or above the rest of the world, but understood as part of it and that ethics requires of humans the capacity to consider animals and nature as a part of our ethical systems. The principle of *equal consideration* means to give equal consideration to the well-being of all beings affected by human actions. This helps us to balance the flourishing of humans and animals. It does not, however, mean sameness of treatment. It is obvious that humans have different capacities than ants or woodpeckers and do not necessarily require the same treatment, but they do require our consideration when and if our actions may impact them. The principle of *hard cases* asks us, when cases arrive pitting animals against humans, that we, as humans, do not automatically privilege humans but seek solutions/compromises that acknowledge and protect the well-being of animals. Finally, the principle of *moral carrying capacity* means that humans should live within the carrying capacity of the planet in a manner that allows the entire geographical community to flourish. For example, some ecosystems are more fragile than others so natural resource extraction and development in these places comes at too high of a cost to human and other animal lives. In summary, geoethics provides a way for us to examine the cost and limits of particular actions on Earth's living (and non-living) parts and systems.

Now that we have described the field of animal geography and the guiding principles of geoethics, it should hopefully be clear why animal geographers have an interest in engineering. Not only are we interested in examining the range of interactions between engineers and animals; the types of places where engineering research takes place and the animals in those places; and how engineering attends to, or neglects, animals when producing the human-built environment across scales, but we are also interested in understanding how engineers conceptually place different animals as subjects in their own right (or not) and the types of ethical boundaries engineers might be drawing between humans and animals. In the next section, we survey the different branches of engineering to show how engagements are occurring in all five categories of the human–animal interaction types studied by animal geographers. This survey will allow us to reflect on how geoethics can be brought more fully into engineering in the last section of the chapter.

3.3 Mapping Engineering's Engagement with Nonhuman Animals

It is an understatement to say that the work of engineers has impacted other species and their environments. These impacts are both direct and indirect. Engineers today engage with animals and their habitats in three ways: biomimicry, environmental impacts, and physical manipulation.

Biomimicry refers to an engineering approach that tries to solve human problems and design novel products by the study and 'mimicry' of animal technologies, anatomy, and physiology. It is based on the premise that nature itself has much to teach engineers. Through the stress-testing of millions of years of evolution and geological history, the natural world offers engineers efficient designs and ready-made, sustainable solutions for problems. To operate using a biomimicry approach, engineers frame their questions in natural science terms, study the problem in the field, then work in the lab to imitate and apply nature's insights. This bio-inspiration approach is most strongly associated with the work of biologist Janine Benyus, of the Biomimicry Institute [6]. Implied in a biomimicry approach is an ethics of primary authorship: if Nature itself is the original creator of superior engineering designs and solutions, then humans should be respectful and deferential towards Nature. For some people, the practice of biomimicry may even have a sacred or religious quality to it.

Environmental impacts refers to any impacts that engineering has on animal worlds. These impacts range from lived encounters between specific animals, engineers, and their products, to the environmental impacts and effects of engineering on animal habitats and living conditions, from resource extraction to the built environment. Thanks to scientific and technological innovation driven by engineers, humans on average are living longer and better lives. With better education, agricultural yields, and public health advances, many people now enjoy higher incomes, the ability to travel long distances quickly, increased purchasing power in bigger living spaces within safer, cleaner cities. They have easy access to medicines, goods, and foods produced halfway around the world. However, this realized progress of modernity—made possible in large part by engineers—has also led to larger per-person environmental footprints and concerns about sustainability. The resulting extraction and overconsumption of renewable (e.g., trees) and nonrenewable (e.g., oil) natural resources on a global scale has had negative effects on wildlife; including pollution, wildfires, habitat loss, deforestation, desertification, climate change, and mass extinction.

Physical manipulation refers to the direct, purposeful physical modification, manipulation, enhancement, and/or control of animal bodies, in whole or part. At the genetic scale, bioengineers create new animal and plant breeds, lineages, and hybrids by removing, adding, or crossing genetic material within and across species. They manipulate RNA and DNA in the development of new vaccines, and they influence the evolution of animal species by creating and releasing genetically altered animals into the world. At the scale of the individual body, engineers design and provide physical augmentations or reductions for scientific experimentation, veterinary, or human medical purposes. To control animal bodies, they design enclosures, tracking devices, collars, (electrified) fences, prods, nets, cages, and confinement mechanisms to prevent their free movements. At a global scale, information about animal bodies is monitored within computer networks and databases that allow scientists to locate animal bodies anywhere in the world. These are just some examples of the enormous influence engineers have on animal bodies and worlds.

In what follows, we share additional examples that illustrate these three modes of engagement, drawn from a representative sample of all the different branches of engineering. The survey also will demonstrate that engineers engage with all five of the categories of animals studied by animal geographers described above.

3.3.1 Mechanical Engineering

Mechanical engineering is a diverse and foundational discipline which intersects with all of the other engineering fields listed below, showcasing its adaptability and innovative potential. These engineers draw from nature's patterns to enhance the functionality and sustainability of machines and structures. Using a biomimicry approach, mechanical engineers imitate the arrangement of nautilus shells when designing pumps, drains, and fans. The nautilus shell and many flowers exhibit the Fibonacci sequence (Fig. 3.1), where each number in the sequence is the sum of the two previous numbers. Mechanical engineers design machines that manipulate animal bodies such as milking machines, squeeze chutes in agriculture and slaughterhouses that facilitate efficient movement

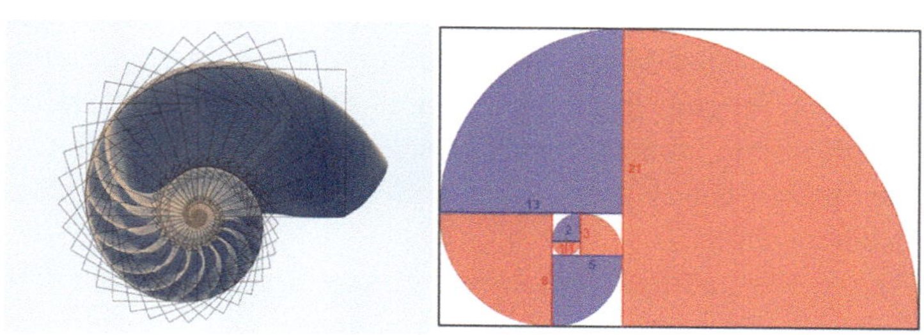

Figure 3.1 An example of studying biomimicry. On the left is a nautilus shell exhibiting the Fibonacci sequence (image by Roberta Conti, Creative Commons license [7]). On the right is a demonstration of the Fibonacci sequence (image by Mabit1, Creative Commons license [8]).

of animals to killing and/or transport floors, fishing gear, nets and traps, as well as harnesses, or (materials for) prosthetics to help animals regain mobility. These engineers also play a pivotal role in addressing environmental challenges; for instance, through the design and development of machines that pull plastics out of the ocean and address climate change by pulling carbon dioxide out of the air. Because it informs all other engineering disciplines, mechanical engineering is the primary discipline for engineers to address the well-being of animals and the environment.

3.3.2 Aeronautics and Astronautics

Aeronautical and astronautical engineering have achieved astonishing feats with respect to flight and space travel. However, they have, and still, depend upon the labor, manipulation, and sacrifice of animal bodies. Perhaps no animal to leave Earth is more famous than Laika, a dog picked up off the streets of Moscow (Fig. 3.2). She was shot into space as the celebrity "Muttnik" in the Russian satellite Sputnik 2 on November 3, 1957. To prepare for her journey, she was spun in centrifuges and forced to live in progressively smaller and smaller confines, surviving only on jellied food. The Russian government, which rushed the launch for political reasons, falsely reported that Laika had died by merciful food poisoning after 7 to 8 days in space, before her oxygen supply was to run out. Fearing public outrage, the authorities concealed the truth—that she quickly died from overheating in a poorly engineered capsule, and perhaps to panic as her pulse shot up to three times higher than normal. Laika can be understood as the face of an army of untold number of animals, ranging from fruit flies to mice, cats, frogs, rabbits, fish, guinea pigs, and chimpanzees used in flying and space travel experiments—nearly all of whom suffered physical and/or mental pain before dying.

Today, aeronautical engineers continue to push the boundaries of flight efficiency and sustainability by harnessing principles from the animal kingdom. Using a biomimicry approach, engineers in aeronautics have long used inspiration from animals to improve the performance of aircraft; for example, the v-shape formation observed in migratory birds has been replicated for planes flying

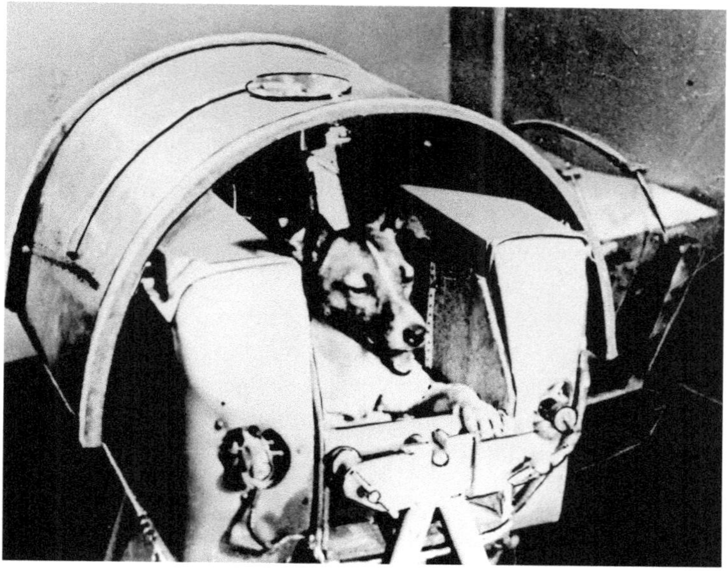

Figure 3.2 Laika, the first living being in space (image by Robert Lewis, Creative Commons license [9].

together, optimizing fuel efficiency. Additionally, the aerodynamic principles of bumps found on the edge of whale fins have been applied to airplane wings to reduce drag and enhance the efficiency of wind turbine blades. Drawing inspiration from beaver pelts, engineers have developed textured surfaces in spacesuits and wetsuits, while NASA has designed space shoes, climbing robots, and wall adhesives for space stations using gecko feet as a model for optimal traction.

Animal worlds are disrupted by aerospace engineering in many ways. Compared to other modes of transportation, airplanes and rockets expel the most carbon emissions, greatly contributing to climate change. Animal habitats and atmospheres are locally disrupted by the noise pollution and habitat loss associated with airports and rocket launch pads; these disturbances can alter mating and migratory behaviors. Additionally, each year, birds collide with airplanes in deadly 'birdstrikes' that present a danger to both wildlife and people. For example, a mid-air collision with a flock of geese resulted in an engine failure that forced a water landing of commercial US Airways Flight 1549 in New York's Hudson River in 2009.

The inside of planes and rockets are not safe for animals: there have been many cases where companion and other service animals have died in transport or as passengers in cargo holds. Planes are an easy way for wildlife to be smuggled or transferred from one place to another, sometimes upsetting local ecologies; for example, a military plane brought hitchhiking brown tree snakes to the island of Guam which escaped and caused devastation among the native bird populations. If viruses may be considered animals (which is an ongoing debate in biology), the fast-moving COVID-19 pandemic was made possible by efficient global airline networks, spreading the disease around the world in a few weeks, with millions of people and animals infected as a result.

3.3.3 Chemical and Nuclear Engineering

While these two disciplines exemplify humanity's striking command over energy and the elements, the projects of these engineers have benefited humans but also wrecked habitats, disrupted entire ecological communities, and threatened the ability of the Earth to support life in the future. For these reasons we have grouped them together here.

Chemical engineers are responsible for bringing plastics, fuels, fibers and fabrics, electronics, cement, paints, foods, perfumes, dyes, water treatments, medications, fertilizers, pesticides, and poisons into the world. Manufactured and distributed on a mass scale since the early 20th century, and often with little regard for environmental impact, chemical engineering products have produced enormous waste and pollution that builds up in animal bodies and ecosystems. For example, plastics are synthetic polymers designed to resist decomposition; they never fully break down, but rather just turn into smaller 'microplastics' (less than five millimeters or 0.2 inches in diameter) that build up in oceans, such as in the Great Pacific Garbage Patch, or in our bodies—with 80% of Americans recently surveyed presenting with microplastics in their bloodstream [10].

Likewise, pesticides and chemicals have been set loose into the environment without proper (or sometimes any) regulation.

Their biological and systemic environmental effects are often unknown before being introduced. For instance, until legislation was passed under Toxic Substances Control Act (1976) and the Superfund Law—or the Comprehensive Environmental Response, Compensation, and Liability Act (1980) in the United States, there were no regulations prohibiting the release of hazardous pollution into animal habitats. Pesticides and insecticides are indiscriminate, systemic, and non-selective, harming a host of plants and animals beyond the target species; for example, eagles became an endangered species as a result of DDT, which fatally softened their egg shells (Fig. 3.3). The effects of DDT were brought to public attention by biologist Rachel Carson in her best-selling book, *Silent Spring* [11] —credited for launching the modern environmental movement by

Figure 3.3 Eagles found dead due to pesticides at the Seney National Wildlife Refuge in 1963 (image courtesy of Seney Natural History Association, Creative Commons license [12].

raising public awareness to the harms of pesticides on animals. At that time, the chemical industry assured a suspicious public that pesticides were safe to people, animals, and the environment; when this proved false, it created doubt about the safety of all chemicals. A similar rift around nuclear power emerged between industry, engineers, scientists, and the public.

Nuclear engineers research and solve problems related to the use, control, and waste associated with nuclear energy production and nuclear weapons. Considered an efficient, 'clean' and 'green' technology because it produces zero greenhouse gas emissions, nuclear power plants are an important alternative to burning fossil fuels in the fight against climate change, which threatens most animal species. The devastating harm nuclear power can unleash, however, has made them controversial and limited their development. On April 26, 1986, an explosion within the reactor of the Chernobyl nuclear power plant in Ukraine killed workers and released more than 100,000 lbs. of radioactive material into the environment. No longer suitable for human settlement, the area was abandoned, and a 1400-mile exclusion zone was set up around the site. Today the area and its wildlife serve as a dystopian science experiment for researchers studying the long-term effects of remnant ionizing radiation on bodies and environments. Interestingly, instead of being a dead zone, scientists have observed that wildlife have taken back the area absent of people. For example, wolves have been found within the exclusion zone at seven times the density in surrounding reserves. These radioactive wolves and other animals found nearby exhibit novel genetic mutations and evolutionary adaptations that researchers and engineers are studying to better understand how tissues and organs respond to radiation exposure.

For decades, animal bodies and genetic lines have been purposefully manipulated or harmed in nuclear engineering tests to study the impacts of radiation on living organisms. As in aerospace and astronautics, scientists exposed animals to radiation in experiments meant to test the physiological effects of nuclear weapons, nuclear accidents, and radiation on patients. By testing on animals, cancer and military researchers have been able to study and determine safe radiation dosing levels, radiation-induced pathologies, efficacy of radiation shielding drugs, and possible side effects of cancer treatments. Other nuclear engineering applications and techniques include radioactive tracing in imaging technologies to detect problems in the body; irradiating technologies to sterilize hospitals, foods, and invasive species; validate vaccines; identify the extent and source of environmental pollution; and induce mutations to produce and select for new, hardier genetic lines of plants and animals.

3.3.4 Biological Engineering, Medical Engineering, Materials Science

Science has long been a target of animal advocates. The modern animal rights and animal welfare movement in the West began in England at the turn of the 19th century as public outcry grew in response to cruel, live dissections of animals (i.e., 'vivisection') without anesthesia purportedly for scientific purposes. Claiming the goal to better understand the human body for the development of effective therapeutics, anatomists such as François Magendie became famous for torturous experiments on animals in front of live audiences, such as in 1824 when Magendie nailed a greyhound dog to the dissecting-table before cutting it open. Legislation was eventually passed in Britain and the United States to ban and regulate such practices on ethical and moral grounds. Today, all bio/medical engineers in the United States who manipulate animal bodies in universities, or who receive federal funding in private laboratories, must operate according to regulations and protocols to protect animals from unnecessary pain and suffering, unless the need for the research is proven to be urgent and essential for human health. Fortunately, in a reversal, engineers have recently made wonderful breakthroughs in the design of virtual dissection technologies that allow for medical training without the need for animal suffering and death.

Vaccine research is one area of biomedical engineering that continues to rely on animal bodies and manipulation. A striking example comes from the case of horseshoe crabs, whose blood contains lysate, a key ingredient in the COVID-19 vaccine and many others. As a threatened species, horseshoe crabs cannot be killed when harvested for scientific reasons; instead, they are caught live, transported to laboratories where they are drained of their blood, and then are returned back to the ocean. About 20–30% die in this violent, invasive process. However, worldwide, there is a growing effort to develop recombinant technologies that can eventually replace the use of animals for testing purposes. In the late 1990s, biologists at the University of Singapore embarked on a groundbreaking journey after realizing that a synthetic alternative to horseshoe crab's blood clotting agent could be engineered in a laboratory setting. This breakthrough marked the birth of Recombinant Factor C, abbreviated as rFC, a genetically engineered protein that serves as a sustainable and ethical alternative to lysate; however, horseshoe crabs are still being harvested and bled today, as activists lobby the pharmaceutical industry to adopt alternatives [13].

Biological engineering navigates the intricate world of genetic manipulation. Some examples involving the manipulation of animal genes by bioengineers include the cloning of entire organisms such as Dolly the Sheep, in vitro testing and fertilization, the manipulation of animal genomes in 'gene drives' designed to eradicate pests and invasives by altering their genetic material to sterilize them or augmenting their DNA to produce heartier organisms, such as the bioengineering of "supercoral" that will be able to better withstand a warmer ocean, or the genetic modification of goats and rabbits to produce therapeutic proteins in their milk. Genetically engineered pig organs have now been successfully transplanted to human and primate recipients. Forensic scientists can apply the tools of engineering to identify causes of animal death, verify the locational sources of ivory and rhino horn in criminal poaching investigations, and determine whether products come from real endangered species or are synthetic. These applications not only improve animal health but also foster a safe and sanitary environment. In the future, these technologies may be involved in the de-extinction of whole species (such as passenger pigeons or dinosaurs) or programming the DNA of animals to make them more resistant to aging and disease.

Taking a biomimicry approach, biomedical engineers are working at the cutting edge of medical research to extend and improve the quality of both human and animal lives. Materials scientists have developed shark-skin-inspired surfaces, modeled off sharks' dermal denticles, in hospitals to resist bacteria growth. Fine particulate matter filters to syphon pollution have been based off manta ray gill designs, and gecko-inspired adhesives are popular in hospitals. Finally, engineers are also exploring the design of self-healing materials that can repair damage autonomously, without scarring, by modeling the ability of axolotls and salamanders to regenerate their limbs.

3.3.5 Computer Science and Electrical Engineering

On March 15, 2000, sixteen dead beaked whales washed ashore in the Bahamas [15]. They showed signs of acute head trauma and were bleeding from the ears. Marine mammal scientists, having seen this phenomenon before, hypothesized that these whales had most likely been exposed to powerful naval sonar technology, which causes marine mammals to suffer, panic, and beach themselves in distress. It was soon confirmed by the U.S. government that they had been testing advanced naval sonar designed to detect submarines in the vicinity of the whales less than a day before. Naval sonar technology is a biomimicry design product that imitates the sonar sensing abilities of whales, dolphins, and other animals who experience their world through sound and echolocation. The deaths shed light on how naval sonar negatively affects the brains of marine mammals, which through necropsies helped scientists to better understand sonar sensing (Fig. 3.4). For conservation, the mass beaching incident in the Bahamas is remembered as a turning point in public awareness and concern about the dangers of sonar to whales and other marine mammals who also (first) used this technology to 'see' objects underwater by pinging them with sounds. Numerous lawsuits against the U.S. Navy have been filed after many more subsequent mass whale strandings events on United States beaches (e.g., Hawaii, Virginia, North Carolina). While this technology is disruptive when used for military or human commercial purposes, it also holds the promise of unlocking marine mammal communication when used for animal behavior research. Through biomimicry of whale sensing abilities, and the

Figure 3.4 Beached whale (Mesoplodon gray) in 2011 (image by Avenue, Creative Commons license [14]).

study of whale communications using algorithmic analysis, humans are gaining a better understanding of what whales need and want and may one day be able to 'talk' to them [16].

Computer science and electrical engineering can also leverage technology to understand, monitor, communicate with, and protect animals. These disciplines work with artificial intelligence, satellite imaging, and algorithms to decode animal language and enhance our understanding of animal behavior. This includes sensing animal positions, tracing migratory patterns, analyzing population dynamics, and even assessing the social or physical environment of the animal through bodily modification. An example would be the surveying, tracking, and remote sensing of endangered species/habitats to determine possible locations where they may be present or actively under threat.

Computer programs and electrical devices such as interactive video games designed for pigs, or apps that allow cats to play through touching the screen, or electrical button boards dogs can learn to press in sequence can allow humans to appreciate what animals understand and are able to do. In addition, these products can also provide physical enrichment, mental stimulation, and promote well-being of animals, especially in captive settings. While tracking companion pets using devices such as Apple Airtags and microchips can increase their safety, other products such as shock collars can cause physical and mental harm.

3.3.6 Civil and Environmental Engineering

Considered the worst fish in the United States, Asian Carp or 'Copi' are not native to the country. Common carp (from Europe), today established in every U.S. state, were first introduced by the federal government in the mid-1800s to feed residents. However, carp populations quickly grew out of control after environmental managers intentionally brought four new carp species (black, silver, grass, and bighead - collectively known as 'Asian Carp') from China in the 1960s and 70s to serve as 'biocontrols' which is the strategic use of one species to 'naturally' control and limit another species (Fig. 3.5). Problems arose when the Asian carp species quickly escaped the waters they were meant for and outcompeted native fish for resources; this resulted in the creation of federal and state task forces to respond to the problem of this 'nuisance fish.' Since the 1970s, biomedical, civil, mechanical, and environmental engineers have been called in to devise technological solutions to eradicate the Asian carp or engineer them to be more benign. To manage hydrilla grass overgrowth in the Florida Everglades, bioengineers have succeeded in limiting carp numbers but effectively use them as biocontrols through a sterilization process of mixing heat, cold, and shock techniques that produce a third set of chromosomes before the fish are released [18]. By all estimates though, feral and fertile

Figure 3.5 Silver carp jumping out of water (image by Great Lakes Restoration Initiative, public domain [17]).

Asian carp are now established in the U.S. and the battle has already been lost. This is evident in the case of the Great Lakes; in 1900, civil engineers, tasked with preventing pollution from the Chicago River from contaminating Lake Michigan, dug a canal to reverse the flow of the river and divert it to the Mississippi Basin and Gulf of Mexico, connecting these two great watersheds for the first time. However, after carp became a problem in the Mississippi Basin and now threaten the Great Lakes, engineers have tried poison, bubble barriers, and electric underwater fence barriers on the Chicago River, while fishermen apply assaulting shock and sonic techniques to stun and corral the fish. The case illustrates how the work of environmental managers and engineers must be done with caution and the ability to undo potentially negative effects.

Civil and environmental engineers have also played important roles in helping animals flourish by building wildlife-friendly infrastructure and rewilding corridors to facilitate the safe and free movement of animals. One example is the design and construction of green overpasses and underpasses strategically located on highways to allow crossings by highly mobile animals such as bison, mountain lions, elephants, tigers, elk, crabs and terrapins (Fig. 3.6). Engineers have also designed bird-proof glass to prevent the deaths of millions

Figure 3.6 Wildlife overpass on Trans-Canada highway between Banff and Lake Louise, Alberta (image by Wikipedant, Creative Commons license [19]).

of birds each year as they strike reflective glass buildings. Engineers have built 'fish ladders' to allow migrating species like salmon to maneuver up and over large dams to reach their spawning grounds. Finally, governments are in necessary conversations with engineers to design and build structures that can protect important marsh and coastal habitats from sea level rise and the more intense storms connected to climate change. With the window of opportunity closing, installations such as seawalls, dikes, and jetty systems become increasingly necessary to ensure that there is habitat available for geographically-restricted species, such as manatees, to live and/or raise their young.

3.4 Geoethical Reflections on Engineering's Impact on Nonhuman Animals

We realize our brief review of engineering and animals is just that—we have only touched the surface of all the ways the branches of engineering have both helped advance our understanding of other animal species and promoted their well-being, yet also impacted them as individuals and communities in negative ways. As animal geographers we ask engineers to consider not *whether* engineering impacts animals, but rather "*how ought* engineering impact animals?" We would like to return to the four principles of geoethics to reflect on potential ethical gaps we found across the different branches of engineering.

The first principle, **geocentrism**, is the recognition of the moral value of animals, humans, and the rest of nature. Remember, this principle says humans are neither placed outside of, nor above, the rest of the world, but understood as part of it. Whether it is Laika, the dog sent into space, wild horseshoe crabs taken and milked for their blood, or pigs genetically altered to grow faster so they can be slaughtered at younger and younger ages, all branches of engineering have treated animals as means to human ends, instruments and resources for human use. This view that humans have the right to treat other animals as they want because humans are seen as above other animals is, today, understood as human supremacy or speciesism [20, 21]. This worldview organizes all animals within an imagined vertical, taxonomic hierarchy of species. In this hierarchy, humans (ranked as exceptional compared to other species) are at the top, with a few species (primates, dolphins, parrots, etc.) ranked as 'higher', more advanced, or more important than others (such as rats or goats), with the least respected animals located at the bottom (e.g., insects, bacteria, viruses). This hierarchy carries an ethical weight because animals ranked at the bottom of the hierarchy are of least concern and are therefore considered disposable, expendable, and ungrievable; i.e., they can be killed or mistreated without much ethical cost. Engineers can problematically reinforce human supremacy by participating in the design and engineering of new ways to confine, manipulate, slaughter, subjugate, exploit, or harvest (the organs, skins, and other products of) animals, alter and degrade their habitats, and/or imitate animal ingenuity solely for human benefit and/or financial profit. These enormous life-and-death consequences demand critical/ethical examination, with the costs in animal lives and suffering made clear and transparent to the public. Engineers, because of their complex understanding of the world, have an opportunity to help lead society in promoting the moral value of humans *and* animals *and* the natural world.

The principle of **equal consideration** refers to giving equal consideration to the well-being of all animals affected by humans. As we said earlier, this does not mean a sameness of treatment, but it means that engineers need to actively consider animals as both individuals and as members of ecological or social communities. One way we found this consideration lacking across engineering is through the issue of consent. Animals are being used in a variety of ways as test subjects or through harvesting their parts or their entire bodies as stand-ins for human beings. The need for physicians and scientists to secure prior consent before treating or operating on a human patient is a foundational principle in modern medicine. In the examples above, it is evident that scientists and engineers benefit from the use of animals in research; however, animal involvement in these experiments may be considered unethical because the individual animals did not and could not give their *consent,* meaning the needs, desires, and freedom of the animal were not considered at all. Speciesism is evident in the fact that if even just a couple of humans were used in the experiments above, the public would be outraged. Humans recognize that animals are close enough physiologically to stand in as proxies, but we do not grant them mercy from suffering in those experiments. In other words, engineers acknowledge human-nonhuman animal likeness but deny these animals the same consideration. For this reason, engineers must do more to reckon with the fact that billions of animals each year are violently made into sacrifices for human progress using their tools and technologies.

When cases arrive pitting animals against humans, the principle of **hard cases** asks that we do not automatically privilege human needs but seek solutions and compromises that acknowledge and protect the well-being of animals. If engineers design a product that we learn harms animals, even if that was not the original intent, engineers have an ethical obligation to participate in designing solutions; for example, the case of the whales and other marine animals negatively impacted by military sonar. It is not going to work to ask the military to stop defending its territory, but it is also unethical to allow sonar to be used in and around mobile, migrating marine animals. Engineers can be at the forefront of developing solutions in such hard cases because of their abilities to innovate and adapt.

Finally, the principle of **moral carrying capacity** asks us to live in a way that allows the entire geographical community to flourish. When chemical engineers create and release chemicals into the environment, this principle asks that they do so only when they can ensure the safety of the entire ecosystem—which includes humans. The pesticides accumulating in nonhuman animal tissue are also accumulating in human tissues. Chemical and nuclear pollution harms humans and animals. Building highways through animal habitats without regard for their needs causes needless human deaths in the way of car accidents as well as prevents nonhuman animals from sharing our world safely. Engineers understand that bridges and buildings must be designed to balance their loads and use their training in solving complex problems to create long-lasting and safe structures. The principle of moral carrying capacity provides engineers with a framework within which to apply their extensive problem-solving abilities to not only determining how human innovations can do the least harm, but how our technological and built infrastructure might promote the well-being of all animals' lives.

3.4.1 Core Questions for Engineers to Address the Ethical Aspects of Their Impact on Animals

We offer this list of questions as a supplementary guide to the principles of geoethics as you continue to move through your engineering education or your career as a practicing engineer:

1. How might other animals, other species, plants, and the environment itself be better considered and protected across all stages of the engineering process?
2. How can engineers, either as individuals or as a professional group, be held responsible for the tools and technologies they invent, as well as address the problem of using animals without their consent?
3. What processes are in place, or need to be created, for engineers to evaluate whether their work is ethical now as well as in the future?
4. At the earliest stages of design, what processes are in place, or need to be created, for engineers to ethically consider if a particular project or product will contribute to or mitigate animal exploitation/harm for purely human wants and needs?
5. What processes are in place, or need to be created, for engineers to consider the impacts of their work on other animals and their ecological and social communities across all the stages of a product's life (extraction of raw materials, refinement/manufacturing, distribution, consumption/use/deployment, waste/disposal/recycling)?

3.5 Conclusion: Engineering for a Multispecies World

As we have shown from both an animal geography and an engineering perspective, the ethical question of *how we ought to live with and treat our fellow animal species* is a very complicated one. Engineers, like all members of human society, must continuously grapple with how the actions we take and the products we make may help or harm animals and the planet's health. Over time, especially as science has facilitated better human understanding and appreciation of animal abilities and experiences, there has been expanding moral consideration for animals across many societies. This shift has been a direct result of decades of public protests, investigations, whistle-blowing, and direct actions by concerned citizens who have convinced policymakers to change the rules and regulations regarding how animals should be treated and kept. This shift has also been the result of the oftentimes incredible work of scientists and engineers who have created products that help us understand and support the myriad animal species with whom we share our world. By incorporating animal geography's focus on human–animal interactions in places, across scales, and between cultures along with the four principles of geoethics into their work, we believe that engineers can act and lead responsibly as the makers of our shared multispecies future, helping create a world where all beings can flourish to the best of their capabilities.

References

1. Glacken, C. J. 1967. *Traces on the Rhodian Shore: Nature and Culture in Western Thought from Ancient Times to the End of the Eighteenth Century*. University of California Press. https://doi.org/10.1525/9780520343054.
2. Livingstone, D. N. 1992. *The Geographical Tradition: Episodes in the History of a Contested Discipline*. Basil: Blackwell.
3. Urbanik, J. 2012. *Placing Animals: An Introduction to the Geography of Human-Animal Relations*. Lanham, MD: Rowman & Littlefield.
4. Lynn, W., B. Liv, F. Santiago-Ávila, and K. Stewart. 2023. Ethics, Wellbeing and Wild Lives. In *The Routledge Handbook on Human-Animal Interactions, Interventions and Anthrozoology*, A. H. Fine, M. Mueller, Z. Ng, J. M. Peralta, and A. Beck, 438–452. London and New York: Routledge. https://doi.org/10.4324/9781032153346-30.

5. Lynn, W. 1998. Animals, Ethics and Geography. In *Animal Geographies: Place, Politics, and Identity in the Nature-Culture Borderlands*, J. Wolch and J. Emel, 280–298. London: Verso.
6. The Biomimicry Institute — Inspired by Nature. 2024. *Biomimicry Institute*. https://biomimicry.org/. Accessed March 10.
7. Conti, R. 2015. *The Golden Number*. http://blog.illustraciencia.cat/2015/05/phi-nautilus-roberta-conti.html.
8. Mabit1. 2020. *Deutsch: Goldene Spirale Mit Fibonacci-Zahlen*. Own work.
9. Lewis, R. 2023. *The First Living Being in Space*. https://www.britannica.com/topic/Laika.
10. Leslie, H. A., M. J. M. van Velzen, S. H. Brandsma, A. D. Vethaak, J. J. Garcia-Vallejo, and M. H. Lamoree. 2022. Discovery and Quantification of Plastic Particle Pollution in Human Blood. *Environment International* 163: 107199. https://doi.org/10.1016/j.envint.2022.107199.
11. Carson, R. 1962. *Silent Spring*. New York: Fawcett Crest.
12. Association, Seney Natural History. 2010. *1963: Two Eagles Found Dead Were Sent to Patuxent for Pesticide Residue Studies. Probably Some of the First Birds Studied after the Release of Rachel Carson's Book "Silent Spring" Which Points the Finger at DDT Causing the Decline in Predatory Bird Numbers.* Seney National Wildlife Refuge—1963. Uploaded by AlbertHerring.
13. Pavid, K. 2020. Horseshoe Crab Blood: The Miracle Vaccine Ingredient That's
Saved Millions of Lives. https://www.nhm.ac.uk/discover/horseshoe-crab-blood-miracle-vaccine-ingredient.html.
14. Avenue. 2011. *An Adult Female Whale, One of a Group of Five Who Stranded and Died (with the Exception of One Juvenile, Who Was Successfully Refloated) on Sunset Beach at Port Waikato on 13 January 2011. They Are Believed to Be Gray's Beaked Whales (Mesoplodon Grayi), although DNA Testing Is Required to Confirm This. See e.g. Single Whale Survives Stranding, Waikato Times, 13 January 2011. The End of This Whale's Beak Was Buried in the Sand, So It Was Longer Than It Appears Here.* Own work.
15. Keim, B. 2024. The Epic Fight to Protect Whales From the U.S. Navy |WIRED. https://www.wired.com/2014/07/war-of-the-whales/. Accessed March 10.
16. Whale Acoustic Library. 2021. *Ocean Alliance*. https://whale.org/whale-acoustic-library/.

17. Silver Carp Jumping Out of Water | U.S. Geological Survey. 2024. https:// www.usgs.gov/media/images/silver-carp-jumping-out-water. Accessed March 10.
18. Sutton, D. L., V. V. V., Jr., and J. E. Hill. 2013. Grass Carp: A Fish for Biological Management of Hydrilla and Other Aquatic Weeds in Florida. *University of Florida Bulletin 867* 2012. https://doi.org/10.32473/edis-fa043-2012.
19. WikiPedant. 2014. *English: One of Several Wildlife Overpasses on the Trans-Canada Highway between Banff and Lake Louise, Alberta*. Own work.
20. Crist, E. 2019. *Abundant Earth: Toward an Ecological Civilization*. Athens: University of Georgia Press.
21. Singer, P. 1975. *Animal Liberation*. Harper Collins.

Chapter 4

Missing Voices: Non-Human Animals as Stakeholders in Engineering Design

Gary Halada
Department of Materials Science and Chemical Engineering, Stony Brook University Stony Brook, New York 11794-2275, USA
gary.halada@stonybrook.edu

4.1 The Nature of Engineering Risk

The role of the engineer is to respond to a need by building or creating something along a certain set of guidelines (or specifications) which performs a given function. Just as importantly, that device, plan, or creation should perform its function without fail. Of course, in discussing what success means in terms of the engineered world, we must consider for whom an engineered device or system is made, and, perhaps even more importantly, who (or even what) it might affect. Even if it seems to function as intended, an engineered system might be considered a failure (or even a disaster) if it causes irreparable harm to people, society, or the environment.

Animals, Ethics, and Engineering: Intersections and Implications
Edited by Rosalyn W. Berne

ISBN 978-981-5129-56-4 (Hardcover), 978-1-003-65231-1 (eBook)
www.jennystanford.com

Engineering disasters are usually catastrophic failures of a human-made system, structure, process, or material, which are perceived to result in an outcome with a high cost to human life or health, the environment, our communities and societal structures, our industry, or the economy. The potential for disaster is often judged based on the probability or likelihood of failure, the vulnerability of a community, ecosystem or business to failure, and the likely severity of such a failure should it happen. The role of engineers (and others) is to design, create and maintain human-made systems so that the likelihood of failure (which causes a system to not perform its intended function, or to perform its function but with unintended disbenefits or byproducts)—or in other words, risk—is as low as reasonably achievable and so that any failure that would occur would have as benign an impact as possible. To do so, it is critical to understand not only a failure in the traditional engineering sense—such as a collapse of a bridge or crash of a plane—but also how a failure results from the missing voices of stakeholders during the design process or neglect of attention to societal, environmental or cultural values in design, implementation or maintenance of technology [1].

Our discussion of "missing voices" in the design and deployment of engineering designs and solutions has often been considered to be sufficiently ethical and value-rich if we take into account the voices of those in disadvantaged or marginalized communities who are significant stakeholders. For example, it is ethical to consider the needs of native populations in considering the potential negative impact of building a dam or a waste treatment facility near tribal lands or within a rainforest. In this case, we are primarily concerned with the impact on native populations. But what if our values require us to also consider the impact of design on non-human animal species present in the affected zone? How would this broader view of "missing voices" impact our design process, and how we choose to solve engineering problems? This is the focus of this chapter.

4.1.1 Probabilistic Risk Assessment (PRA) in Engineering

In order to better understand the impact of consideration of non-human stakeholders, we can approach the problem from the perspective of probabilistic risk assessment, or PRA. PRA has been a guiding principle in design and deployment of engineered systems since 1975 when it was used in the US nuclear industry to improve the safety and reliability in the fledgling nuclear reactor industry [2]. This makes sense as the catastrophic failure (or leakage or any of a number of less than catastrophic failures) of a technology as potentially risky and dangerous as nuclear power required a refined and rigorous approach.

In essence, we can express the probability of failure (or risk) of a technological design or development as:

$$\text{Risk, or conseqences of risk} = \frac{\begin{array}{l}[\text{Vulnerability} \times \text{Probability of failure} \\ \times \text{Cost (in health, lives, economic and} \\ \text{environmental impact)}]\end{array}}{\begin{array}{l}\text{Mitigation (strategies or technologies} \\ \text{to reduce risk)}\end{array}} \tag{4.1}$$

Clearly, risk evaluated in this sense is largely dependent on not only the nature of the engineering design, including the likelihood of failure of individual or interdependent components [3], but also expresses the central role of vulnerability and cost of failure [4]. The vulnerability of the environment, the economy, or human health to potential failure modes requires a clear understanding of the sensitivities of each of these to the potential outcomes of failure. However, it is not just failure which must be considered in this case. A properly operating and performing engineered system, if the potential harm to stakeholders is not fully taken into account in the designed outputs of the technology, may be equally dangerous and carries the potential to cause irreparable harm. This is clearly the case if we consider the impact of engineered systems and their operations on non-human animals as stakeholders.

Equally important is the "cost" term in the PRA equation. Traditionally, the cost considered is the cost to human life and livelihood (which includes financial costs of failure). In recent years, the role of the impact on environmental costs has been included in risk considerations. When we focus on the cost to the health, procreation, and indeed very existence of non-human species, we ensure the incorporation of risk to non-human animals in the

PRA assessment. Finally, we come to the denominator of the PRA equation, mitigation of risk. For nuclear power plants or other potentially hazardous facilities or machines, mitigation usually takes the form of redundancy, incorporation of fail-safe systems, use of safety factors in design (which prevent ever-present variability in material properties, for example, from leading to catastrophic failure), safety interlocks and warning labeling, "worst-case" design methods, and extensive simulation and testing. To prevent a non-observant operator or bystander from walking too close to a high-temperature piece of equipment, systems such as fencing, warning lights and light curtains may be used. However, this would do little to prevent electromagnetic radiation, including that which would not normally be hazardous to a human being, from disrupting the well-being of nearby wildlife. If we are to consider the risk to non-human animals as stakeholders in design, it is their vulnerability, the cost to them of engineered hazards, and the necessity of broadening our view of mitigation which engineers must consider.

This chapter will explore these aspects through specific case studies focused on the need to reduce vulnerability, lower potential costs, and find forms of mitigation for non-human animal stakeholders.

4.1.2 Factors Critical to Reducing Vulnerability and Enhancing Mitigation

When considering the vulnerability of non-human animal species, it is critical to understand the differences between these species and humans which exacerbate the potential for harm. In particular, we need to focus on differences in how humans and other animals perceive the world, and further understand how perceptions of their environment impact the survival characteristics of non-human species. For example, animals rely on their ability to forage for food, migrate, mate and procreate in order to establish a sustainable population. Furthermore, animals often rely on their interdependence, as predator and prey, as well as on symbiotic relationships, in order to exist. In general, animals are extremely sensitive to changes in their natural environment and habitat. While we are already keenly aware, in many cases, of the impact

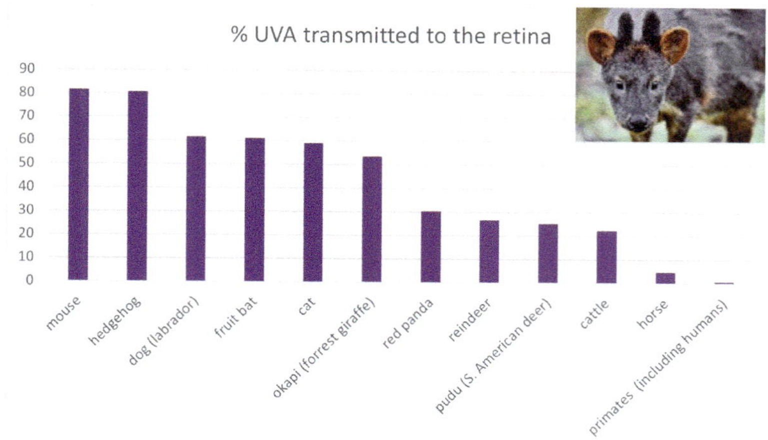

Figure 4.1 Percent transmission of UVA (315–400 nm wavelength) light by the eye of animals, based on data presented by Douglas and Jeffery, 2014 [6]. The figure given for primates, 0.5%, represents an average of seven different species, all of which have eyes which can detect less than 1% of UVA light. Inset shows a Southern Pudu (second smallest deer in the world, about 40 cm tall).

of chemical pollutants and encroaching development (with the concomitant removal of natural resources and potential habitats) on disruptions to animal life, there are subtler or at least lesser known sensitivities which may adversely impact the ability on non-human animals to thrive.

As extensively discussed by Ed Yong in his best-selling book, *The Immense World*, animal species often perceive their environment in very different ways as compared to humans [5]. For example, as shown in Fig. 4.1, based on a recent study of the properties of structures that make up the eyes of a number of animal species, the optical range of animals often differs significantly as compared to humans.

Overall, human eyes can detect about 0.5–0.6% of UVA, but this also can vary with age and conditions with degrade the lens of the eye, where, with the vitreous humor, most of the blockage occurs [7]. Various theories have been proposed for this difference, with a number of different reasons why such differences evolved.

For certain animals who hunt nocturnally, UV sensitivity helps in identifying potential prey. For reindeer, UV sensitivity enables them to better detect predators, such as polar bears.

As almost all terrestrial animals depend on their vision (which, surprisingly, may be less acute than that of humans in many cases) to detect food (prey or plants), possible mates, members of their own communities, possible nesting or hiding places, tell-tale signs to follow during migratory routes, and possible danger from predators or competitors, bright UV light from engineered sources, whether intentional or not, will be potentially disruptive. In general, unusual signals, colors or radiations may frighten animals, resulting in subtle changes to lines of travel and resting.

Likewise, many animals are more sensitive than humans to other aspects of electromagnetic fields, subtle movements in their physical environment, air or water movement, and other factors which human designers of engineered structures and systems may easily not notice and hence will not consider in their engineering designs. A recent OES report on potential hazards associated with marine renewable energy systems (MRE) considered these factors in terms of "stressors and receptors." Stressors are those aspects of engineered systems which can cause harm or disturbance to marine animals and their habitats (mostly considered in terms of vibrations from moving turbine blades, or physical obstructions such as mooring lines, or—as we shall look at more extensively below—electromagnetic radiation from cables that transmit the electricity generated to the shore or other collection points). The receptors are the animals themselves—whether they live primarily in the vicinity of stressors or pass through areas which are impacted by stressors. Receptors may also be aspects of the environment itself, such as ocean currents, which may be impacted by engineered structures and changes to which become additional stressors on marine life. As we will discuss below, mitigation strategies may be focused directly on the removal of stressors from sensitive animal habitats, though where that is not practical, designing methods to reduce the impact of stressors on receptors (for example by placing physical separations or by reducing the stressor effect generated) may represent an acceptable alternative course of action.

4.2 Case Studies

4.2.1 Corona Discharge in High-Voltage Transmission Wires

As described above, a wide variety of animals have visual acuity which extends into the ultraviolet range. In fact, as Yong states, the ability to see in the UV spectrum may be more the norm than not being able to, as is the case for humans.

The extent of high-voltage power transmission lines in the United States is approximately 1 million km, most of which is three-phase alternative current (AC) [8]. Suffice to say that the existing network of these cables traverses nearly every animal habitat and ecosystem in the country, though more dispersed in sections of the American West. There is also a growing need for significant increases in the amount of high-voltage direct current (DC) lines, especially to service the growing number of dispersed solar and wind energy generating sources. Overall, the National Academies predict a tripling of the electrical power grid would be required to meet carbon net-zero needs for alternative energy sources [9].

High-voltage and high-current-capacity transmission lines, such as those which traverse much of the United States, often generate strong emanations in the UV range through a process called corona discharge. Corona discharge occurs around high-tension wires or more likely the connections between wires and insulators when the voltage gradient exceeds a critical value, leading to the ionization of air molecules and partial electrical discharge [10]. Besides the many other damaging effects of corona around high-tension lines (including radio interference, possible damage to insulating materials and nearby conductors, and formation of hazardous chemical species such as ozone and nitrogen oxides as well as other ionized air-borne pollutants), the radiative spectrum produced by corona is primarily in the UV part of the spectrum. Most of the UV light created by corona discharge occurs with a wavelength between 295 and 405 nm, peaking at about 340 nm which places the great majority of emanation in the UVA range [11]. As pointed out above and illustrated in Fig. 4.1, this range is essentially invisible to human eyes (except perhaps to the very young and to those with particular types of damage to the lens of the eye [12]) but is quite visible to the majority of non-human animal species.

The incidence of coronal discharge from high-tension lines is increased by humidity and by increased wind speed [13]. However, other aerosols or particulate matter (i.e., dust) can also impact corona. This can be due to surface roughening due to the accumulation of dust and dirt (a rougher surface can result in increased corona due to the concentration of charge at high aspect ratio sites) or surface imperfections formed by degradation due to the electrical discharge itself. In any case, it is certainly possible that changing weather patterns and extreme weather may result in increased incidents of corona discharge in existing power transmission systems. As many of these high-voltage wires and support infrastructure cross sensitive ecological regions, it is likely that animals conducting normal activity will be impacted by their sensitivity to UV emanations generated by corona. A number of researchers have recently reported on cases where animal life cycles and habits have been adversely impacted by corona discharge and their visual sensitivity to the resulting UV radiation [14–16].

4.2.1.1 Reducing risk: vulnerability

As stated above, vulnerability depends on the exposure of impacted stakeholders, in this case UV-sensitive animal species, to the source of risk. Hence, it is possible to reduce vulnerability with more careful considerations of placement of new high-voltage systems. However, considering the need for more grid capacity as carbon neutrality in energy generation is pursued, it will be increasingly difficult, even if the political will is present and an expanded concept of environmental impact statements is adopted which considers these broader impacts as a high priority. Wireless transmission as well as subsurface buried high-voltage lines are options, but are disruptive in other ways (and the impact on non-human animals of extensive wireless power transmission itself is not well understood). Burying high-voltage lines is particularly expensive, and would result in disruption of habitat in many parts of the country.

4.2.1.2 Reducing risk: mitigation

If moving or segregating high-voltage transmission lines from sensitive habitats is impractical, at least without a yet unknown transformative technology, then strategies for mitigation would be necessary. Several of these methods are currently available (corona rings, coatings, remote monitoring), deployed primarily due to the cost of lost transmission line current and efficiency (rather than to protect the sensitivities of non-human species).

A corona ring (see Fig. 4.2) can be affixed to critical locations (near ends of insulators) on high-voltage transmission lines. Its purpose is to spread out voltage (with a rounded shape so as not to provide sharp points or edges which concentrate charge) so as to reduce the voltage gradient below a critical threshold for corona discharge (as shown in Fig. 4.2). Of course, anything that disrupts the smoothness of the surface of the ring can increase discharge from the ring itself [17]. While these methods can protect the transmission line itself, it also helps to limit the generation of UV light which disrupts animals' activities.

Of course, the cost of adding corona ring (and especially optimized corona rings) to the network of high-voltage transmission lines that cross sensitive areas would be significant, as would

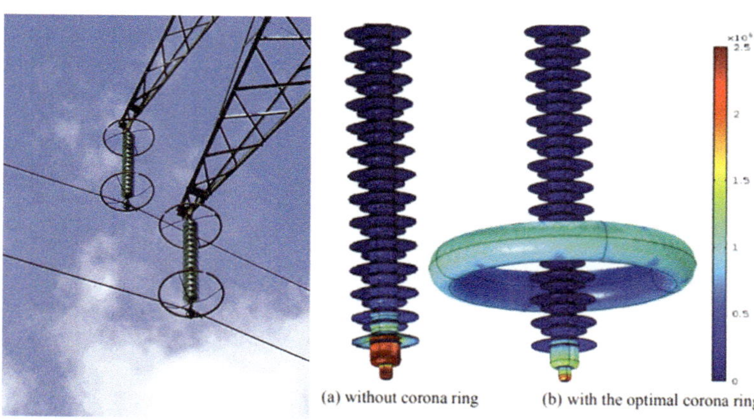

Figure 4.2 Picture of corona rings (left, from Wikipedia Commons) and computational model showing the impact of corona ring on electric charge (right, reprinted with permission from [18]).

repairing, cleaning or replacing connectors where they are found to be likely to produce corona discharge. However, one potential mitigation would be to use drones equipped with UV and thermal detection equipment to identify specific faults along HV transmission infrastructure, allowing for targeted repairs and adaptation to limit UV radiation [19]. Such equipment exists, and may be purposed for these applications.

Other methods for designing HV transmission lines to better distribute voltage density and limit UV radiation include bundling cables, with periodically spaced separators, and removal and replacement of sharp-edges connectors, clamps, and other components on or near HV lines. Other engineering fixes have been suggested, including anti-corona coatings (e.g., room temperature vulcanizing (RTV) silicone coating [20]), which can delay the onset of corona discharge with increasing voltage, but not stop it entirely, especially if the coating weathers or becomes cracked. Coatings can lose their hydrophobicity (necessary to maintain properties) with time or become brittle due to electrochemical processes [21]. It is also possible that corona discharge occurring near a coating will significantly damage nearby coated areas. Any of these effects results in a need for cleaning and recoating, again a labor-intensive process and potentially expensive process.

Overall, careful siting of high-voltage lines, redesign (including addition of corona-preventing rings and coatings) and careful inspection and maintenance would be necessary to limit the impact on non-human animal species.

4.2.2 Underwater Electric Fields from Power Transmission Cables

As another example of how emanations from engineered systems can impact non-human stakeholders, let us consider the impact of power cables. In particular, while we could easily discuss land-based power transmission infrastructure as in the previous example, let us instead consider power cables which run on or near the ocean floor.

As of 2019, there were over 1.2 million kilometers of undersea power cables installed worldwide (there are far more communication cables, but as these do not emit much of an electromagnetic field, we will not be discussing those [22]), and the total length predicted to be in use by 2030 is 2.5 million km [23]. A large percentage of this growth is driven by the offshore wind industry. As wind turbine networks are built off the coast of countries throughout the world, power cable networks will be built on the ocean floor to interconnect turbines, tie turbines to the power grid, and transport power to land-based battery storage facilities. As wind power needs grow, turbines will be placed not only near shoreline locations, anchored to the ocean floor, but will more likely be placed on floating platforms, ballasts and moorings, allowing for denser construction and optimization of location further out at sea where winds may be stronger and more constant [24]. This will require subsea power cables that are longer and have a greater current-carrying capacity.

Cables can be buried beneath the ocean floor or be tethered to it (and rest on the bottom). Positioning is critical for the determination of the impact of electric fields emitted by cables on sea life. Burying cables, while safer for marine life, can result in potentially damaging temperatures as heat dissipation is limited by soil properties. Most cables are actually bundles of conductors (usually three main sub-bundles) surrounded by insulating material, which makes heat dissipation critical.

4.2.2.1 Sensitivity of marine animals to electric fields

A number of marine organisms are very sensitive to electromagnetic radiation. For example, members of *Elasmobranchii*, a subclass of *Chondrichthyes* or cartilaginous fish, including modern sharks, rays, skates, and sawfish, are electroreceptive in that they can detect the subtle electromagnetic fields created by their prey, as well as by the movement of seawater itself through the earth's magnetic field. One study showed that dogfish and blue sharks can detect, and in fact orient their direction of motion, to signals with voltage gradients of only 5 nanovolts per centimeter [25]. Sharks and other cartilaginous fish have hundreds of receptors, named Ampullae of Lorenzini (Fig. 4.3), on their heads and jaws which allow them to detect prey and orient themselves when biting [26]. These are tiny glycoprotein filled sacs or bulbs with semiconductor-like properties. Differences

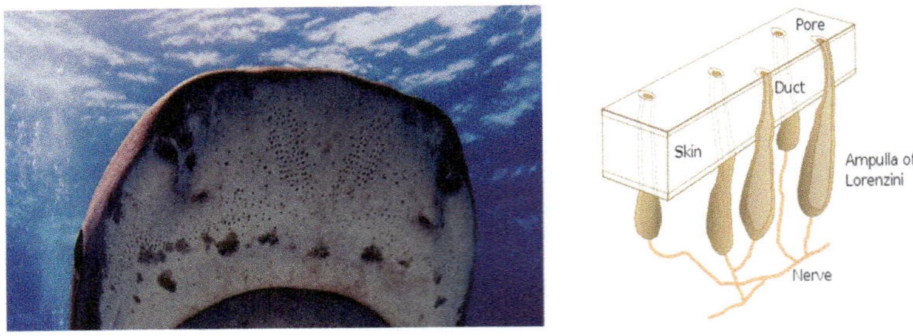

Figure 4.3 Pores of Ampullae of Lorenzini (illustrated, right) in snout of tiger shark (from Albert Kok, CC BY-SA 3.0 <https://creativecommons.org/licenses/by-sa/3.0>, via Wikimedia). Illustration on left shows gel-filled pores in skin with nerve bundles at base (from Chiswick Chap—Own work, CC BY-SA 4.0, https://commons.wikimedia.org/w/index.php?curid=117476439).

in voltage between the pore opening at the interface with water and the base of the bulb is relayed to specialized nerves via polarization of a membrane caused by movement of calcium and potassium ions. This is somewhat akin to the way our nerve cells transmit signals in the body.

Studies have shown that the electric field sensitivity of the pores on the juvenile scalloped hammerhead sharks *Sphyrna lewini* and sandbar sharks *Carcharhinus plumbeus* as they orient to dipole fields (and subsequently try to bite the source of the signal) in the water may be as low as 0.5–1 nV/cm [27]. For perspective, this is an approximately 100 million times weaker signal than that produced by a typical cell phone when you make a call. It is no surprise then that electric fields emanating from subsea cables can distract and in fact disrupt the lifecycle of marine animals. Other creatures besides the aforementioned sharks also are quite sensitive to electric fields and hence may also experience disruptions or misleading signals as a result of the presence of power cables or other electric field sources. These include turtles, mollusks, crustaceans, some freshwater fish such as catfish, lamprey and sturgeon, and even the duck-billed platypus, sensitive to low electric fields such as that produced by tail flips in freshwater shrimp, due to its need to find buried food in murky waters [28]. In addition to detecting the bioelectric field generated by prey (and indeed all living organisms), electroreceptors are used by sharks and other marine life for navigation. Electric fields are generated in the ocean by water flowing through the vertical component of the earth's magnetic field, resulting in electric fields on the order of 5 to 500 nV/cm, which marine animals can use for orientation [29].

While there are a limited number of studies specifically analyzing the impact of electric fields from subsea cables on marine life, the ones which have been published show a definite effect which may be deleterious. Crayfish have been shown to be impacted by electric fields near the sea floor, and burrowing invertebrates in general may be adversely impacted by marine electric fields [30]. In addition, the construction and placement of cables may also disrupt marine life, as does chemical pollution and noise generated, as well as the magnetic fields which generally accompany electric fields [31]. One study showed that the benthic shark avoided electric fields of 10 microvolt/cm (close to 1000 times the sensitivity of the Ampule of Lorenzi pores), which is the maximum predicted field emitted from typical 3-core undersea 150 kV, 600 A cables [32]. Dogfish were attracted to an electric field of 0.1 μV/cm, which is consistent with the predicted bioelectric field emitted by prey species. In any case, whether attracted or repelled by the electric field near subsea power cables, the disruption to the normal behavior of marine animals is an important concern.

4.2.2.2 Reducing risk: vulnerability

The growth in offshore wind (as well as other power transmission systems) will clearly cause an impact on marine life which uses electroreceptors for hunting, navigation, mating, communication, and orienting. If we consider these marine animals as stakeholders in the design process, certain considerations in placement and installation may become important. This may be a doable mitigation in sensitive habitats where new systems are being installed, but it may be cost prohibitive to retrofit cabling in existing installations. Where existing installations are located, the use of sensing and monitoring equipment may at least be able to indicate where, due to lack of insulation or movement of cables, electric field leakage is greatest.

Figure 4.4 Cross-section of typical subsea three-phase power cables [34](left) and typical array of offshore wind turbines (from NSW Submarine power).

4.2.2.3 Reducing risk: mitigation

Most subsea power cables associated with offshore wind turbines are AC current, three-phase systems using three conductors and heavily insulated lines (see Fig. 4.4). An important effect of alternating current is the induction of electric fields in salt water outside of the cable due to the varying magnetic fields generated by high current capacity lines. While typical insulation as shown can effectively block electric fields generated by internal conductors, the magnetic fields cannot be effectively blocked, resulting in the generation of electric fields (much like the electric fields generated by seawater moving relative to the earth's magnetic field) [33].

The strength of the magnetic field is inversely proportional to the distance from the cable, meaning that the strength of the induced electric field also reduces with distance to the cable, though can reach detectable levels some distance (at least 10 m) from a cable. Interestingly, the strength of the magnetic field (and hence induced electric field) increases proportionally to the flow of seawater past the cable. These induced fields may be reduced by burying the cables at least 1 meter beneath the ocean floor, though burial is not possible in rocky areas, and some section of the cable may not be buried for some part of its length due to the need to connect cables to junction boxes and wind turbines. Even at a distance of 10 meters from a 3-conductor, 100 A carrying cable of standard construction, the induced electric field in typical-salinity seawater (and not taking water flow into account) has been calculated to be approximately 350 nV/cm, easily within the range of the sensitivity of a shark, for example [35]. Higher current (and most cables are designed to carry 400 A or greater) will generate larger magnetic, and hence electric, fields.

Mitigation of the impact of the electric field induced by subsea power cables typically takes the form of burying cables more deeply. However, while burying cables at a depth of 10 m, (which is not typical as most cables are closer to the surface due to cost and construction difficulties) can reduce the induced electric field somewhat, it still may not be effective, especially as the current carrying capacity of cables increases. It is also possible to reduce the distance between conductors in the cable, but this can create other problems with heat dissipation and potential shorting due to damage. Another solution may be to use magnetic shields, similar to that used in buried cables on land. This usually takes the form of aluminum, copper or ferromagnetic shielding, with a variety of box-shape configurations, optimized via computational simulation [36].

Finally, it may also be advisable to find ways to keep marine organisms at a greater distance from subsea power cables. A metallic or composite cage could be placed around cables, but installation of sufficiently large cages or barriers (keeping marine animals at least 10 or more meters from cables, including those buried beneath the ocean floor) would be potentially expensive. Yet this level of separation may be the best choice until some additional research on the design of power transmission cables provides a better solution.

4.3 The Path Forward

The challenge facing engineers today (as well as scientists, business leaders, politicians, government regulators, policy makers, economists, scientists, and the public) is to develop methodologies and policies which recognize the needs of all stakeholders, balance those needs with the need for sustainable and societally healthy progress in technology and human development, and implement procedures to create a beneficial future for all. This is clearly a broad and demanding mandate, and one with many layers of complexity. While we cannot address this level of complexity in a brief summary, there are some methods of ethical engineering which may point the way toward protecting our non-human stakeholders in a technologically advanced engineered future.

4.3.1 Value Sensitive Engineering Design

Value-sensitive engineering design (Fig. 4.4) is a process for considering the role the values of engineers play in identifying a project or problem to be solved, clearly analyzing potential negative byproducts or disbenefits of a proposed solution or design, and determining what defines success for the outcome of that design or problem-solving task. Batya Friedman has written extensively on value-sensitive design, in particular for software development [37]. More recently, for example, Ron Sandler at Northeastern University has discussed value-sensitive design specifically for the development of nanotechnology [38]. He expanded on the three guiding principles (see Fig. 4.5) to include specific questions an engineer should ask themselves when considering the ethical basis of their work.

If we apply this reasoning to the development of technological systems and solutions which take into account the needs of and potential harm to non-human animals, we can use it as a guideline for potential solutions. For example, as we consider the nature of the problem to be solved and our choices as engineers, some will choose to directly address the design of systems which avoid the production of radiation or fields which might not impact humans directly but which do limit the ability of animal species to thrive. Likewise, engineers can uphold in their designs and work the value of not generating damaging byproducts which would harm animals or disrupt their lives. When an engineer considers how they define success in engineering design, creating solutions which consider, as best as possible, the needs of all stakeholders, human and non-human alike, could be a crucial part of that definition. If this sort of thinking is kept in focus during the problem-solving process, reducing vulnerability of animals and mitigating potential negative

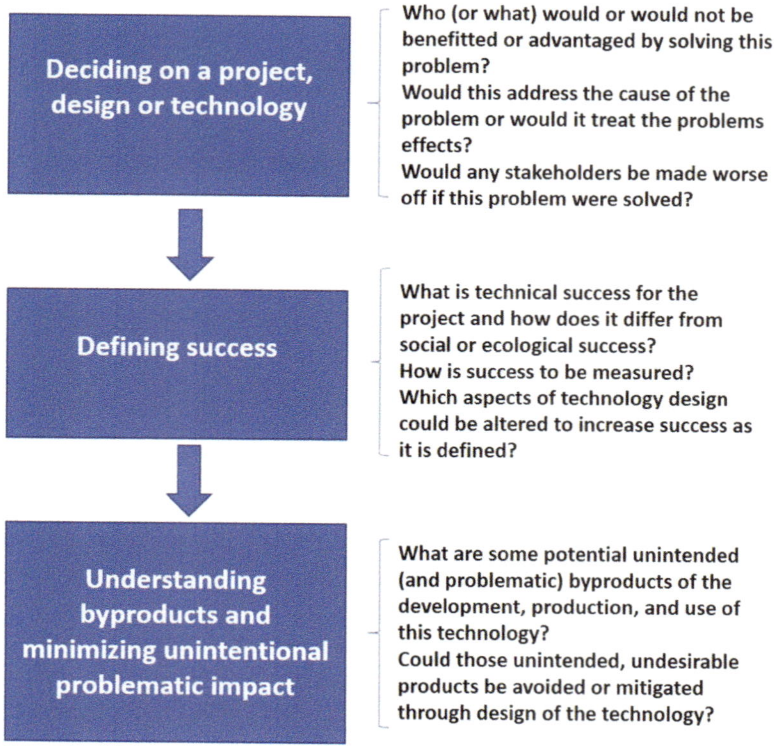

Figure 4.5 Value sensitive design process, adapted from R. Sandler [38].

impacts and hence reducing engineering risk will be well-integrated into the engineering design process.

4.3.2 Adaptive Management

Adaptive management is a process developed for conservation and resources management and is especially valuable when mitigation of an engineering risk is required but significant uncertainties exist concerning the impact and possible sources of risk from engineering systems under development. It may be especially valuable for the reduction of risk to non-human animals from emerging threats with poorly defined or controlled sources (such as UV emissions from powerlines or electric fields near subsea

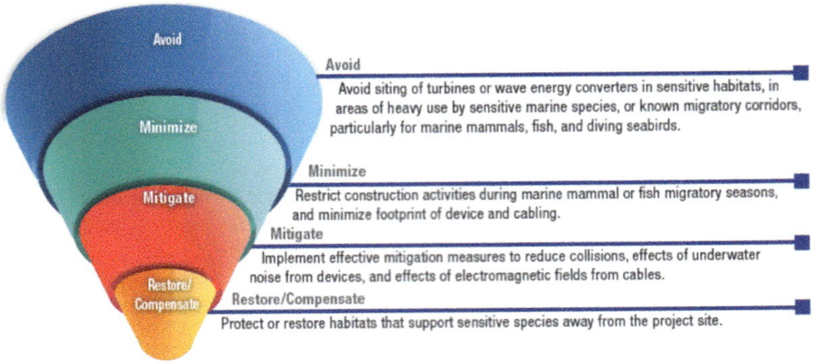

Figure 4.6 An example of a mitigation hierarchy used to avoid impacts when possible, minimize remaining impacts, mitigate to diminish impacts, and provide compensation for unavoidable impacts [39].

cables). This approach involves a hierarchal approach to mitigation which considers avoidance of the risk (for example through the consideration of non-human stakeholders in the determination of where to construct high-voltage transmission towers); minimizing exposure to UV light or electric fields through modifications of power lines and subsea cables or use of shielding; and restoration of habitats where it is possible to move or modify existing structures. Figure 4.6 below shows an example of this process proposed for the marine renewable energy industry, specifically considering the impact of offshore wind turbines.

4.3.3 Engineering Education

*********Clearly, the process of educating the next generation of engineers requires training that is ethical, value-conscious, and inclusive in consideration of the voices of all stakeholders in engineering design and problem solving. This is true whether we consider the needs of marginalized communities, socioeconomically diverse populations, and non-human animals as well as the ecosystem as a whole. This can take a number of forms. For example, faculty can modify existing coursework to include teaching of values through case studies and open-ended problem solving which incorporates risks to non-human animals. Capstone design courses can emphasize the need to consider the full range of stakeholders and the broader needs of an ecosystem facing growing risk and challenges. In some cases, it may be possible to build engineering coursework around study of engineering risk, vulnerability and mitigation strategies. For example, the author of this chapter has developed and taught for a number of years an online course entitled "Learning from Engineering Disaster" at his home institution [40]. While the course initially focused on forensic engineering and utilized "classic" engineering disasters as case studies (e.g., the Hindenburg, the Challenger), the course has been redesigned to include a focus on engineering failure which results from missing voices in the design process—from failure due to a lack of gender considerations in design of automobiles and voice assistants to the impact of racial bias on urban design and software used by the justice system and siting of hazardous chemical processing facilities in marginalized communities [41]. The course content has recently been further expanded to consider the impact of climate change as an engineering threat multiplier, especially in disadvantaged communities and developing nations, and most recently to include the case studies described in this chapter.

In conclusion, when considering the voices of non-human stakeholders (as in the case of climate change impacts), competing interests, such as the driving forces of economics and global competition, as well as the uncertainty surrounding the impact of byproducts of engineered systems should not prevent implementation of mitigating solutions and corrective actions. The ability of scientific research on sensitivity on non-human species; ever more sensitive technology capable of remote data collection for monitoring the health of flora and fauna; deep machine learning algorithms to detect impacts of technology on animals; and the will of an informed citizenry and enlightened educators will hopefully lead the way to an engineered world which fully considers the needs of the animal kingdom. As famously stated by the Harvard biologist E.O. Wilson, "The problem is that we have paleolithic emotions, medieval institutions, and god-like technology." It is in large part up to the current and next generation of engineers to determine what kind of god we will be.

References

1. Halada, G.P. 2010. Teaching by Disaster: The Ethical, Legal and Societal Implications of Engineering Disaster, *Proceedings, American Society for Engineering Education, Middle Atlantic section ASEE Annual Conference,* October 15–17, Villanova University, PA., accessed at https://peer.asee.org/teaching-by-disaster-the-ethical-legal-and-societal-implications-of-engineering-disaster.pdf
2. Keller, W., and M. Modarres. 2005. A Historical Overview of Probabilistic Risk Assessment Development and Its Use in the Nuclear Power Industry: A Tribute to the Late Professor Norman Carl Rasmussen. *Reliability Engineering & System Safety* 89: 271–285. https://doi.org/10.1016/j.ress.2004.08.022.
3. Armen, H., S. Flumerfelt, G. P. Halada, and F.-J. Kahlen. 2011. Complexity and Consequence. *Mechanical Engineering* 133: 46–49. https://doi.org/10.1115/1.2011-DEC-6.
4. Flumerfelt, S., G. Halada, and F.-J. Kahlen. 2012. Complexity by Design. *Mechanical Engineering* 134: 29–33. https://doi.org/10.1115/1.2012-MAR-1.
5. Yong, E. 2022. *An Immense World: How Animal Senses Reveal the Hidden Realms Around Us*. Random House.
6. Douglas, R. H., and G. Jeffery. 2014. The Spectral Transmission of Ocular Media Suggests Ultraviolet Sensitivity Is Widespread among Mammals. *Proceedings of the Royal Society B: Biological Sciences* 281. Royal Society: 20132995. https://doi.org/10.1098/rspb.2013.2995.
7. Nunomura, C. Y., and S. Julio de Faria E. Sousa. 2022. Ultraviolet Radiation and the Human Eye *Arquivos Brasileiros De Oftalmologia*: S0004-27492022005010214. https://doi.org/10.5935/0004-2749.2021-0354.
8. Transmission—NEMA Electrification Infographic. 2024. https://energytransition.nema.org/transmission/#. Accessed February 20.
9. *Accelerating Decarbonization of the U.S. Energy System*. 2021. Washington, D.C.: National Academies Press. https://doi.org/10.17226/25932.
10. Yahaya, E. A., T. Jacob, M. N. Nwohu, and A. A. Sadiq. 2013. Power loss due to Corona on High Voltage Transmission Lines. *IOSR Journal of Electrical and Electronics Engineering* 8: 14–19. https://doi.org/10.9790/1676-0831419.
11. Riba, J.-R. 2022. Spectrum of Corona Discharges and Electric Arcs in Air under Aeronautical Pressure Conditions. *Aerospace* 9: 524. https://doi.org/10.3390/aerospace9090524.

12. Boettner, E. A., and J. Reimer Wolter. 1962. Transmission of the Ocular Media. *Investigative Ophthalmology & Visual Science* 1: 776–783.
13. Matthews, J. C. 2012. The Effect of Weather on Corona Ion Emission from AC High Voltage Power Lines. *Atmospheric Research* 113: 68–79. https://doi.org/10.1016/j.atmosres.2012.03.016.
14. Tyler, N., K.-A. Stokkan, C. Hogg, C. Nellemann, A.-I. Vistnes, and G. Jeffery. 2014. Ultraviolet Vision and Avoidance of Power Lines in Birds and Mammals. *Conservation Biology* 28: 630–631. https://doi.org/ 10.1111/cobi.12262.
15. Copping, A., and L. Hemery. 2020. *OES-Environmental 2020 State of the Science Report: Environmental Effects of Marine Renewable Energy Development Around the World. Report for Ocean Energy Systems (OES).* PNNL–29976, 1632878. https://doi.org/10.2172/1632878.
16. Gill, A., and M. Desender. 2020. *2020 State of the Science Report, Chapter 5: Risk to Animals from Electromagnetic Fields Emitted by Electric Cables and Marine Renewable Energy Devices.* PNNL–29976CHPT5, 1633088. https://doi.org/10.2172/1633088.
17. Chatterjee, S., and Udayakumar. 2012. Design of Corona Control Rings for EHV/UHV Line Hardwares and Substations. In *2012 IEEE 10th International Conference on the Properties and Applications of Dielectric Materials*, 1–4. https://doi.org/10.1109/ICPADM.2012.6318996.
18. M'hamdi, B., M. Teguar, and A. Mekhaldi. 2016. Optimal Design of Corona Ring on HV Composite Insulator Using PSO Approach with Dynamic Population Size. *IEEE Transactions on Dielectrics and Electrical Insulation* 23: 1048–1057. https://doi.org/10.1109/TDEI.2015.005383.
19. Brady, J. 2024. Corona, Utility Practice & Experience, *INMR Weekly Technical Review*, https://www.inmr.com/infrared-uv-inspection-of-overhead-lines/, published July 5, 2024.
20. Xu, Z., and R. Li. 2013. Research on the Anti-Corona Coating of the Power Transmission Line Conductor. *Energy and Power Engineering* 05: 148–150. https://doi.org/10.4236/epe.2013.54B028.
21. Gorur, R. S., and B. S. Bernstein. 1998. Field and Laboratory Aging of Polymeric Cable Terminations. II. Laboratory Aging. *IEEE Transactions on Power Delivery* 13: 323–329. https://doi.org/10.1109/61.660897.
22. Meisner, K., H. Schabelon, J. Bellebaum, and H. Sordyl. 2006. Impacts of Submarine Cables on the Marine Environment: A Literature Review. *Federal Agency of Nature Conservation/Institute of Applied Ecology Ltd.*
23. Submarine Power Cables Market Forecast 2020-2030: Visiongain. 2024. *https://www.visiongain.com/.* https://www.visiongain.com/report/

global-submarine-power-cable-market-report-2019-2029/. Accessed March 9.

24. Christopher, B. 2023. Powering Offshore Wind Farms With Numerical Modeling of Subsea Cables, *IEEE Spectrum*, https://spectrum.ieee.org/hellenic-cables-offshore-wind, published Jan 25, 2023.
25. Kalmijn, A. J. 1982. Electric and Magnetic Field Detection in Elasmobranch Fishes. *Science* 218. American Association for the Advancement of Science: 916–918. https://doi.org/10.1126/science.7134985.
26. Hofmann, M.H., and A. Ferrell. 2011. Physiology of Ampullary Electrosensory Systems. In *Encyclopedia of Fish Physiology: From Genome to Environment*. Academic Press.
27. Kajiura, S. M., and K. N. Holland. 2002. Electroreception in Juvenile Scalloped Hammerhead and Sandbar Sharks. *Journal of Experimental Biology* 205: 3609–3621. https://doi.org/10.1242/jeb.205.23.3609.
28. Langner, G., and H. Scheich. 2009. Electric Senses in Monotremes: Electroreception and Electrolocation in the Platypus and the Echidna. In *Encyclopedia of Neuroscience*, ed. M. D. Binder, N. Hirokawa, and U. Windhorst, 1056–1060. Berlin, Heidelberg: Springer. https://doi.org/10.1007/978-3-540-29678-2_2919.
29. Kramer, B. 1996. *Electroreception and Communication in Fishes*: Progress in Zoology, Vol 42, Gustav Fischer.
30. Albert, L., F. Deschamps, A. Jolivet, F. Olivier, L. Chauvaud, and S. Chauvaud. 2020. A Current Synthesis on the Effects of Electric and Magnetic Fields Emitted by Submarine Power Cables on Invertebrates. *Marine Environmental Research* 159: 104958. https://doi.org/10.1016/j.marenvres.2020.104958.
31. Taormina, B., J. Bald, A. Want, G. Thouzeau, M. Lejart, N. Desroy, and A. Carlier. 2018. A Review of Potential Impacts of Submarine Power Cables on the Marine Environment: Knowledge Gaps, Recommendations and Future Directions. *Renewable and Sustainable Energy Reviews* 96: 380–391. https://doi.org/10.1016/j.rser.2018.07.026.
32. Gill, A. B., and H. Taylor. 2001. The Potential Effects of Electromagnetic Fields Generated by Cabling between Offshore Wind Turbines upon Elasmobranch Fishes, *Tethys.pnnl.gov*, report 488, UK: Countryside Council for Wales, 2001, accessed at https://tethys.pnnl.gov/sites/default/files/publications/Gill-and-Taylor-2001.pdf.
33. Gill, A. B., Y. Huang, J. Spencer, and I. Gloyne-Philips. 2012. The Potential Effects of Electromagnetic Fields Generated by Cabling Between Offshore Wind Turbines upon Elasmobranch Fishes |Tethys. In . London, UK: COWIRE.

34. El Fahem, M. 2017. Key considerations in designing subsea power cables, *EMWorks*, https://www.emworks.com/blog/high-voltage/3-phase-high-voltage-submarine-power-cables, published June 8, 2017.
35. Olsson, T., A. Larsson, P. Bergsten, and J. Nissen 2010. Impact of Electric and Magnetic Fields from Submarine Cables on Marine Organisms: The Current State of Knowledge. *Sustainable Energy Authority of Ireland, Vattenfall Power Consultant*, accessed at https://www.seai.ie/sites/default/files/renewable-energy/ocean-energy/foreshore-lease/Appendix-4-Impact-of-electric-and-magnetic-fields.pdf.
36. Machado, V. M. 2012. Magnetic Field Mitigation Shielding of Underground Power Cables. *IEEE Transactions on Magnetics* 48: 707–710. https://doi.org/10.1109/TMAG.2011.2174775.
37. Friedman, B., and P. H. Kahn. 1992. Human Agency and Responsible Computing: Implications for Computer System Design. *Journal of Systems and Software* 17. Computer Ethics: 7–14. https://doi.org/10.1016/0164-1212(92)90075-U.
38. Sandler, R. 2012. Value Sensitive Design and Nanotechnology. In *Debating Science: Deliberation, Values, and the Common Good*, 205–225. Amherst, NY: Prometheus Books.
39. Lievre, L., and Celia. 2020. *Chapter 12 Supplement. Adaptive Management Related to Marine Renewable Energy*. PNNL-29976CHPT12SUPP. Pacific Northwest National Laboratory (PNNL), Richland, WA (United States). https://doi.org/10.2172/1633207.
40. Halada, G. 2017. Learning from Engineering Disasters: A Multidisciplinary Online Course. In *2017 ASEE Annual Conference & Exposition Proceedings*, 28612. Columbus, Ohio: ASEE Conferences. https://doi.org/10.18260/1-2–28612.
41. Halada, G. 2023. Redesign of an Engineering Failure Course to Incorporate Learning Objectives in Diversity, Ethics and Inclusivity, *Proceedings of the 2023 ASEE Annual Conference & Exposition*, accessed at https://peer.asee.org/44045.

The Doorbell for Fish

In the springtime, many species of fish flock through the city of Utrecht, located in the Netherlands. Perch, bream, roach, and many other species of fish are traversing these waters to find a safe place to reproduce and lay their eggs. On their journey, the fish run into a major obstacle: humans. To control water levels and travel in the city canals, the Weerdsluis lock was put into place. It successfully performs its job, isolating boats and changing the water level to match the boats' direction of travel, but the fish are prevented from continuing to their spawning ground. Locals recognize the importance of this marine migration and have come up with a very creative way to help the animals pass.

A camera, craftily placed in the water along the wall of the Weerdsluis, is angled to capture any fish that might be waiting at the lock. The footage from this camera is linked to a website, which can be accessed by anyone in the world. Locals can watch this camera, and if they see a fish, can press the "doorbell" next to the camera livestream on the website. Pressing the doorbell alerts the lock operator to a fish sighting, and he can investigate further. If fish have collected to pass, he can open the lock, allowing them to pursue fresher waters and reproduce.

Not only does this create a working relationship between technology, humans, and non-human animals, but it promotes awareness of the effect we have on animals and how we can create an ecosystem of balance. Additionally, the "doorbell" serves to collect data on the species of fish in the waters of Utrecht. Over the years, the data from thousands of picture submissions from the doorbell can be analyzed to watch for major fluctuations in fish population. It is the perfect example of how intersections of engineering and wildlife can restore equilibrium between humans and non-human animals.

Citation

1. Visdeurbel. 2024. About the fish doorbell. *visdeurbel.* https://visdeurbel.nl/en/about-the-fish-doorbell/. Accessed May 9.

Chapter 5

Bridging the Divide: Toward a Holistic Approach to Animal Ethics

Philipp Aerni
Center for Corporate Responsibility and Sustainability (CCRS), School of Management Fribourg, HES-SO, 8005 Zurich, Switzerland
philipp.aerni@ccrs.ch

The contemporary debate on animal rights has been shaped by ethicists who apply the principles of utilitarian and deontological ethics to non-human animals. Even though principle-based ethics may provide guidance to citizens concerned about animal rights, it is not very helpful for practitioners who are committed to improving animal welfare in the livestock industry. They resort instead to a pragmatic approach to ethics recognizing that normativity is implicit in practice-oriented human interactions rather than a separate form of reasoning. Spinoza's virtue ethics is based on this insight. It also provides valuable ethical guidance for citizens in search of a more holistic understanding of ethics that is rooted in an adequate understanding of human nature.

Spinoza was aware that human as well as non-human animals are sentient beings. Yet, he was opposed to extending human rights to animals since this would amount to a futile attempt to moralize the amoral order of nature. In this context he may have agreed

Animals, Ethics, and Engineering: Intersections and Implications
Edited by Rosalyn W. Berne

ISBN 978-981-5129-56-4 (Hardcover), 978-1-003-65231-1 (eBook)
www.jennystanford.com

with many contemporary ethicists who believe that a commitment to improving animal welfare does not require an "anti-speciesism" mindset but must take advantage of the human ability to jointly address concrete problems. Such an ethic of meliorism aims at concrete steps to minimize animal suffering in the short run while enabling institutional and technological change designed to enhance animal welfare in the long run.

5.1 Introduction

In a recent publication, the ethicist Per Bauhn [1], who grew up on a pig farm, explains in his foreword how he had to reconcile his strong commitment to be kind to animals with the fact that they live on the farm for a certain purpose. He kept a piglet as a pet and eventually realized that neither the piglet nor his family were really happy about it. Bauhn noticed that the emotional attachment was real, but his moral attitude proved to be primarily a human projection that came into conflict with other moral responsibilities of a farm family. This experience led him to question whether non-human animals are agents who should have the same rights as humans. Even though he regards wanton cruelty against non-human animals as morally wrong for reasons related to the concept of justice, he does not see the need to assign rights to non-human animals themselves.

This self-critical form of reasoning has a long tradition in ethics in general and animal ethics in particular, as will be outlined in this chapter. Yet, it tends to be discarded in the polarized contemporary debate on animal rights. Bauhn's attempt to reconcile the practical with the ideal world tends to be denounced by contemporary animal rights advocates as an uncritical endorsement of the preferential moral treatment of humans over other species. This bias in moral consideration is called "speciesism" and compared by many ethicists as being on par with racism and sexism [2–4]. In other words, research on animal rights in general and the field of critical animal studies (CAS), in particular, tend to frame the discourse by starting from an alleged unchallenged certainty that speciesism is used as a means of justification of the "oppressor" (human animal) to exploit the "oppressed" (non-human animal). The next step is then to explore how this "wrong" can be "righted" (positive critique) from a normative standpoint [5]. This approach skips, however, the core of any scientific method, which starts with a commitment to critical inquiry that may result in insights that challenge one's personal views or an established hypothesis that was assumed to be true. John Dewey [6–8] and more recent representatives of pragmatist ethics and normative naturalism argue that ethics must be understood primarily as a process of critical inquiry [9–12]. The goal of this inquiry is to discover solutions to problems. Identifying a problem is all about a perceived mismatch between the real situation (bad) and the desirable situation (good), which already contains a normative claim. The second step is to find out how to achieve the desirable end effectively. If there are established practices that proved empirically to be effective in achieving the end, then they represent a prudential norm in the sense that you ought to do things this way rather than that way to get the problem solved. In this context, it is important to note that this practical reasoning is no longer made by single individuals but is reflected in professional norms and institutions associated with particular practices [13]. Such prudential norms are primarily what Kant meant with the Hypothetical Imperative, motivated by the desire to achieve a particular purpose. But what about the means applied to achieve the end and is the end really what society as a whole considers good and desirable? These may be the real normative questions, yet, again, the best approach to ensure that the applied means produce as minimal harm to third parties as possible and that as many people as possible benefit from the ends achieved is again based on inquiry. That is why normative claims share the same epistemology as empirical ones, both obtain their warrant or justification from empirical outcomes [9]. As for desirable ends there is a large societal consensus that covering essential human material (food security, shelter, human health, personal safety) and immaterial needs (justice, fairness, inclusiveness, respect, etc.), on this planet is a "good" end. After all, the opposite is linked to human suffering and therefore considered to be a problem or a "bad." In addition, there is also the critical question to what extent fulfilling essential human material needs creates unnecessary suffering and death of non-human animals and how this undesirable impact may be minimized or avoided through the use of innovative practices and new technologies but also through awareness building on the side of producers as well as consumers.

But translating these good desires into reality is only possible through the world of practice involving scientists, veterinarians, livestock farmers and decision makers across the value chain, who must be incentivized to embark on collective action to address a desirable public end in an effective way and to secure the financial means to cover for the costs. At the same time, they also have a moral responsibility towards the people working in the value chain who must earn a decent salary that allows them to meet their moral duties to secure a living for their families.

This pragmatist approach takes into account these multiple responsibilities and the hard trade-offs to be solved. This is the case neither with the binary moral categories that frame the moral reasoning of postcolonial studies (oppressor/oppressed, perpetrator/victim, etc.), nor with principle-based non-empirical ethics that develops its rules, categories, and distinctions without considering the context of the respective moral problem [14]. The first part of this chapter is designed to outline the basic reasoning behind the various forms of principle-based ethics and why they all end in the conclusion that speciesism is morally wrong. The second part then discusses why context matters in ethics: one needs to have an understanding of the history of animal domestication as well as the situation of animal welfare in technology-deprived low-income countries. There is also a need to cast a critical look at the claims of ethical consumerism and to create greater awareness of the real global sustainability challenges among the large majority of the population who are not involved anymore in the production of food in general and animal rearing in particular. It will be shown that principle-based ethics largely addresses concerns related to ethical consumerism but not the concerns of those who are immersed in the practice of animal welfare, such as veterinarians. Finally, it highlights the fact that practical ethics on the consumer as well as on the producer side run the risk of committing what G. E. Moore called naturalistic fallacy or, the other side of the coin, metaphysical fallacy [15].

The third part of this chapter focuses on the need to integrate and to some extent reconcile the ethical perspectives on the producer and the consumer side. For that purpose, a more holistic approach to ethics is proposed, which is able to reconcile ethical theory with the real nature of human beings. This requires exploring the real trade-offs found in the world of practice while responding to the growing need in society to have a dependable moral compass in life that is able to provide meaning, orientation and identity in life. In this chapter, it is argued that Spinoza's opus magnum *Ethics: Proved in Geometrical Order* provides such a holistic ethical approach. Even though the work was published more than 250 years ago, its basic insights have been validated by the latest science in the fields of neurology, biology and moral psychology. Moreover, it avoids the two types of fallacies associated with ethical theory as described by G. E. Moore.

Spinoza's development-oriented approach to virtue ethics is grounded in the proper understanding of human nature while also offering a moral compass that is free of moral hypocrisy and the application of moral double standards. Finally, it can also be integrated into an ethics of meliorism that matters to animal scientists and engineers who are committed to ensure that unnecessary suffering of non-human animals in the process of mobilizing the non-human means (production of animal-based resources) for human ends (food security, health, clothing, transport, companionship) can be avoided to the greatest extent possible. The final part provides a brief synopsis and a suggestion on how to move forward.

5.2 Speciesism or the Problem with Principle-Based Ethics and Postcolonial Studies

5.2.1 Speciesism from Peter Singer's Utilitarian Point of View

The term "speciesism" was primarily derived from principle-based animal ethics. It gained early prominence in the 1970s with *Animal Liberation*, a seminal publication by Peter Singer [16] that denounced the disregard of animal suffering in intensive livestock management. The book strongly contributed to the rise to the animal rights movement and continues to remain one of the most cited books on animals rights to date [17].

As a utilitarian ethicist, Singer follows Jeremy Bentham's definition of the right action as that which produces the greatest pleasure for the greatest number. Bentham's utility principle says that actions should be judged right or wrong to the extent that they increase or decrease well-being or "utility." Singer combines this principle with the principle of equal consideration of interests, which states that one should include all affected interests, including the interests of non-human animals, when calculating the rightness of an action. In this context, he considers those who deny that non-human animals deserve equal consideration of interest as advocates of speciesism understood as "prejudice or attitude of bias in favour of the interests of members of one's own species and against those of members of other species" [17]. While there is a debate about the proper understanding of Speciesism within the academic community of ethicists [18, 19], the term has also been embraced in many other research disciplines such as political science, law and even ecology.

The interdisciplinary discussions revolve around the rights of animals [4, 20] as well as the negative implications of speciesism in the livestock industry for the global management of ecosystems [21].

5.2.2 Critical Animal Studies

The emerging field of critical animal studies (CAS) stands for the alleged posthumanist turn in animal ethics [22] advocating for the dissolution of the clear distinction between human and non-human animals that would still be maintained by earlier animal ethicists referring to the works of Peter Singer [17, 23] as well as Tom Reagan, who argued in favor of animal rights mainly from a justice point of view considering that sentient species have an inherent value [24]. CAS advocates support the view that "speciesism" is a crime on par with racism and sexism; concepts that allow to reduce complexity in moral debates by dividing the world into oppressed (people of color, women, non-human animals) and the oppressors (white, male, human animals). CAS has been criticized for its tendency to only look at multiple forms of oppression (intersectionality) while shunning research on complex overlaps and a commitment for practical solutions to address the animal welfare challenge [25].

The alienation between Peter Singer, the original opponent of speciesism, and the new animal rights activists, inspired by CAS, has also become manifest in Singer's recent publication *Animal Liberation Now* [26]. He points out that equal consideration does not mean equal treatment. According to Singer, the focus should remain on effective action designed to minimize animal suffering on this planet rather than the liberation of animals meant in the literal sense [16].

5.2.3 The Problem of the Principle of Impartiality in Moral Consideration

Singer's approach and its emphasis on the application of impartiality in moral consideration has been questioned especially in view of emotions that humans share with non-human animals such as compassion, sympathy, grief, or empathy [27]. For some ethicists, such emotions provide sufficient evidence for the claim that animals are moral beings [28]. Singer may have implicitly admitted the limits of impartiality when pursuing impartial altruism in his personal life. He donates a large share of his income to charities that ensure anonymity of the human and non-human beneficiaries. Yet, when his own mother was affected by Alzheimer's disease, he felt morally obliged to particularly take care of her, independent of strangers who suffered equally from the disease. This made him admit that "....it is different when it's your mother" [29, 30].

5.2.4 Principle-Based Ethics as Applied by Christine Korsgaard and Martha Nussbaum

Critical views about Singer's utilitarian approach have also come from ethicists who emphasize the intrinsic value of animals. Christine Korsgaard, a well-known animal ethicist who is largely following Kant's deontological approach in her ethical deliberations but also finds fault with his exclusion of non-human animals in the demand that individuals should be treated as ends on themselves rather than means to an end. Her justification to treat sentient animals as ends in themselves is based on the argument that our similarities to non-human animals are morally more important than the differences that exist between them and us. One such similarity is the claim that animals also form an identity, a self, based on experiencing their conditions and things in the world as good or bad and then act upon it in order to pursue the final good of well-being [31].

In this context, Christine Korsgaard's ethical approach to animal treatment overlaps with the one of the philosopher Martha Nussbaum, who is primarily influenced by Aristotle's emphasis on the importance of providing the conditions in which a good human life can be actively chosen and pursued [32]. Nussbaum extends this responsibility to non-human animals by pointing at the intrinsic value of preserving the inviolability of each individual animal. She believes that this can be achieved by applying the "capabilities approach" touched upon by Aristotle and further developed by the Nobel Prize–winning economist Amartya Sen to non-human animals. It is based on the assumption that all sentient beings should have an enabling environment that allows them to flourish [33] based on the foundations of the experience of emotions as "elements of our common animality" [34]. She shares the view with Korsgaard that animals have an intrinsic value even though Korsgaard's argumentation strictly follows the de-ontological approach with its appeal to use reason rather than affect in moral judgment. Her latest book on animal ethics [35] is, however, dedicated to her cats, which indicates that empathy may have also been a major motivation to focus on the topic of animal ethics.

As for attributing rights to animals, Nussbaum and Korsgaard diverge in their conclusions. Korsgaard argues that we should be concerned about legal protection of animals [36], rather than securing a wide range of rights that humans consider relevant human rights in liberal societies. After all, non-human animals strive to live in accordance with their natures, not their presumed values [35]. Nussbaum disagrees and is a supporter of the animal rights movement. Even though they both consider Peter Singer's "Animal Liberation" an important milestone due to its inclusion of all sentient beings in the demand to reduce suffering of living beings on the planet, they also have reservations about Singer's strictly utilitarian approach.

Singer's utilitarianism may also rely on the assumption that there are objective principles such as the principle of utility and the principle of equal consideration that should serve as a guide in taking rational moral action. But the focus is on the consequences of the actions and the extrinsic objective to minimize suffering and maximize pleasure among sentient beings. Even though emotions do not play a role in this, Singer acknowledges the intrinsic value of sentient beings insofar as that they are not merely objects for our use but have lives of their own that need to be respected. Yet, what primarily counts in his ethical consideration are the consequences: the numbers of animals that suffer from instrumental use by humans. That is why his focus is on action against the livestock industry to minimize the suffering of sentient beings [37].

Despite the disagreements among animal ethicists that pursue either a utilitarian or a deontological approach, they all tend to agree that "Speciesism" is wrong, albeit for different reasons.

5.2.5 Inconsistencies of Anti-Speciesism Inherent in Principle-Based Ethics

As someone experienced with the world of practice in animal farming as well as the field of practical ethics, Per Bauhn points at some inconsistencies in how deliberations in principle-based ethics lead to the conclusion that speciesism is to be condemned. He challenges the view that non-human animals can be considered to a certain degree as moral agents because we share with them certain moral emotions. According to Bauhn, moral action is also about the capability of using a moral concept and to act upon it [1]. Reagan [24] agrees to some extent and suggests that non-human animals must be considered as moral patients rather than moral agents. Nevertheless, they would have to be treated equally because they satisfy the "subject-of-a-life criterion" and therefore have inherent value. Yet, in Reagan's view, only moral agents who have the cognitive and other abilities necessary for being held morally accountable have moral duties. In other words, while humans have duties and rights, non-human animals would only have rights, such as the right not to be killed. However, this right is also violated if a sheep is killed by a wolf, but since a wolf is considered to be a moral patient rather than a moral agent, there is no duty to respect the rights of the sheep. This makes Bauhn wonder why a wolf can kill and eat a sheep without doing anything morally wrong, while it would be morally wrong for a human to kill and eat a sheep? From the perspective of the sheep, it is equally bad whether it is killed by a wolf or by a human [1].

Reagan's view, nevertheless, tends to be endorsed by Nussbaum and Korsgaard who emphasize that non-human animals possess equal moral worth and therefore merit equal protection of their basic rights. But since sentient creatures are unable to stand in for their basic rights, who is really authorized to represent them? At this point in Bauhn's argument, animal rights activists may inadvertently turn into speciest paternalists as self-appointed representatives of non-human animals. After all, non-human animals are incapable of speaking for themselves and defending their basic rights as "moral patients." Consequently, humans must "distribute rights and duties in accordance with a belief that one species—the human one—is superior to others when it comes to moral reasoning" [1]. Martha Nussbaum takes this to the extreme by suggesting that as humans capable of moral reasoning we must balance predator interests (e.g., lions) against prey interests (e.g., gazelles) to find out what could possibly be best to accommodate the interests of both parties. Ironically, if animal rights advocates would be in charge of deciding which animal interest should be allowed to override which other animal interest, it would amount to a human-centered species imperialism or species colonialism [1]. This reasoning is obviously inconsistent with the principle of animal equality.

Bauhn suggests that more consistency in animal ethics discussions can be achieved by accepting the fact that each species—including the species of humans—act in accordance with their own natural interests. This would also explain why the evolution of humankind is strongly tied to the domestication and breeding of animals for meat, milk, fur, and other products that are still essential to human well-being in many regions of the world. In view of the human capacity for species-transcending sociality, it is also too reductionist to denounce humans simply as "oppressors" when it comes to their relationship with non-human animals. In fact, keeping and pampering pet animals as companions, caring for the well-being of domesticated animals and protecting them against dangerous predators is observed in all cultures. If we therefore speak out in favor of equal consideration, we have to first take into account the specific human nature with its complex needs and inherent capacity to use reason to transcend merely instinctual behavior [1]. With his suggestions that ethicists should pay more attention to the complex nature of humans and how their relationship toward non-human animals has evolved over time, Bauhn seems to invite his guild to look beyond their own discipline and to revisit some of the baseline assumptions underlying the different ethical approaches that rely on principles rather than human experience.

Yet, Bauhn mainly focuses on other well-known ethical theorists, such as Immanuel Kant and Alan Gewirth in challenging the moral argumentation against speciesism. Kant argued in his work *Metaphysics of Morality* originally published in 1785 [38] that non-human animals may not be considered as ends in themselves but that we would have an indirect duty to refrain from applying wanton cruelty against them. Gewirth reaches a similar conclusion based on a particular understanding of moral agency. Moral agency is constituted by normative structure comprising the setting of goals, realizing them through action and subjecting them to continuous evaluation in regard to their effectiveness. In this context, there is a clear justification for claiming rights that secure equal access to the necessary goods that enable us to live as moral agents.

Non-human animals may also be considered as agents in the sense that they respond to good and bad things in their pursuit of the final good of well-being [35]. But non-human animals lack the reflexive normative structure to be considered moral agents that justify the claim of generic rights associated with human freedom and well-being. These rights are deemed necessary to pursue one's self-interest in consideration of other-regarding interests in life, which explains the essential human dimension of the discussion on "moral rights" which are not to be confused with "possessions" [39]. They rather represent a claim to the non-interference of others with one's possession of freedom and well-being. Even though the explicit and collectively agreed claim to non-interference cannot be applied as such to non-human animals, Gewirth argues that humans and non-humans share the feeling of pain and their desire to avoid it. Consequently, non-human animals also have a right to immunity from wanton infliction of pain [39].

In his work, Bauhn [1] hardly refers to the latest insights in the natural sciences and does not discuss naturalist approaches to ethics based on Spinoza's work that later also contributed to an ethics of meliorism. Instead, he attempts to reconcile principle-based ethics with a proper understanding of human nature. It is therefore worth further exploring what that means by diving into evolutionary aspects of moral development that are also related to the history of domestication of animals. This may then help to understand the implications that these empirical insights have on ethical theory in general and on the practical ethical debate on the application of modern biotechnology to animal breeding in particular.

5.3 Minding the History and Context of Animal Domestication and Its Impact on Animal Welfare

Throughout the history of humankind, non-human animals have been killed on a daily basis by humans either for direct human consumption or indirectly through competition for resources. Even crop producers are forced to kill animals that endanger the harvest of their crops [40]. This has nothing to do with humans simply enjoying the killing of non-human beings. It is rather due to the fact that we are part of the global food web that evolves in an ever-changing ecological context. Life, death and decay is all about the cycle of nutrients, no matter who is at the apex of the food web [41].

5.3.1 Domesticated Animals Are Part of the Human-Made Cultural Landscape

Popular writers like Yuval Harari tend to add a moral dimension to this amoral order of nature by portraying the history of humankind as a history of subjecting the natural environment not just to meet essential human needs but also for the sake of accumulating goods to increase its power over non-human animals through oppression [42, 43]. Yet, this Neo-Darwinian view of evolution is increasingly questioned. Since its onset in early neolithic times, human animals share a living with domesticated animals resulting in a culturally managed landscape designed to adapt nature to the needs of humans [44]. This cohabitation may be characterized as Mutualism rather than Darwinism [45–47]: animals are raised and taken care of by human communities in return for making use of their resources, while being alive (dairy products, eggs, manure, wool) and after slaughtering (meat, hides, bone tools, etc.). The mutual dependence has left a genetic imprint on both sides meaning that the survival of humans has become dependent on the reproduction and use of domesticated animals while domesticated animals have become dependent on human care; this also explains why most of them would not be able to survive outside the cultural landscape anymore [48]. Humans are unique insofar as they have the ability to select partner species based on genetic variants, leaving one relationship in favor of another to enhance the benefits for the domesticator. Animal breeding and animal husbandry are therefore linked to countless historical decisions by humans to modify their environment in a way that ensures a more predictable, steady and abundant supply of essential resources for themselves and their community of which they are part of and on which their survival depends [49].

As such, modern animal biotechnology may be considered as the latest sort of technique in the history of animal selection and breeding. An ethical perspective must take into account the history of human and non-human animal suffering and welfare. It is too simplistic to reduce the history of animal domestication to an account of the oppressors (humans) at the expense of the oppressed (non-human animals) [50].

5.3.2 The Popularity of Mythical Accounts Starting from an Original Age of Innocence

Popular accounts that attribute the oppression and destruction of nature to the system of "greed" that humans have created by embracing capitalism tend to follow the mythical structure primarily found in religious accounts [51–54]. It implicitly or explicitly assumes that there must have been a prior age of innocence, a moment that represents the fall of humankind and a possibility to restore the original balance with nature [54–56]. The problem with such narratives is that they fail to take into account that human communities and their relationship to wild and domesticated animals face different socioeconomic and historical circumstances all over the world. And these communities have multiple responsibilities not just toward non-human animals but also fellow-human beings [56].

5.3.3 Minding That Poverty Rather Than Affluence Is Often the Enemy of Animal Welfare

The current discourse on animal rights with its focus on denouncing the animal welfare situation in modern animal production facilities in high-income countries tends to ignore the fact that the situation is often much worse in low-income countries where the companies involved in the rearing and transportation of animals often lack the means to adopt the latest animal-friendly technologies and practices despite laws against the cruelty of animals that exist also in many of these countries [57]. Besides, many of the domesticated animals raised by pastoralists and semi-subsistence farmers in these countries suffer and die either from preventable communicable diseases due to lack of access to adequate medical treatment or from hunger and starvation due to extended periods of drought [58–60].

In this context, it is worth considering the respective historical and geographical context in which non-human animals are raised, used and consumed by humans. For example, prior to the invention of meat refrigeration systems in the late 19th century, cattle was shipped alive across the Atlantic. Mostly, the animals arrived sick and starved at the destination and were then force-fed to increase in weight quickly to fetch a decent price at the local market [61]. In this context, technology that allowed the shipment of processed refrigerated meat definitely generated a massive improvement in animal welfare. It reduced the need for live transports to a significant extent, at least in high-income countries. The same applies to the potential of more recent technologies such as gene-editing to improve animal welfare. The company EggXYt is using it for pre-incubation sex identification of embryonic chickens [62]. It uses the CRISPR-Cas9 technique to cause male embryos to fluoresce at the earliest possible time after egg fertilization. As such, this technique effectively eliminates pain suffered by male chickens that are not meant to hatch. Chicken culling continues to represent one of the most serious ethical challenges in the chicken industry [63].

All these aspects highlight the potential of innovative practices and new technologies to reduce animal suffering. However, they remain largely unaddressed by principle-based ethicists [26, 33, 35]. They tend to neglect the specific context in which animals are raised and bracket out concrete moral dilemmas that force us to balance animal versus essential human material needs [64]. The reason for this omission may be their focus on general questions related to the consumption of animal-based products—an ethical question that primarily concerns those who are not involved in the business of animal husbandry.

5.3.4 Practical Ethical Challenges from a Producer's Perspective

The impact of new technologies on the well-being of domesticated animals needs to be critically assessed on a case-by-case basis and understood in the larger historical and geographical context [42, 64]. The goal of such a practice-oriented ethical approach in the production of animal-based resources is to achieve more real and tangible improvements in animal welfare while ensuring that the needs of a growing and more affluent population can be met as long as commercially viable substitutes for animal-based resources are still missing.

A focus on practical challenges to improve animal welfare must consider the following questions:

- Where do we stand today in terms of animal welfare compared to earlier times, and to what extent can new technologies, such as biotechnology, combined with more strict animal welfare laws, improve the well-being of farmed animals?

- What is the net impact of intensive animal production systems on the welfare of wild as well as domestic animals?
- To what extent is the improved implementation of animal welfare standards linked to the means available in government and industry to enforce such standards?

Veterinarians are practitioners who face the challenge of meeting their ethical duty to act in the best interests of the animal patient while minding the client's financial background and the emotional bond between client and animal [65]. They are committed to find concrete solutions to identify animal welfare problems even though these are first of all associated with animal health rather than animal ethics [66]. In their practical work, animal welfare is described as "the physical and mental state of an animal in relation to the conditions in which it lives and dies" [67]. In this context, the World Organization for Animal Health (WOAH) [68] asks practitioners in the livestock industry to comply with the "Five Freedoms" of animal welfare. They consist of freedom

(1) from hunger, thirst and malnutrition,
(2) from physical and thermal discomfort,
(3) from pain, injury and disease,
(4) from fear and distress, and
(5) to express normal patterns of behavior.

The Five Freedoms have been complemented in some legislative frameworks on animal welfare with the concept of the Five Domains (nutrition, environment, health, behavior, mental state), with the mental state of the animal being an additional critical aspect to be considered [69].

WOAH counts 170 member states that are expected to take these guiding principles into account in national legislation on animal welfare. Overall, the organization emphasizes the fact that animal welfare goes far beyond animal ethics; it also has scientific, economic, cultural, social, religious and political dimensions that must be taken into account.

There is, however, a growing awareness that perceiving animals through a purely positivist lens in veterinary science tends to overlook the fact that animals are more than a set of genetic-based physicochemical functions that may fail and then be treated accordingly. The biological concepts of pain and death cannot capture feelings such as "anxiety," "distress," "grief" and "sorrow" and it may be one of the core tasks of ethics to accommodate what cannot be captured in purely biological terms. This would require a practical ethics of care that is able to provide meaningful guidance to veterinarians [64].

However, veterinarians and livestock producers operate in a complex environment with sometimes diverging expectations and scarce means available for animal welfare. In this context, it is often ignored that compliance with more strict standards of animal welfare requires additional means to be invested in animal health and welfare-enhancing management practices and technologies. Since the profit margins in the animal production industry tend to be quite narrow, there is a reluctance to adopt such measures unless government or the retail industry are willing to contribute [70].

5.3.5 Ethical Concerns about Intensive Animal Farm Practices from the Consumer's Perspective

Contemporary urban consumers in high-income countries have mostly ceased to raise their own animals for personal use. Instead, they rely on already processed animal-based products from retail stores. Due to this detachment from the production side, many consumers feel great anxiety about the origin of these animal-based products and the fact that animals may have been killed in large-scale abattoirs. In order to attenuate their feeling of guilt [52], they consider various choices: One is to change their consumption habits moving toward a vegan lifestyle [71]. Another option is to register for a so-called box scheme where the trusted rural producer in the region builds up a relationship of trust by delivering the sustainably produced animal-based food products directly to the urban consumer who then also visits the farm for family holidays [61]. Finally, many affluent urban consumers focus on ethical labels that claim to have been produced with the highest standards of animal welfare [61, 72, 73].

However, such premium products and alternative lifestyles tend to be more expensive and time-consuming when it comes to the careful selection of the best choice. Moreover, urban consumers who care about ethical labels are often not aware that there may be a gap between the promise and the reality concealed by such labels [74]. After all, the retail industry is primarily concerned about marketing [75, 76].

It may also be problematic if the purchase of a product with an ethical label is mostly meant to signal virtue and distinguish oneself from others [77, 78]. It induces such consumers to look down upon the average price-conscious middle-class urban consumer who does not strive for a higher ideal but merely a better life. It may not be fair to "pity" them as being guided by a "false consciousness" and therefore in a certain way "pathological" [79].

Less affluent consumers who struggle to feed their families due to budget constraints may care primarily about the affordability of a food product. However, this does not necessarily imply that poor consumers act more selfishly, whereas the affluent consumer is devoid of egoistic motives. In fact, the anthropologist Daniel Miller highlighted the fact that ordinary price-conscious consumers tend to think more about others (family members, guests to be invited) rather than themselves when they shop in retail stores [73]. He calls this a "warm," human-centered type of ethics as opposed to a "cold" type of ethics that is focused on more abstract ethical concepts represented by well-known ethical premium labels such as fair trade and organic. While Miller does not play out one type against the other, he highlights the fact that humans tend to face multiple moral responsibilities and that they tend to prioritize local engagements based on the principles of fairness and reciprocity to more abstract global engagements, such as commitment toward animal welfare and environment justice. Consumers who care more about the latter tend to link their commitments more to abstract ethical concepts and rely in their argumentation primarily on stories told by advocacy groups and other trusted intermediaries that reveal a tendency toward moral absolutism [80]. Since advocacy groups provide moral guidance and are able to reduce complexity in consumer decision making, there is often a reluctance to question ideological motives and economic interests that may also guide the agenda of advocacy groups. Critical inquiries tend to increase ambiguity and with it, complexity, which stands in conflict with moral absolutism [81].

The moral hazard represented by the gap between claim and delivery is usually more effectively addressed on the local level where an effective system of personal monitoring in place [82]. This would again explain why most people care more about their local moral responsibilities where they are more likely to be held accountable.

5.3.6 The Blind Spot of Ethical Consumerism in Postmaterial Societies

The political scientist Ronald Inglehart predicted in his book *The Silent Revolution* [83] that increasing prosperity experienced by the post-war generations in North America and Europe may eventually induce the coming generations to take their material security for granted and to increasingly emphasize the importance on non-material value orientation expressed in concerns about the environment, social justice, cultural identity, and self-expression. In this context, animal welfare and livestock management is primarily regarded as a matter of justice toward animals and to some extent also of environmental concern considering the substantial amount of greenhouse gas emissions released by the livestock industry.

Interestingly, social justice questions related to access to essential material goods in low-income countries seem to be less of a concern since it is often implied that excessive consumption rather than material deprivation and poverty is the main problem on this planet [84]. This assumption stands in strong contrast to the current real challenges related to food security, especially in low-income countries: the latest UN report on the state of food security and nutrition in the world [85] estimates that almost a third of the global population, equivalent to 2.4 billion people, did not have constant access to food. Pastoralists in semi-arid environments who largely depend on livestock for food security constitute a significant share of this vulnerable population [86].

In this context, one has to also become aware that people most vulnerable to hunger and starvation may actually be those who are not integrated into the global formal economy, usually denounced as the global capitalist system [84, 87]. They tend to be ignored because they cannot be part of the great anti-globalization narrative [56].

5.3.7 Principle-Based Ethics, the Culture of Consumerism and the Risk of Metaphysical Fallacy

Contemporary academic literature on the ethics toward non-human animals hardly addresses these concerns related to ethical consumerism, multiple responsibilities in the daily lives of middle-class families and the unmet material needs of 2.4 million people on this planet.

One reason for this detachment of principle-based ethical theory may be related to the argument that empirically validated insights on how things are (facts) do not necessarily provide a higher insight on how things should be (values). David Hume already observed this so-called fact/value gap in his Treatise on Human Nature. Yet, it eventually became known as the Naturalistic Fallacy described by the philosopher G. E. Moore in his book *Prinicipia Ethica* published in 1904 [88].

Moore maintained that the "Good" (what we ought to do) cannot be deduced from its natural properties. The reason for that is the unique nature of "Good" which cannot be defined because it is simple and has no parts. Its elusive nature is due to fact that "good" is an adjective that always refers to a thing. But what is good about the thing is not found in its property. In other words, although we may experience many good things, that which is good about these things, is not found in their properties. "Good" therefore cannot be divorced from an individual moral judgment in regard to a particular thing. Consequently, the attempt to define it through its natural properties amounts to naturalistic fallacy.

But according to Moore, "Good" cannot be a super-sensible property either. Any reference to something supernatural or divine that is beyond human experience is therefore a manifestation of metaphysical fallacy, which he regards as the other side of the coin of naturalist fallacy. Moore is therefore also critical about the proposition to follow universal ethical utilitarian or deontological principles that are allegedly derived from pure reason.

Moreover "being rational" is often understood in utilitarian as well as de-ontological ethics as the ability to adopt the standpoint of an impartial spectator assigned with the task to reach a universalizable reason-based moral judgment largely independent of the latest empirical insights in other fields of research. This stands, however, in strong contrast to findings in neuroscience [89] and moral psychology [90] highlighting the fact that emotion and unconsciousness are always prior to reason and consciousness. Moreover, comparative studies in anthropology [13] clearly indicate that normativity is implicit in practice-oriented human interactions rather than a separate form of reasoning.

5.4 Spinoza's "Ethics" as a Guide for a More Holistic Approach to Animal Welfare Challenges

As illustrated in the prior part, the contemporary public debate on animal rights with its focus on the denunciation of speciesism tends to ignore the complex trade-offs we face as citizens of different countries and individuals operating in very different socioeconomic circumstances. Even though a strong commitment to improve animal welfare is important and commendable, it may also conflict with other moral responsibilities in our more immediate social and natural environment.

What most people look for is a moral compass that enables them to address the multiple ethical challenges and trade-offs in life in a humane and pragmatic way. This moral compass, also regarding the treatment of non-human animals, is, however, not "for sale" in the sense that someone would own and reproduce it to meet a market demand. Instead, we "ought" to acquire such a compass as self-interested but also self-critical human beings.

5.4.1 Spinoza's "Conatus Doctrine"

Baruch Spinoza's work *Ethics: Proved in Geometrical Order*, shortly published after his death in 1677 [91], may serve as one of the best guides in developing a moral compass. Spinoza counts as one of the great philosophers of the early age of enlightenment who inspired diverse German philosophers in the 19th century such as Schelling, Hegel, Marx and Feuerbach [92] as well as Anglo-Saxon scientists and scholars concerned with the mind-body problem [93, 94].

In *Ethics*, Spinoza discusses how a moral compass may emerge from an active human life that makes use of reason in the pursuit of self-interest to gain insight and achieve virtue (EIVP24),[1] understood as "the power of bringing about certain things, which can be understood through the laws of his nature alone" (EIVP8).

Unlike Aristotelian virtue ethics that has influenced Jewish, Islamic and Christian moral theory as well as various forms of stoicisms, Spinoza's virtue ethics is based on the conatus doctrine [95], the egoistic drive to preserve in being (EIIIp6). "Conatus," understood as the pursuit to preserve in being, is observed by all living beings independent of the duration of their derived existence (natura naturata) (EIIIp8). It is therefore "natura naturans," following from the eternal necessity of "Nature or God" (Deus sive Natura > EIVp4) (EIIp45).

As Natura Naturata, all bodies of human and non-human animals age and eventually die despite the best efforts to persevere in being. That is why we cannot have an adequate idea of the duration of our lives (EIIp31). The reason for that is that there is always something different and more powerful that is able to destroy us as individual beings (EIVa). In other words, living beings are subject to the laws of entropy and this explains why they do not live forever [96].

Some ethicists presume that Spinoza's Conatus doctrine would represent an example of naturalistic fallacy according to Moore's *Principia Ethica*, because the "Good" (what we ought to do) cannot be deduced from its natural properties. Spinoza does, however, not deduce "the Good" simply from natural properties. Instead, he starts

[1] Spinoza's *Ethics* consists of five books: Part I (*of God*), Part II (*Of the Nature & Origin of the Mind*), Part III (*Of the Origin & Nature of Emotions*, Part IV (*Of the Servitude of Humanity, or the Strength of the Emotions*) and Part V (*Of the Power of the Intellect, or the Liberty of Humanity*). Each part consists of propositions (p), Definitions (d) and Annotations or Scholium (s), Auxiliary Set or Lemma (l) and Axioms (a). The Abbreviations Refer to the book *Ethics* ('E'), the Part (I, II, III, IV, V) and the particular type ('p', 'd', 's', 'l', 'a').

from the fact that humans equate the good with what they desire or find useful in their urge to satisfy their needs. The path toward virtue is, however, characterized by developing an internal desire to go beyond such an understanding of the Good.

In the preface of Part IV, Spinoza suggests that this happens through a normative orientation toward achieving a model of human nature that is guided by insight and therefore worth striving for. Consequently the "Good" reflects an approximation to a model of human nature that we are to set before ourselves [97]. At the same time, Spinoza avoids Moore's metaphysical fallacy, the tendency to generalize by attributing ends to transcendent or supernatural principles, laws and beings. This rejection of the teleological conception of human agency distinguishes Spinoza also from Aristotle [98].

5.4.2 Spinoza's Differentiation between Human and Non-Human Animals

In addition, Spinoza admits that non-human animals are also "animated" beings in different degrees "for of everything there is necessarily an idea in God, of which God is likewise the cause..." (EIIp13s). But "apart from the joy and desire, which are passions that we may share with non-human animals, there are other affects of joy and desire which are related to us insofar as we act" (IIIP58). This ability to act is related to the human ability to "understand" and thus gain insight into adequate ideas that enable us to develop the desire to strive for virtue. This desire enables us to free ourselves from the bondage of passions, lead an active life (to act rather than to be acted upon) and join forces with other human beings to achieve a particular common objective—which indirectly also serves our self-interest. The ability to cooperate is common to all humans who live in accordance with their nature (4p18s); and that is why Spinoza argues that "there is no individual thing in nature, which is more useful to man, than a man who lives according to the guidance of reason" (4p35c1). It illustrates why it makes sense for a human being to focus on achieving a common goal that serves a particular common need and enhances my freedom to act in the community. In order to be perceived as a useful and trustworthy member of this community, on which my well-being ultimately depends. I need to restrain the passive passions generated by external causes that are falsely felt to serve the egoistic drive to preserve in being [99].

5.4.3 Virtue: Pursuing a Non-Rival Good While Making the World a Better Place

Every human life starts with basic egoistic drives primarily linked to tangled, inadequate ideas. Children primarily associate "the good" with things that tend to benefit them immediately (idea of joy) and "the bad" with things that may prevent us from enjoying such immediate benefits (idea of sadness) (E4p8). But in the course of life, we learn that short-term benefits may eventually turn out to be harmful for us and that what we believed to be bad could benefit us in the long run. The insights gained through experience and self-questioning may enable us to replace inadequate ideas, dominated by affects generated by external causes, with adequate ideas of which we are the cause (E3d2). As grown-up agents in life, we may be guided by such adequate causes, which would indicate that we were able to acquire a moral compass and become virtuous through an active life.

Since virtue is a non-rival good, a good that "can be enjoyed by all equally" (EIVp36), we may pursue virtue in life out of pure self-interest without harming anyone: if I acquire virtue, there will not be less virtue for others. Once we realize that "virtue" is a good that becomes more useful to us the more it is shared, we desire it not just for ourselves but also for others (EIVp36, 37). As a consequence, society may become better as a whole. This line of argumentation essentially points out that the pursuit of self-interest must not go at the expense of ethical behavior but can actually reinforce it. Pursuing a socially desirable good in my own self-interest also reduces the risk of moral hypocrisy since publicly declared motives and personal motives are more aligned and acceptance of moral pluralism is higher [100].

Since it is impossible for a human being to exist, seek virtue and thrive without external support from other human beings, Spinoza implicitly endorses the so-called Golden Rule, the principle of treating others as one would want to be treated by them, since it is well-aligned with the pursuit of self-interest on the long run (EIVp37). Nadler [99] also illustrates this by means of Spinoza's example of deception: acting deceptively can never be in accordance with the nature of a virtuous human being, who has an interest in the design of common rules of human exchange that encourage collaboration (EIVp72). Deception would destroy trust and therefore discourage collaboration.

5.4.4 How to Treat Other Sentient Beings Who Do Not Share Our Nature

When ethicists discuss Spinoza's view on how to treat non-human animals, they tend to point at two crucial statements in ethics. The first, found in part III of Ethics (EIIIp57s), makes Spinoza look like an early advocate of Singer's principle of equal consideration of interest; in opposition to Descartes's "beast-machine" approach that merely recognizes mechanical response to stimuli in non-human animals, Spinoza recognizes that animals are sentient beings like us. Yet, Spinoza also points out that affects generated by non-human animals "differ from men's affects as much as their nature differs from human nature" (EIIIp57s). Both the horse and the man are driven by a lust to procreate; but the one is driven by an equine lust, the other by a human lust" (EIIIp57s).

In part IV, Spinoza disappoints potential hopes that he could be seen as a pioneer of anti-speciesism when he denounces the law against killing animals as being based on empty superstition rather than sound reason arguing that "we have the same right against them (non-human animals) that they have against us". In the same paragraph he writes "Not that I deny that the lower animals have sensations. But I do deny that we are therefore not permitted to consider our own advantage, use them at our pleasure, and treat them as is most convenient for us. For they do not agree in nature with us, and their affects are different in nature from human affects" (EIVp37s1).

This latter sentence would be considered outrageous if uttered in the contemporary public debate on animal welfare. But would

Spinoza have endorsed cruelty toward animals if it would trigger pleasure in certain humans—as some authors claim [101]? Clearly not, since applying cruelty is not of any individual use beyond satisfying such base feelings as revenge. Like deception, cruelty, if considered as acceptable in society, cannot be in one's individual interest. Moreover, it cannot be based on an adequate idea since applying cruelty is always based on an external cause that affects us in a certain way (EIIIp41c, s).

Nevertheless, Spinoza's view that we do not share the same nature with animals and therefore should make use of them to meet our needs would undoubtedly earn him the label "speciest" in the contemporary ethics debate. Does Spinoza indeed contradict himself when attributing a privileged position to humanity while also admitting that the natures of distinct humans may also diverge [102–104]? For example, people who are affected by a mental disease may lose their ability to reason and eventually barely survive without permanent assistance from others. As a consequence, they cease to be rational in their pursuit of self-interest, and consequently do no more agree in nature with us. Why then treat them any different from non-human animals [103, 105]? One of the main arguments to justify a clear distinction is that "the affects of other animals differ from men's affects as much as their nature differs from human nature" (EIVp57). In other words, the joy I feel petting my dog cannot be the same joy experienced by the dog for that joy is specific to its nature. In this context, the view that modes are always caused by and understood through the things of which they are modifications (E Id5) serves as a conceptual, and hence causal, barrier to the transmission of affects [106].

Spinoza addresses the question from another angle in part IV when referring to the infinitively more powerful external force that eventually makes us age, change in nature and die (EIVp3). But as regards change in nature, Spinoza makes clear in the preface of Part IV that an animal (e.g., a horse) cannot turn into a human and reversed for that would not alter but actually destroy its respective nature [107]. Instead, he illustrates how the enduring nature of a living human individual can accommodate radical but not unlimited changes in the way it strives to persevere in its being (EIVp39s).

For example, the nature-changing effect of a disease can clearly be observed in a person who is affected by Alzheimer's disease. Spinoza describes it as "an external force that alters the bodily rules of motion and rest" (EIVp39). It deprives humans of their ability to reason and engage with their environment. As a consequence, the elderly may turn into children that need our support and assistance. Yet, Spinoza points out that "a composite individual can be affected in many ways and still preserve its nature" (EII l7s/EIIP13).

In this regard, Spinoza makes it clear that solidarity with humans is always linked to the perception of direct or indirect usefulness. After all, only human bodies and minds are able to join forces to achieve a common objective that is in the common as well as the self-interest. In this context, Spinoza argues that only humans have the ability "to become one mind and one body striving together, as far as they can, to preserve their being; and that all, together, should seek for themselves the common advantage of all" (EIVp18s). The self-interest in human solidarity is also expressed in Spinoza's insight that "... to man, then, there is nothing more useful than man." (EIVp18s). Yet, solidarity goes beyond immediate self-interest insofar as it is also extended to human beings that are affected by an external force that makes them less "useful." For virtuous human beings strive toward social unity not just by caring only for their own needs but also the needs of those who have been (e.g., parents) or will be (e.g., children) enablers of our own ability to preserve in existence and strive toward virtue (Ethics IV/p40).

In sum, Spinoza recognizes the fact that we are transient beings, like non-human animals, who are born with a different nature and end up with a different nature. But we would never be able to change our fundamental human nature [107]. He also agrees that human and non-human animals share basic instincts and affects linked to survival and procreation and that they all have a natural right to persevere in being in any way their nature dictates (EIVp37s2). This also involves the right to eat other organisms that sustain our lives, and reversed, other organisms having the right to eat us for their need to persevere in their being (EIVp37s1). After all, we are as much part of nature as they are (EIVp4) and nature itself does not follow a moral order.

5.4.5 Making a Distinction between Wild and Domesticated Animals

When Spinoza makes a comparison to non-human animals he does not distinguish between wild and domesticated animals even though they may indeed have diverged in nature to a considerable extent. One could even argue that domesticated animals may have more in common with human than non-human animals in the sense that they cannot live without human care and protection, and humans cannot dispense with the resources provided by the domesticated non-human animals. As pointed out before, the resulting mutual relationship has even left a genetic imprint on both parties [49]. Domesticated animals are part of the cultural space that humans must protect against predators and other forces of nature, and, unlike their wild counterparts, domesticated animals learned to cooperate with humans in shaping the cultural space. Since domesticated animals are clearly useful to humans, solidarity with them manifests itself also in the fact that we care for them and defend them against predators. They join the human effort "to become one mind and one body striving together", as Spinoza formulates it (EIVp18s).

Treating domesticated animals well may, nevertheless, be considered a means to an end, because the whole purpose of domesticated animals throughout history was and continues to be the sustainable provision of animal-based resources. But this does not mean that domesticated animals are per se worse off compared to their wild counterparts who are also part of a food chain and as such may starve or get killed [40, 108].

5.5 Concluding Remarks

The primary criteria whether domesticated animals are better or worse off compared to their counterparts is, on the one hand, related to the sense of moral responsibility animal producers have toward their animals. On the other hand, it is linked to the institutional framework conditions that either encourage or discourage cooperation in an effort to develop innovative solutions to improve animal welfare [108–110]. However, the contemporary debate on the ethics of non-human animals hardly touches upon the ethical dilemmas on the producer side and how to solve them. Instead, it is shaped by advocacy groups that denounce the conscious or unconscious bias in moral treatment against non-human animals. As such, they ask policy makers to assign rights to non-human animals that would protect them from the instrumental use by humans and consumers to stop consuming animal-based products for the sake of animal welfare and the climate.

The ethicist Per Bauhn, who is very familiar with the ethical challenges on the supply side of animal-based products but also shows great understanding for the ethical concerns on the demand side, aims with his recent publication *Animal Suffering, Human Rights, and the Virtue of Justice* [1] to bridge the divide by resorting to the arguments of the principle-based ethicist Alan Gewirth [39]. Gewirth argues that humans are moral agents that should act in accord with their generic rights to freedom and well-being while also ensuring that all others are able to claim these rights too (Principle of Generic Consistency). However, there is very little in his book that concerns the real ethical challenges that actors in the values chain of animal-based products face and how to address them effectively.

5.5.1 Animal Welfare in the World of Practice

Generally, family-owned farms may care more about their animals than employees of large corporate-owned animal production sites [111, 112]. One could argue in this context that normativity is more likely to be implicit in practice-oriented human-animal interactions on family-owned farms. At the same time, large sites may have more means available to comply with prescribed standards of animal welfare in the value chain [113].

There is, however, no doubt that cruelty toward animals takes place on farms, in research institutions and in private households with companion animals [114], and laws are already in place in most countries in Europe and North America to treat animal cruelty as a crime. In this context, the numerous societies against cruelty toward animals are pushing for a worldwide coverage of laws that criminalize cruelty toward animals [115].

Animal suffering on farms continues to be a matter of serious concern, but it is less related to anecdotes widely covered by the media and documentaries of deliberate cruel treatment committed by indifferent actors in factory farming than the lack of commitment by actors in the value chain as well as by governments to develop and jointly fund practical solutions to improve animal welfare and detect and punish abuse in more efficient ways [116].

Even though veterinarians play a core role in this, they face a precarious situation: torn between their commitment to patient care and their dependence on clients that are not always willing or capable to pay the adequate price for the service, they often experience moral distress [117]. The reason for the scarce resources available may be related to the unequal capture of margins in the value chains [118]. While retailers tend to obtain by far the highest margin of the final product, they are seldom willing to spend their profits on investment in animal research designed to improve animal welfare or to co-share expenses for the work of veterinarians.

Retailers may advocate for more strict private animal welfare standards in the supply chain, which certainly also have a positive impact on animal welfare. But the problem is that these private standards primarily reflect consumer perception (their ultimate clients) rather than the latest insights from animal welfare science [75]. As gatekeepers in the value chain, they have the power to set the standards while not being obliged to cover the costs of implementation [61, 119, 120]. Since consumers tend to distrust the use of modern technology in animal breeding and animal husbandry, retailers are unlikely to promote new technologies even if they would have a great potential to improve animal welfare [42]. In a way, these are the real ethical problems of modern animal production systems. They are related to a misalignment of incentives, scarce financial resources available for animal welfare and a preference among large retailers to regard animal welfare primarily as a reputation management issue which they aim to address by signaling to consumers that they care about the issue. Increasing the choice of premium vegan products is one way of signaling their objective to support consumers who have decided to refrain from consuming animal-based products because they oppose speciesism and would like to align their political concerns with their consumer choice. In other words, it is about ethical principles of affluent consumers in postmaterial societies who must not worry anymore about access to essential material needs. Even though their conscious ethical choices are commendable, there is also the risk of being oblivious about the fact that the majority of the population on this planet still struggles to cover material needs. It is not that poor people do not care about animals, but their primary responsibility is to ensure that their families and communities are safe and have access to essential material needs. In other words, the socioeconomic, historical and geographical context matters in moral judgment, and this makes it so difficult to resort to ethical principles as a universal guide for morally responsible action.

5.5.2 The Gap between Theory and Practice as Well as Activism and Concrete Action

The different forms of principle-based ethics that are advocated by well-known animal ethicists such as Kosgaard, Singer and Nussbaum are primarily concerned with general questions concerning the ethical treatment of non-human animals. In this context, they all agree that "Speciesism" either in its weak form (disregarding equal consideration) or its strong form (disregarding equal treatmment of non-human animals), is to be condemned and, consequently, the consumption of animal-based resources to be questioned.

If one takes into account the historical, geographical and socioeconomic context, the ethical situation becomes more complicated. Historically, non-human animals have been killed on a daily basis by humans either for direct human consumption or indirectly through competition for resources. Why this happened and continues to happen has nothing to do with humans simply enjoying the killing of non-human beings. In fact, the history of animal domestication can also be characterized as a history of caring for and protecting animals from predators and diseases. What remains is, however, the undeniable existence of a global food web that is not just about life and well-being but also decay and death, independent of the existence of human beings.

Geographically, animal welfare challenges in affluent high-income countries diverge considerably from animal welfare challenges in resource-constrained low-income countries, where poverty is often the main obstacle to improved animal protection. Yet, in the contemporary ethical debate dominated by actors from high-income countries, the implicit assumption that affluence and its cause (capitalism) are the main enemy of animal welfare is hardly ever challenged.

From a socioeconomic point of view, the crucial challenge is providing access to nutritious food for those who cannot afford it. Cultured meat may once become a perfect substitute for real meat and initiate the decline of the livestock industry. The decline of the livestock industry may then generate new social problems since it generates income and employment of millions of less privileged people. Yet, for the moment, there is plenty of evidence that the decline of the livestock industry is not imminent: meat consumption has been increasing on a global level over the past three decades and continues to increase. There is no evidence so far that appeals to reduce meat consumption has an effect on revealed consumer behavior on a global level [121].

What is sorely needed is a holistic approach to animal welfare that integrates the ethical perspectives on the supply (production) and the demand side (consumption) of animal-based resources, takes into account the context within which the ethical debate takes places and is able to provide meaning, orientation and identity to citizens who are concerned about animal welfare. In this chapter, we proposed that Spinoza's opus magnum *Ethics* may be the best candidate for such a holistic approach because it combines an ethic of meliorism advocated by pragmatist ethicists with a virtue ethic that reconciles self-interest with moral responsibility as well as affect with reason.

5.5.3 The Importance of Embracing a Holistic Approach in Animal Ethics

As mentioned in the introduction to this chapter, pragmatist ethicists associate ethics primarily with the ability to solve problems in an effective way. Based on trial and error, solutions are developed and further refined while new problems may emerge from solving old problems. The accumulation of experience and knowledge may then eventually lead to established practices that help achieve shared objectives in an effective way [6–8]. It then becomes a prudential norm that practitioners ought to follow. In other words, practices are implicitly normative in the sense that they prescribe and proscribe right ways to attain the ends, no matter whether these ends are good in a moral or merely a practical sense [11–13]. In this context, it is important to note that this practical reasoning is no more about deliberations of single individuals but becomes incorporated in professional norms and institutions associated with particular practices. This is a clear indication that human moral reasoning is essentially "social" in character [13, 122] and that human social interaction and organization are cooperative in ways not observed with non-human animals [123].

Spinoza's ethics is very much in line with the pragmatist approach in ethics. It is reflected in his appreciation of the ability of humans to achieve a common purpose, reflected in his observation that "To man, then, there is nothing more useful than man" (EIVp18s). After all, it is through cooperation that humans align their self-interest with the interest of others and become "one mind and one body" seeking for themselves the common advantage of all (EIVp18s). In this chapter, we argued that the term "all" may even include domesticated animals that have become dependent on us while we have become dependent on them. In other words, they are part of the human-made cultural landscape rather than the natural landscape.

Pragmatist ethics provides important guidance to practitioners in the production of animal-based resources who tend to follow an ethics of meliorism, constantly exploring and testing new practices and technologies that may help solve problems in animal welfare, provided that costumers and society as a whole are willing to share the costs.

But is pragmatist ethics also able to provide meaning, orientation and identity to non-practitioners who are very concerned about the instrumental use of non-human animals and denounce the unequal moral treatment of animals in society? Many of them tend to embrace principle-based ethics with its advocacy against "Speciesism" because it suggests moral clarity provided by a clear distinction between "wrong" and "right" as well as "the bad" and "the good."

Spinoza questions this type of moral reasoning since it tends to resort to universal principles or beliefs that cannot be empirically validated and suggests an overarching purpose or end toward which humanity or nature as a whole are striving (teleology). Yet, can there be moral meaning and orientation without the belief in a higher purpose? Spinoza's work is designed to prove that it is possible through self-critical moral inquiry designed to achieve virtue in life.

His virtue ethics is designed to provide moral individual guidance in life while avoiding moral hypocrisy through a better alignment of self-interest with other-regarding interests as well as the admittance that affect rather than reason tends to shape our moral outlook on life. Striving for virtue is then reflected in the desire to gain insight and to lead a fulfilled and self-directed life by learning to master our passions. It must be understood as a journey of critical moral inquiry, a tenacious ethical commitment to increasing understanding about how things work, and to live an active and generous life to achieve human flourishing not just for ourselves but also others [97].

Julia Annas, the great contemporary philosopher of virtue ethics, very much shares this understanding even though she does not explicitly refer to Spinoza. Annas argues that striving for virtue is structurally similar to acquiring particular practical skills [124]. It requires a willingness to learn from others and to eventually learn to think for ourselves. But whereas a person who acquires practical skills is focused on achieving a particular and well-defined end, a person who pursues virtue has only a vague idea of the end associated with human flourishing within the constraints of given life circumstances. In this context, Annas emphasizes the importance that ethical action cannot be detached from the agent, as utilitarian and deontological ethics suggest, but is rather determined by the agent's willingness to pursue an active and self-determined life seeking knowledge and showing a commitment to realize the personal potential in life. The motive to strive for virtue is, however, beyond egoism and altruism. Instead, it is about becoming more useful to others by making something of ourselves [124].

Spinoza's as well as Annas's ethics start from what we all have in common rather than what divides us, namely the pursuit of self-interest, the desire to persevere in being observed in all living organisms, and the awareness that we can achieve nothing in life without the support of others. They both also make a clear distinction between the nature of human and non-human animals pointing at the fact that nature was never meant to please human beings. Instead, it follows its own laws in which all living beings have the same right to live but are also exposed to the same natural forces that cause decay and death. In this context, humankind was able to gain insight into these natural laws and harness them to develop more effective survival strategies.

In the age of the Anthropocene, planet earth may be threatened by the very success of humankind to convert nature, that is not primarily designed to cater to human needs, into culture that has the primary purpose to address human needs; yet, human-made threats to the global ecosystem cannot be addressed effectively by calls for "a return to nature" or by appeals to humanity to abandon individually defined goals of human flourishing for the sake of saving the planet. What is required instead is an awareness that there is a problem as well as a sufficient sense of urgency to solve the problem that poses a threat to human flourishing in the long run. However, these problems need to be well-defined in order to develop solutions that really work. Throughout the history of animal domestication, there are many examples of how humans were able to find solutions to animal welfare problems, for example—be it better ways to protect them from predators or communicable diseases or, more recently, to decrease levels of stress by changing their environment in captivity to make it more conducive to their original natural environment. These problems were solved because of the human dependence on animal-based resources. Yet, by solving certain problems, new problems may emerge in addition to the challenge of feeding more people that have become more affluent. But again, there is no way to reverse the clock and return to old times of self-sufficiency.

What could potentially solve animal welfare problems as well as problems related to greenhouse emissions from the livestock industry altogether are perfect substitutes for meat, leather, dairy products and other animal-based resources that are not just equal in taste but also lower in price. Assuming that consumer acceptance would be high, many domesticated animals would indeed eventually disappear from the cultural landscape. But would that solve all the problems? Probably not, because substitutes for animal-based resources are technological solutions that may bring their own problems in terms of energy consumption and ingredients. It would also generate a significant societal challenge since millions of most underprivileged people who are employed in the livestock industry would lose their jobs. Moreover, semi-subsistence farmers who are not integrated into the formal global economy of food production are unlikely to have access to these substitutes since they are simply not available in remote rural regions.

In other words, solving problems is essentially what humans do in cooperation with other humans. It is a process of learning, understanding and acquiring new knowledge and skills to make things better. This ethics of meliorism does not contain any salvation promise and that makes it less apt to provide meaning, orientation and identity for humans who long for ideals worth fighting for.

But learning how to solve problems, whether they are institutional or technical in nature, ultimately makes humanity better off—and not necessarily at the expense of the natural environment.

Usually, we are very good at solving local problems but very bad at solving global problems. This has a lot to do with the fact that collective action on the local level involves less transaction costs because individual interests are better aligned with the collective interest. Problems on the global level are, however, more likely to get solved if there is a willingness to move from ideology and vested interests to preserve the status quo towards enabling change through practical solutions that proved to work somewhere and have the potential to be scaled up. This requires, however, an inclusive governance approach that ensures that innovative ideas can spread as widely as possible and are easily adaptable to the local context elsewhere. So what do such innovative ideas have to do with virtue? One could argue that innovative ideas help to solve trade-offs between the use of natural resources and the striving for human flourishing. So does virtue; striving for virtue provides meaning, orientation and identity in a positive sense, because it is not based on a negative view of humankind as the root of all problems, but rather a positive one that shows confidence in the ability of humankind to solve problems. What innovative ideas and virtue further have in common is the fact that they are both non-rival goods. If I use an innovative idea to solve a problem, it does not mean that it is no more available for others. The same is true for virtue, if I become virtuous through an active life in pursuit of human flourishing, that does not mean that the availability of "virtue" for others has been diminished. In addition, whereas scarce and rival resources such as concrete material goods and natural resources diminish in value the more they are used, innovative ideas and virtue actually increase in value the more they are used. As such they help make the world a better place for more people and thus increase the probability of human flourishing. Since human flourishing is not possible without a natural environment that provides the very foundation of living, innovative ideas have already been employed effectively to restore ecosystem services. But despite all these positive aspects of the natural environment, we should not forget the fact that the laws of nature were never meant to please humankind, and therefore nature continues to be a threat for many humans and domesticated animals who live in resource-constrained rural tropical areas. Farmers in these regions have no means to protect their animals and crops from the forces of nature that are likely to aggravate due to climate change. These are basic anthropological insights that should be kept in mind in any form of ethical reasoning and animal ethics in particular.

References

1. Bauhn, P. 2023. Duties Regarding Animals and the Virtue of Justice. In *Animal Suffering, Human Rights, and the Virtue of Justice*, ed. P. Bauhn, 73–80. Cham: Springer Nature Switzerland. https://doi.org/10.1007/978-3-031-27048-2_9.
2. Pluhar, E. 1987. Speciesism: A Form of Bigotry or A Justified View? *Between the Species* 4. https://doi.org/10.15368/bts.1988v4n2.1.

3. Milburn, J., and A. Cochrane. 2021. Should We Protect Animals from Hate Speech? *Oxford Journal of Legal Studies* 41: 1149–1172. https://doi.org/10.1093/ojls/gqab013.
4. Stucki, S. 2023. *One Rights: Human and Animal Rights in the Anthropocene*. SpringerBriefs in Law. Cham: Springer International Publishing. https://doi.org/10.1007/978-3-031-19201-2.
5. Fairclough, N. 2013. *Critical Discourse Analysis: The Critical Study of Language*. 2nd ed. London: Routledge. https://doi.org/10.4324/9781315834368.
6. Dewey, J. 1986. *John Dewey: The Later Works, 1925-1953: Vol. 7: Ethics 1933*, ed. J. A. Boydston. 1st ed. Carbondale/Edwardsville: Southern Illinois University Press.
7. Dewey, J. 1986. *John Dewey: The Later Works, 1925–1953. Vol. 8: Essays and How We Think 1933*. ed. J. A. Boydston. 1st ed. Carbondale/Edwardsville: Southern Illinois University Press.
8. Dewey, J. 1986. *John Dewey: The Later Works, 1925–1953. Vol. 12: Logic: The Theory of Inquiry*, ed. J. A. Boydston. 1st ed. Carbondale/Edwardsville: Southern Illinois University Press.
9. Laudan, L. 1990. Normative Naturalism. *Philosophy of Science* 57. Cambridge University Press: 44–59. https://doi.org/10.1086/289530.
10. Kitcher, P. 2024. The Ethical Project. *Harvard University Press*. https://www.hup.harvard.edu/books/9780674284289. Accessed February 24.
11. Liszka, J. J. 2021. *Pragmatist Ethics: A Problem-Based Approach to What Matters*. State University of New York Pr.
12. Wallace, J. 2009. *Norms and Practice*. Cornell University Press.
13. Tomasello, M. 2014. *A Natural History of Human Morality*. Cambridge (Mass.): Harvard University Press.
14. Preti, A. A. 2018. Developing Habits of Moral Reflection: Dewey, Moral Inquiry, and Practical Ethics. In *Ethics across the Curriculum—Pedagogical Perspectives*, ed. E. E. Englehardt and M. S. Pritchard, 147–163. Cham: Springer International Publishing. https://doi.org/10.1007/978-3-319-78939-2_10.
15. Teehan, J., and C. diCarlo. 2004. On the Naturalistic Fallacy: A Conceptual Basis for Evolutionary Ethics. *Evolutionary Psychology* 2. SAGE Publications Inc: 147470490400200108. https://doi.org/10.1177/147470490400200108.
16. Singer, P. 1979. *Practical Ethics*. Cambridge University Press.

17. Barber, E. 2023. What Would It Mean to Treat Animals Fairly? *The New Yorker*, December 16.
18. Hopster, J. 2019. The Speciesism Debate: Intuition, Method, and Empirical Advances. *Animals* 9: 1054. https://doi.org/10.3390/ani9121054.
19. Albersmeier, F. 2021. Speciesism and Speciescentrism. *Ethical Theory and Moral Practice* 24: 511–527. https://doi.org/10.1007/s10677-021-10168-6.
20. Cochrane, A., R. Garner, and S. O'Sullivan. 2018. Animal ethics and the political. *Critical Review of International Social and Political Philosophy* 21: 261–277. https://doi.org/10.1080/13698230.2016.1194583.
21. Swartz, B., and B. D. Mishler. 2022. Speciesism, Science, and Society. In *Speciesism in Biology and Culture: How Human Exceptionalism is Pushing Planetary Boundaries*, ed. B. Swartz and B. D. Mishler, 3–31. Cham: Springer International Publishing. https://doi.org/10.1007/978-3-030-99031-2_1.
22. Calarco, M. 2015. *Thinking through Animals: Identity, Difference, Indistinction*. 1 ed. Stanford, California: Stanford University Press.
23. Singer, P. 1975. *Animal Liberation*. Harper Collins.
24. Reagan, T. 1984. *The Case for Animal Rights*. London, Melbourne and Henley: Routledge and Kegan Paul.
25. Humphreys, R. 2023. *Animals, Ethics, and Language: The Philosophy of Meaningful Communication in the Lives of Animals*. Springer Nature.
26. Singer, P. 2023. *Animal Liberation Now*. Random House.
27. Meyza, K. Z., I. Ben-Ami Bartal, M. H. Monfils, J. B. Panksepp, and E. Knapska. 2017. The Roots of Empathy: Through the Lens of Rodent Models. *Neuroscience & Biobehavioral Reviews* 76. SI:IBNS-2015: 216–234. https://doi.org/10.1016/j.neubiorev.2016.10.028.
28. Rowlands, M. 2012. *Can Animals Be Moral?* Oxford University Press. https://doi.org/10.1093/acprof:oso/9780199842001.001.0001.
29. MacFarquhar, L. 2016. *Strangers Drowning: Impossible Idealism, Drastic Choices, and the Urge to Help*. Reprint edition. New York, New York: Penguin Books.
30. McManus, R. M., M. Kleiman-Weiner, and L. Young. 2020. What We Owe to Family: The Impact of Special Obligations on Moral Judgment. *Psychological Science* 31: 227–242. https://doi.org/10.1177/0956797619900321.
31. Korsgaard, C. M. 2023. Amicus Brief. *Perspectives in Biology and Medicine* 66: 29–37.

32. Nussbaum, M. 1988. Nature, Function, and Capability: Aristotle on Political Distribution. *Oxford Studies in Ancient Philosophy: Supplementary Volume*.
33. Nussbaum, M. C. 2023. *Justice for Animals: Our Collective Responsibility*. New York (N.Y.) London: Simon & Schuster.
34. Nussbaum, M. C. 2003. *Upheavals of Thought: The Intelligence of Emotions*. 1 ed. Cambridge: Cambridge University Press.
35. Korsgaard, C. M. 2018. *Fellow Creatures: Our Obligations to the Other Animals*. Oxford University Press. https://doi.org/10.1093/oso/9780198753858.001.0001.
36. Korsgaard, C. M. 2012. A Kantian Case for Animal Rights. In *Animal Law—Tier and Recht: Developments and Perspectives in the 21st Century*, M. Michael and D. Kuhne, ed. J. Hanni, 3–27. Zurich: Dike Verla. https://doi.org/10.1093/acprof:oso/9780199396078.003.0010.
37. Russell, D., and P. Singer. 1997. An Interview with Professor Peter Singer. *Animal Issues* 1.
38. Kant, I. 1996. *The Metaphysics of Morals*. Cambridge: Cambridge University Press.
39. Gewirth, A. 1978. *Reason and Morality*. Chicago, IL: University of Chicago Press.
40. Singleton, G. R., P. R. Brown, J. Jacob, and K. P. Aplin. 2007. Unwanted and Unintended Effects of Culling: A Case for Ecologically-Based Rodent Management. *Integrative Zoology* 2: 247–259. https://doi.org/10.1111/j.1749-4877.2007.00067.x.
41. Allen, B. L., C. Bobier, S. Dawson, P. J. S. Fleming, J. Hampton, D. Jachowski, G. I. H. Kerley, et al. 2023. Why Humans Kill Animals and Why We Cannot Avoid It. *Science of the Total Environment* 896: 165283. https://doi.org/10.1016/j.scitotenv.2023.165283.
42. Aerni, P. 2021. The Ethics of Farm Animal Biotechnology from an Anthropological Perspective. *Sustainability* 13: 3674. https://doi.org/10.3390/su13073674.
43. Harari, Y. N. 2015. Industrial Farming Is One of the Worst Crimes in History. *The Guardian*, September 25, sec. Books.
44. Zeder, M. A., and B. D. Smith. 2009. A Conversation on Agricultural Origins: Talking Past Each Other in a Crowded Room. *Current Anthropology* 50: 681–690. https://doi.org/10.1086/605553.
45. Odling-Smee, J. 2010. Niche Inheritance. In *Evolution: The Extended Synthesis*, ed. M. Pigliucci and G. B. Müller. The MIT Press. https://doi.org/10.7551/mitpress/9780262513678.003.0008.

46. Bowles, S., and H. Gintis. 2011. A Cooperative Species: Human Reciprocity and Its Evolution. In *A Cooperative Species: Human Reciprocity and Its Evolution*. https://doi.org/10.1515/9781400838837.
47. Chomicki, G., E. T. Kiers, and S. S. Renner. 2020. The Evolution of Mutualistic Dependence. *Annual Review of Ecology, Evolution, and Systematics* 51: 409–432. https://doi.org/10.1146/annurev-ecolsys-110218-024629.
48. Ammerman, S. 2022. Animal Agents in the Human Environment. In *The Power of Nature: Archaeology and Human-Environmental Dynamics*. ed. M. L. Smith, 213–232. University Press of Colorado.
49. Zeder, M. A. 2015. Core Questions in Domestication Research. *Proceedings of the National Academy of Sciences* 112: 3191–3198. https://doi.org/10.1073/pnas.1501711112.
50. Taylor, C., and K. S. Montford, ed. 2020. *Colonialism and Animality: Anti-Colonial Perspectives in Critical Animal Studies*. London: Routledge. https://doi.org/10.4324/9781003013891.
51. Billet, M. I., A. Baimel, S. S. Sahakari, M. Schaller, and A. Norenzayan. 2023. Ecospirituality: The Psychology of Moral Concern for Nature. *Journal of Environmental Psychology* 87: 102001. https://doi.org/10.1016/j.jenvp.2023.102001.
52. Eom, K., T. Qian H. Tok, C. S. Saad, and H. S. Kim. 2021. Religion, Environmental Guilt, and Pro-Environmental Support: The Opposing Pathways of Stewardship Belief and Belief in a Controlling God. *Journal of Environmental Psychology* 78: 101717. https://doi.org/10.1016/j.jenvp.2021.101717.
53. Thompson, C. J. 2004. Marketplace Mythology and Discourses of Power. *Journal of Consumer Research* 31: 162–180. https://doi.org/10.1086/383432.
54. Rangan, H. 2000. *Of Myth and Movements: Rewriting Chipko into Himalayan History*. 1st published by Verso, 2000 edition. London; New York: Verso Books.
55. Barthes, R. 1957. *Mythologies*. Editions du Seuil.
56. Aerni, P. 2018. *Global Business in Local Culture: The Impact of Embedded Multinational Enterprises*. SpringerBriefs in Economics. Cham: Springer International Publishing. https://doi.org/10.1007/978-3-030-03798-7.
57. Asebe, G. 2016. The General Status of Animal Welfare in Developing Countries: The Case of Ethiopia. *Journal of Veterinary Science and Technology* 7. https://doi.org/10.4172/2157-7579.1000332.

58. Janssen-Tapken, U. 2009. Developing Optimized Cattle Breeding Schemes Based on the Demands and Opportunities of Poor Livestock Keepers in Eastern Africa with a Special Focus on Trypanosomiasis. Doctoral Thesis, ETH Zurich. https://doi.org/10.3929/ethz-a-006031104.
59. Pearson, R. A., and R. C. Krecek. 2006. Delivery of Health and Husbandry Improvements to Working Animals in Africa. *Tropical Animal Health and Production* 38: 93–101. https://doi.org/10.1007/s11250-006-4363-y.
60. Perry, B. D., T. P. Robinson, and D. C. Grace. 2018. Review: Animal Health and Sustainable Global Livestock Systems. *Animal* 12: 1699–1708. https://doi.org/10.1017/S1751731118000630.
61. Freidberg, S. 2010. Perspective and Power in the Ethical Foodscape. *Environment and Planning A* 42: 1868–1874. https://doi.org/10.1068/a43217.
62. eggXYt—Count Your Chickens Before They Hatch. 2024. *eggxyt*. https://www.eggxyt.com. Accessed March 5.
63. Coppola, F., G. Paci, M. Profeti, and S. Mancini. 2023. Stop Culling Male Layer-Type Chick: An Overview of the Alternatives and Public Perspective. *World's Poultry Science Journal* 0: 1–21. https://doi.org/10.1080/00439339.2023.2288315.
64. Devienne, P. 2023. Reflections on the Ethics of Veterinary Medicine. In *Animal Suffering*, 161–184. John Wiley & Sons, Ltd. https://doi.org/10.1002/9781394228935.ch9.
65. Springer, S., P. Sandøe, H. Grimm, S. A. Corr, A. T. Kristensen, and T. Bøker Lund. 2021. Managing Conflicting Ethical Concerns in Modern Small Animal Practice—A Comparative Study of Veterinarian's Decision Ethics in Austria, Denmark and the UK. *PLOS ONE* 16: e0253420. https://doi.org/10.1371/journal.pone.0253420.
66. Hernandez, E., P. Llonch, and P. V. Turner. 2022. Applied Animal Ethics in Industrial Food Animal Production: Exploring the Role of the Veterinarian. *Animals: An Open Access Journal from MDPI* 12: 678. https://doi.org/10.3390/ani12060678.
67. Animal W. 2024. *WOAH—World Organisation for Animal Health*. https://www.woah.org/en/what-we-do/animal-health-and-welfare/animal-welfare/. Accessed March 5.
68. Home. 2024. *WOAH—World Organisation for Animal Health*. https://www.woah.org/en/home/. Accessed March 2.
69. Mellor, D. J., N. J. Beausoleil, K. E. Littlewood, A. N. McLean, P. D. McGreevy, B. Jones, and C. Wilkins. 2020. The 2020 Five Domains

Model: Including Human-Animal Interactions in Assessments of Animal Welfare. *Animals: An Open Access Journal from MDPI* 10: 1870. https://doi.org/10.3390/ani10101870.

70. Grandin, T. 2014. Animal Welfare and Society Concerns Finding the Missing Link. *Meat Science* 98. Meat Science, Sustainability & Innovation: '60th International Congress of Meat Science and Technology 17–22 August 2014, Punta Del Este, Uruguay': 461–469. https://doi.org/10.1016/j.meatsci.2014.05.011.
71. Vieira, L. C. 2022. Urban Approaches to Sustainability: Food. In *The Palgrave Handbook of Global Sustainability*, ed. R. Brinkmann, 1411–1427. Cham: Springer International Publishing. https://doi.org/10.1007/978-3-031-01949-4_89.
72. Carolan, M. 2016. *The Sociology of Food and Agriculture*. 2nd ed. London: Routledge. https://doi.org/10.4324/9781315670935.
73. Miller, D. 2012. *Consumption and Its Consequences*. 1st ed. Cambridge: Polity.
74. Coral, C., and D. Mithöfer. 2023. Contemporary Narratives about Asymmetries in Responsibility in Global Agri-Food Value Chains: The Case of the Ecuadorian Stakeholders in the Banana Value Chain. *Agriculture and Human Values*: 1–20. https://doi.org/10.1007/s10460-022-10405-3.
75. Aerni, P. 2013. Resistance to Agricultural Biotechnology: The Importance of Distinguishing between Weak and Strong Public Attitudes. *Biotechnology Journal* 8. https://doi.org/10.1002/biot.201300188.
76. Aerni, P. 2018. The Use and Abuse of the Term 'GMO' in the 'Common Weal Rhetoric' Against the Application of Modern Biotechnology in Agriculture. 39–52. https://doi.org/10.1079/9781786394644.0039.
77. Brick, C., D. K. Sherman, and H. S. Kim. 2017. "Green to Be Seen" and "Brown to Keep Down": Visibility Moderates the Effect of Identity on Pro-Environmental Behavior. *Journal of Environmental Psychology* 51: 226–238. https://doi.org/10.1016/j.jenvp.2017.04.004.
78. Cisek, S. Z., C. Sedikides, C. M. Hart, H. J. Godwin, V. Benson, and S. P. Liversedge. 2014. Narcissism and Consumer Behaviour: A Review and Preliminary Findings. *Frontiers in Psychology* 5: 232. https://doi.org/10.3389/fpsyg.2014.00232.
79. Miller, D. 2023. *The Good Enough Life*. John Wiley & Sons.
80. Jauernig, J., M. Uhl, and G. Waldhof. 2023. Genetically Engineered Foods and Moral Absolutism: A Representative Study from Germany. *Science and Engineering Ethics* 29: 34. https://doi.org/10.1007/s11948-023-00454-0.

81. Aerni, P., and K.-J. Grün. 2011. Moral und Angst: Erkenntnisse aus Moralpsychologie und Politischer Theologie.
82. Aerni, P. 2021. Decentralized Economic Complexity in Switzerland and Its Contribution to Inclusive and Sustainable Change. *Sustainability* 13. Multidisciplinary Digital Publishing Institute: 4181. https://doi.org/10.3390/su13084181.
83. Inglehart, R. 1977. *The Silent Revolution: Changing Values and Political Styles Among Western Publics*. Princeton University Press.
84. Aerni, P., M. Stavridou, P. Aerni, M. Stavridou, and I. Schluep, ed. 2021. *Transitioning to Decent Work and Economic Growth*. MDPI—Multidisciplinary Digital Publishing Institute. https://doi.org/10.3390/books978-3-03897-779-7.
85. FAO, Unicef, IFAD, WFP, and WHO. 2022. The State of Food Security and Nutrition in the World 2022. https://doi.org/10.4060/cc0639en.
86. Niamir-Fuller, M., and E. Huber-Sannwald. 2020. Pastoralism and Achievement of the 2030 Agenda for Sustainable Development: A Missing Piece of the Global Puzzle. In *Stewardship of Future Drylands and Climate Change in the Global South: Challenges and Opportunities for the Agenda 2030*, ed. Simone Lucatello, Elisabeth Huber-Sannwald, Ileana Espejel, and Natalia Martínez-Tagüeña, 41–55. Springer Climate. Cham: Springer International Publishing. https://doi.org/10.1007/978-3-030-22464-6_3.
87. Hollander, J. M. 2003. *The Real Environmental Crisis: Why Poverty, Not Affluence, Is the Environment's Number One Enemy*. 1 ed. Berkeley, Calif.: University of California Press.
88. Moore, G. E., ed. *Principia Ethica*. Cambridge University Press.
89. Damasio, A. R. 2018. *The Strange Order of Things: Life, Feeling, and the Making of Cultures*. Vintage.
90. Haidt, J. 2007. The New Synthesis in Moral Psychology. *Science* 316. American Association for the Advancement of Science: 998–1002. https://doi.org/10.1126/science.1137651.
91. Kisner, M. J., ed. 2018. *Spinoza: Ethics: Proved in Geometrical Order*. Translated by Michael Silverthorne. Translation edition. New York: Cambridge University Press.
92. Löwith, K. 1972. Vermittlung Und Unmittelbarkeit Bei Hegel, Marx Und Feuerbach. *Revue Internationale de Philosophie* 26. Revue Internationale de Philosophie: 308–335.
93. Sacks, O. 1990. *Neurology and the Soul*. Vol. 37(18). The New York Review of Books.

94. Damasio, A. 2003. *Looking for Spinoza: Joy, Sorrow, and the Feeling Brain*. 1 ed. New York: Harvest.

95. Lin, M. 2019. The Conatus Doctrine. In *Being and Reason: An Essay on Spinoza's Metaphysics*, ed. M. Lin, 0. Oxford University Press. https://doi.org/10.1093/oso/9780198834151.003.0006.

96. Hidalgo, C. 2015. *Why Information Grows: The Evolution of Order, from Atoms to Economies*. Illustrated edition. New York: Basic Books.

97. Garrett, D. 2018. *Necessity and Nature in Spinoza's Philosophy*. Oxford University Press. https://doi.org/10.1093/oso/9780195307771.001.0001.

98. Hübner, K. 2014. Spinoza on Being Human and Human Perfection. In *Essays on Spinoza's Ethical Theory*, ed. M. J. Kisner and A. Youpa, 0. Oxford University Press. https://doi.org/10.1093/acprof:oso/9780199657537.003.0007.

99. Nadler, S. 2014. The Lives of Others: Spinoza on Benevolence as a Rational Virtue. In *Essays on Spinoza's Ethical Theory*, ed. M. J. Kisner and A. Youpa, 0. Oxford University Press. https://doi.org/10.1093/acprof:oso/9780199657537.003.0003.

100. Graham, J., P. Meindl, S. Koleva, R. Iyer, and K. M. Johnson. 2015. When Values and Behavior Conflict: Moral Pluralism and Intrapersonal Moral Hypocrisy. *Social and Personality Psychology Compass* 9: 158–170. https://doi.org/10.1111/spc3.12158.

101. Rogers, C. D. 2021. Being Consistently Biocentric: On the (Im) Possibility of Spinozist Animal Ethics. *Journal for Critical Animal Studies*.

102. Wilson, M. D. 1999. Chapter 12. "For They Do Not Agree in Nature with Us": Spinoza on the Lower Animals. In *Ideas and Mechanism: Essays on Early Modern Philosophy*, ed. M. D. Wilson, 178–195. Princeton University Press.

103. Grey, J. 2013. "Use Them at Our Pleasure": Spinoza on Animal Ethics. *History of Philosophy Quarterly* 30: 367–388.

104. Özbey, S. N. 2023. Spinoza's Serpentine Worries: Animal Affects in Spinoza. In *Different Beasts: Humans and Animals in Spinoza and the Zhuangzi*, ed. S. N. Özbey, 0. Oxford University Press. https://doi.org/10.1093/oso/9780197686386.003.0009.

105. Rocca, M. D. 2004. Egoism and the Imitation of Affects in Spinoza. In *Spinoza on Reason and the Free Man*, ed. Yirmiahu Yovel. Little Room Press.

106. Steinberg, J. 2013. Imitation, Representation, and Humanity in Spinoza's Ethics. *Journal of the History of Philosophy* 51: 383–407. https://doi.org/10.1353/hph.2013.0045.

107. James, S. 2021. Spinoza on the Constitution of Animal Species. In *A Companion to Spinoza*, 365–374. John Wiley & Sons, Ltd. https://doi.org/10.1002/9781119538349.ch33.

108. Dawkins, M. S. 2017. Animal Welfare and Efficient Farming: Is Conflict Inevitable? *Animal Production Science* 57: 201. https://doi.org/10.1071/AN15383.

109. Maia de Souza, D., R. Petre, F. Jackson, M. Hadarits, S. Pogue, C. N. Carlyle, E. Bork, and T. McAllister. 2017. A Review of Sustainability Enhancements in the Beef Value Chain: State-of-the-Art and Recommendations for Future Improvements *Animals* 7: 26. https://doi.org/10.3390/ani7030026.

110. Amos, N., and R. Sullivan, ed. 2017. *The Business of Farm Animal Welfare*. 1st ed. Routledge.

111. Devitt, C., P. Kelly, M. Blake, A. Hanlon, and S. J. More. 2013. Dilemmas Experienced by Government Veterinarians When Responding Professionally to Farm Animal Welfare Incidents in Ireland. *Veterinary Record Open* 1: e000003. https://doi.org/10.1136/vropen-2013-000003.

112. Thompson, P. B. 2017. *The Spirit of the Soil: Agriculture and Environmental Ethics*. 2nd ed. New York: Routledge. https://doi.org/10.4324/9781315559971.

113. Fernandes, J. N., P. H. Hemsworth, G. J. Coleman, and A. J. Tilbrook. 2021. Costs and Benefits of Improving Farm Animal Welfare *Agriculture* 11: 104. https://doi.org/10.3390/agriculture11020104.

114. McEwen, B. 2017. Eternally Vulnerable: The Pathology of Abuse in Domestic Animals. *Academic Forensic Pathology* 7: 353–369. https://doi.org/10.23907/2017.032.

115. Williams, J., and G. Ferreira, ed. 2023. *Understanding Animal Abuse and How to Intervene with Children and Young People: A Practical Guide for Professionals Working with People and Animals*. London: Routledge. https://doi.org/10.4324/9781003165552.

116. Van Assche, A., and R. Narula. 2023. Internalization Strikes Back? Global Value Chains, and the Rising Costs of Effective Cascading Compliance. *Journal of Industrial and Business Economics* 50: 161–173. https://doi.org/10.1007/s40812-022-00237-x.

117. Moses, L. 2018. Another Experience in Resolving Veterinary Ethical Dilemmas: Observations From a Veterinarian Performing Ethics

Consultation. *The American Journal of Bioethics* 18: 67–69. https://doi.org/10.1080/15265161.2017.1409825.

118. Ihle, R., O. D. Rubin, Z. Bar-Nahum, and R. Jongeneel. 2020. Imperfect Food Markets in Times of Crisis: Economic Consequences of Supply Chain Disruptions and Fragmentation for Local Market Power and Urban Vulnerability. *Food Security* 12: 727–734. https://doi.org/10.1007/s12571-020-01084-1.
119. Syukron, M., and Y. Su. 2023. Understanding Small- and Very-Small-Scale Size Meat Processors in Missouri to Strengthen the Local Supply Chain. *Journal of Agriculture, Food Systems, and Community Development* 12. https://doi.org/10.5304/jafscd.2023.122.006.
120. Frohlich, X. 2023. *From Label to Table: Regulating Food in America in the Information Age*. 1 ed. Oakland, California: University of California Press.
121. Parlasca, M. C., and M. Qaim. 2022. Meat Consumption and Sustainability. *Annual Review of Resource Economics* 14: 17–41. https://doi.org/10.1146/annurev-resource-111820-032340.
122. Chater, N., H. Zeitoun, and T. Melkonyan. 2019. The Social Character of Moral Reasoning. *The Behavioral and Brain Sciences* 42: e149. https://doi.org/10.1017/S0140525X18002583.
123. Tomasello, M., and A. Vaish. 2013. Origins of Human Cooperation and Morality. *Annual Review of Psychology* 64: 231–255. https://doi.org/10.1146/annurev-psych-113011-143812.
124. Annas, J., and D. Copp. 2007. Virture Ethics. In *The Oxford Handbook of Ethical Theory*. Oxford Handbooks.

Chapter 6

The Value of Life: Resuscitation, Resurrection, and the Ethics of De-Extinction

Kurt Reymers
Division of General Education, School of Arts and Sciences,
SUNY Morrisville, Morrisville, New York 13408, USA
reymers@morrisville.edu

This essay on the ethics of de-extinction evaluates two metaphors used to describe recent interest and attempts at de-extinction, or the process of bringing extinct species back into existence through back breeding, cloning, and gene-editing technologies. The metaphors of resuscitation and resurrection are particularly common in the literature on de-extinction, so they are independently investigated as to the moral and ethical ramifications related to each and then compared to the literature on the ethics of de-extinction. The analysis concludes that at the core of the debate over de-extinction can be found expressions regarding the value of life and how its definition is approached by competing sociological and ethical interests.

Animals, Ethics, and Engineering: Intersections and Implications
Edited by Rosalyn W. Berne

ISBN 978-981-5129-56-4 (Hardcover), 978-1-003-65231-1 (eBook)
www.jennystanford.com

6.1 De-Extinction Metaphors: Resuscitation and Resurrection

"Metaphors have a way of holding the most truth in the least space."

—Orson Scott Card, *Alvin Journeyman* [1]

Scientists estimate that a million species are threatened with extinction currently [2]. Efforts being made to bring back extinct species (de-extinction) are now backed with significant resources and attention. A recent press release header notes that the de-extinction firm *Colossal Biosciences* is a "billion dollar startup" and there has been a great amount of media hype for several years accumulating around projects such as *Colossal* and *Revive and Restore* [3–5]. However, there are few national or international policies and regulations guiding decisions made about such questions, other than that prohibiting human cloning or "heritable" gene-editing (according to the Center for Genetics and Society, "approximately 46 countries have formally banned human cloning. . . represent[ing] less than a quarter of all countries" [6]. As to what species should or should not be resuscitated or whether it will ever be ethical to resurrect from extinction another human species, such as Neanderthals, scientists can rely only on a relatively small number of bioethics and philosophy scholars to guide their decision-making. As some have noted [7–11] the ethical complexities of these goals are significant. Cohen says, "due to its special character, de-extinction tests the limits of our ethical notions from multiple angles" [8]. Rather than rehash the convoluted ethical quandaries of the topic that have been adequately handled by more well-armed philosophers, this essay will examine only two of the metaphors used to draw analogies between de-extinction and the practices and phenomena related to "resuscitation" and "resurrection." These metaphors are frequently used in public dialogue about species de-extinction [12–15] and it may be useful to investigate the ethical constraints surrounding those metaphorical categories to gain insight into what factors might help to guide the development of de-extinction technologies and decisions on how they might be used.

6.1.1 Back-Breeding, Cloning, and CRISPR

There are three technologies that might be used, alone or in conjunction, to bring an extinct species back into existence. The first of these is selective breeding. This is hardly a new technology, having been utilized at least as far back as the einkorn wheat grown in the hills of what is now northern Turkey at the dawn of civilization. Without selective hybridization, humanity would arguably still be in the stone age. Selective breeding that introduces elements of a prior gene pool that has all but disappeared is known as *back breeding*. Currently, one of the more famous back-breeding projects involves the quagga, a creature closely related to the modern zebra. Significant phenotypic (trait) differences are a lack of stripes on the brownish rear quarter of the animal and a narrow snout. While attempts at breeding zebra with features akin to the quagga have been undertaken, scientists recognize that the resultant animal may not be significantly different enough from the modern zebra to classify as a separate sub-species [16].

The second method of de-extinction involves cloning. In this method, the nucleus of a cell captured from an extinct species is implanted into an embryo of the closest living female relative of the extinct species and allowed to come to term. This process is called somatic cell nuclear transfer, or SCNT. For example, to de-extinct the passenger pigeon, which went extinct in 1914, cells from a preexisting passenger pigeon would be harvested, the nucleus of a cell transferred to an embryo to be inserted into the womb of a band-tailed pigeon, the passenger pigeon's closest living relative, and allowed to come to term. This prompts two questions: first, where do cells from extinct species come from? One source is the "Frozen Zoo" in San Diego, California. The Frozen Zoo is a cryogenic facility that houses the stem cells of a diversity of creatures that are, or are in danger of, going extinct. Dr. Oliver Ryder, a geneticist at the Frozen Zoo, says "We have species in the Frozen Zoo that are not from extinct species, but are from species that have a very limited gene pool. And if we can flow the genes from the Frozen Zoo into the living population, we can reduce the impacts of small populations like inbreeding" [2]. The Zoo now boasts a collection of cryogenically frozen tissues representing 1,100 species [17].

The second question about cloning is, how successful is SCNT? Unfortunately, success rates of viable embryos are relatively low, and implantation rates even lower, making the prospects for a living clone tenuous in any one given trial. Also, clones frequently have genetic defects. In the case of the first cloned de-extinction, an SCNT of DNA from the last Pyrenean ibex (named Celia) was implanted into a closely related relative, which came to term. Siipi and Finkelman note, "Using cells cultured from those samples, scientists working for the company Advanced Cell Technology set about cloning Celia. The fruit of their labor was born in 2009 and survived for 7 minutes before succumbing to lung problems" [[11]:428]. Another example is Elizabeth Ann, a cloned black-footed ferret, the first of any native endangered species in North America, who will not be able to reproduce because of genetic abnormalities in her reproductive system that required a hysterectomy [18, 19]. The fact that cloning in this way is prone to genetic abnormalities is a concern for returning extinct species, but scientists believe that with enough time these problems can be surmounted.

The third method of de-extinction involves gene editing. This technique manipulates the genome directly, changing the sequence of amino acids that make up the DNA strand which codes genotypes and relates, in some way, to specific phenotypes. This allows for the use of incomplete DNA fragments to be used, a uniquely useful ability when dealing with species that are "deep" or long extinct. The technology that has come to the fore in this area is CRISPR-Cas9, a genome-editing technique developed in the first two decades of the 21st century by a cohort of biological scientists. The development of the CRISPR-Cas9 genome editing technique was recognized by the Nobel Prize in Chemistry in 2020 which was awarded to Emmanuelle Charpentier and Jennifer Doudna [20] (Royal Swedish Academy of Sciences 2020). CRISPR stands for Clustered Regularly Interspaced Short Palindromic Repeats. A recent variation on the tool adds "Cas9" (CRISPR-associated proteins) to the acronym. Paired with cloning, CRISPR-Cas9 methods offer the ability to support the reintroduction of even deep extinct species, or at least some of their genetic traits, into the contemporary world.

An analysis of the ethical similarities and differences between the metaphors of resuscitation and resurrection proper and de-extinction may help to frame de-extinction. Resuscitation is a long-held practice, taking different forms over the centuries as medical knowledge and technology have changed and as cultural norms and boundaries ebb and flow. Resurrection is even more complicated, bringing deeply entrenched and culturally specific theological, mythological, and moral dimensions into the argument and demanding an accounting for the common portrayal of resurrection-technology as "playing God."

6.2 The Ethics of Resuscitation

"In dark times, the definition of good art would seem to be art that locates and applies CPR to those elements of what's human and magical that still live and glow despite the times' darkness."

—David Foster Wallace, *Conversations with David Foster Wallace* [21]

Medical resuscitation is the process of reviving and stabilizing a person who is in a life-threatening condition, such as cardiac arrest, respiratory failure, or severe trauma. The goal of medical resuscitation is to restore vital functions, such as breathing and circulation, as quickly and efficiently as possible to prevent organ damage and death. This process typically involves a combination of techniques and interventions, including cardiopulmonary resuscitation (CPR), defibrillation, airway management, and administration of medications. Resuscitation practices have existed for thousands of years. The earliest documentation of CPR dates to the Old Testament, which describes how the prophet Elisha resuscitated an apparently dead child by blowing air into his mouth [22].

Medical professionals are trained to perform resuscitation procedures in various healthcare settings, including hospitals, emergency rooms, and pre-hospital environments, like ambulances and public spaces. Early and effective resuscitation can significantly improve the chances of survival and reduce the risk of long-term complications for patients experiencing life-threatening emergencies.

While not exclusively cardiovascular in nature, many cases involve cardio-pulmonary resuscitation, or CPR. The practice of CPR is attempted over 150,000 times a year in the U.S. alone [23]. It involves compression of the chest to simulate a heart rhythm and provision of oxygen to the lungs. When performed properly, it can extend the opportunity for critical medical care to arrive and the patient to be attended to in a hospital setting (if not already there – there are an estimated 370,000–750,000 cases of in-hospital CPR administered in the United States yearly [24].

New technologies and practices related to medical resuscitation have introduced new challenges to parties responsible for the decision-making about and the administering of resuscitation. At the same time, there has been a shift from paternalistic to patient-centered care with an increasing focus on individual rights and values [25]. Ethical issues may arise in the practice of resuscitation when there is a lack of clarity in the wishes of a non-responsive patient.

Four ethical principles dominate the decision-making process regarding resuscitation [26]. They are *autonomy, beneficence, nonmaleficence,* and *justice* [27, 28]. Mentzelopoulos et al. suggest the following definitions for these ethical principles:

- *Autonomy*: respect for the right of self-determination.
- *Beneficence*: selection of beneficial interventions for the patient after assessment of the risk-to-benefit ratio.
- *Non-maleficence*: avoiding harm or inflicting the least possible harm in the course of achieving a beneficial outcome.
- *Justice*: means fair and equal distribution of benefits, risks, and costs; pertains to the equality of rights to healthcare, and the legal obligation of healthcare providers to adhere to appropriate care and allocation of burdens and benefits [25].

6.2.1 Autonomy

In 1976, the *New England Journal of Medicine* (*NEJM*) was among the first to report hospital policies on the process for making and communicating decisions about a patient's resuscitation status [29], developing the infrastructure for what was to become the Do Not Resuscitate (DNR) order, allowing for patients to provide advanced directives regarding their care. This documentation of intent—often included in a living-will, health care proxy, or (more recently) a MOLST form (Medical Orders for Life Sustaining Treatment)—reinforces one of the central values that undergirds the ethical use of CPR today: autonomy. The idea that "persons have intrinsic and unconditional worth, and therefore, should have the power to make rational decisions and moral choices" ([30]: p.19) is a central value of most individualist cultures. It is at the heart of the enlightenment philosophies of Immanuel Kant and John Stuart Mill and of democratic political reform throughout the world over the past 200 years. So, it is no wonder that its reinforcement is a dominant ethical value within modern healthcare.

DNRs and MOLSTs are not the only way in which patients can express or medical practitioners can elicit it. Varkey notes three other ways in which autonomy is reinforced: through informed consent, truth-telling, and confidentiality [30]. While it's hard to see how the first and last of these practices might apply to the case of de-extinction, the central practice of truth-telling might. Honesty, say Mentzelopoulos et al. [[25]: p. 704] is the accurate and transparent communication to the patient/family of the best research evidence, and clinical judgment including uncertainties. This emphasis on truth-telling, as with all science, should be noted as a core value in the practice of resuscitation.

One challenge to autonomy can be paternalism (although *parentalism* is the proper term, I will use this term, as it is already rooted in ethics literature) [30–32]. This occurs in cases where the practitioner or state believe that they have the best interests of the patient at heart, despite what the patient might want. This is particularly the case when autonomy is compromised temporarily or permanently. O'Neill notes that in asking 'What would this patient have chosen in this situation?', "it may be possible to maintain elements of respect for the particular patient as he or she was in former times. But usually this provides only vague indications for medical or other treatment, and respect for absent autonomy can be at best vestigial" [[31]: p. 178]. Rather, O'Neill argues, we should be asking another question: "What would the ideally autonomous choose in this situation?" [[31]: p. 178]. Though the caretaker may strive to satisfy the patient's needs through a perceived beneficence, it may in fact be paternalistic.

6.2.2 Beneficence

"The obligation of physician to act for the benefit of the patient" [[30]: p. 19] describes the ethic of beneficence. This ethic entails a number of moral rules to protect and defend the right of others, for instance, to help persons with disabilities and rescue persons in danger. It also involves aiding the psychological state of the patient, particularly the state of self-worth as it is related to human dignity. Maintaining dignity "regarding resuscitation and post-resuscitation care, means avoiding disproportional interventions and an 'end-of-life' contradicting patient's preferences" [[25]: p. 705]. The issue of human dignity is very much at the core of resuscitation response, but is also culturally, even sub-culturally, specific and extraordinarily difficult to define. Nonetheless, the caring physician will recognize the boundaries for given patients (sometimes reinforced by a living will or health care proxy, other times by family or community) and respect the patient's dignity to provide a "good end" [33]. Death, at some juncture, must be accepted as an inevitable outcome.

This may be cold comfort to a physician. "The primary goal of physicians is traditionally accepted as that of preserving or restoring life to the patient. Because of this traditionally accepted goal, sometimes physicians face significant difficulties accepting death as a natural process" [[27]: p. 610]. Resuscitation provides perhaps the final opportunity for beneficence. "However, in many cases, death is not necessarily unwelcome by the patient and family, and should be accepted as a natural part of life, by the patient, loved ones, and health care providers. . . The physician should address the many non-medical challenges facing patients and loved ones at the end of life, including numerous physical, emotional, social, cultural, and spiritual challenges" [[27]: p. 610–611].

Focus on the practicality of providing a "good" or dignified outcome often requires the physician to manage the equally or more challenging to define concept of futility [22, 25, 27, 30,

34, 35]. Futility is "the use of considerable resources without a reasonable hope that the patient would recover to a state of relative independence or be interactive with their environment" [[25]: p. 710]. "However," Mentzelopoulos et al. continue, "without quantification of 'considerable', 'reasonable', 'relative independence', and 'interactive', beneficial treatment can be arbitrarily/unethically denied to vulnerable population subgroups. . ." [[25]: p. 710]. Under this definition, the balancing of resources might be tied to a more utilitarian ethical outlook on resuscitation. However, Marco notes that "several authors have shown that there is no consistent consensus among physicians about the definition of futility" [[27]: p. 609].

Without question, finding the throughline to the provision of a good quality of life post-resuscitation, or even a good death, is difficult and demands of the care provider a sensitivity that can be hard to summon, especially if unfamiliar with the culture or ethno-religious norms of the patient and the patient's family, and a need to understand and then balance the resources needed to take on resuscitation practices.

6.2.3 Nonmaleficence

Nonmaleficence is the obligation of a physician not to harm the patient [30]. Nonmalificence is tied to beneficence through Hippocrates, whose famous line "to do good or to do no harm" established the so-called Hippocratic Oath, which became a staple of medical school convocations in the early 1700s [36].

The oath to do no harm involves a nuanced understanding of what we mean by harm, and how it relates to actions taken from a caretaker, or even their inaction toward the patient. Active (but unintentional) harm might result from post-resuscitation damage. For instance, brain injury after resuscitation, a common sequela following cardiac arrest, can range in severity from mild impairment to devastating brain injury and brainstem death [37]. Inaction can also lead to harm, as the case of futility outlines above. To avoid the appearance of maleficence, medical practitioners are often guided to "err on the side of caution" and, barring "objective signs of irreversible death and in the absence of known advanced directives against resuscitative efforts, full resuscitation should be offered" [[32]: p. 218]. The dilemmas that often arise with respect to autonomy, beneficence, and nonmaleficence are that "full resuscitation" poses the possibilities of (1) being unaware of (or simply not responding to) a DNR order, compromising the patient's autonomy, (2) denying the patient their dignity in what might be the final moments of living, and (3) making a bad situation worse by saving a patient yet leaving them compromised, for example, by brain trauma occurring as a result of a deficit of oxygen reaching the brain before or during the very resuscitation efforts designed to keep them alive.

6.2.4 Justice

To complicate matters further, when the goal of justice is added to the first three ethical pillars, new conditions emerge. Justice, in the context of clinical healthcare, "means fair and equal distribution of benefits, risks, and costs; pertains to the equality of rights to healthcare, and the legal obligation of healthcare providers to adhere to appropriate care and allocation of burdens and benefits" [25]. This definition describes the ethical category of distributive justice. Fleishacker says, "Distributive justice refers to the fair, equitable, and appropriate distribution of health-care resources determined by justified norms that structure the terms of social cooperation" [30]. In other words, justice is equity in the context of culture and social structure.

Tied to the resource issue related to nonmaleficence, distributive justice requires the medical practitioner to evaluate important resources that relate to resuscitation, including "allotment of scarce resources (equipment, tests, medications, organ transplants), care of uninsured patients, and allotment of time for outpatient visits (equal time for every patient? based on need or complexity? based on social and or economic status?)" [30]. A brazen example of a violation of the principle of distributive justice is when a physician orders an expensive but unnecessary test or prescribes a patented rather than generic pharmaceutical without consideration of the patient's ability to pay and which would benefit the physician financially.

Being sure that the general public is well-informed can advance the cause of distributive justice; it is well-proven that education and knowledge reduces social differences and conflict. Medical associations [38–40] and governments, NGO's, and non-profits [41–43] have developed public campaigns to teach the importance of resuscitation generally and to provide training in the techniques of life-saving CPR. But one area in which the public is grossly misinformed is in the success rates and outcomes of resuscitation.

The American public believes that resuscitation is far more successful than it, in fact, is [34]. This may be because, in the United States at least, accounts of resuscitation are more likely to derive from fictional medical television shows than reality itself. In a study of three popular medical shows–*Chicago Hope*, *ER*, and *Rescue 911*–Diem et al. reveal four misconceptions about resuscitation: first, these three television programs give a misleading impression about the kind of people most commonly given CPR, vastly underrepresenting elderly populations; second, underrepresenting the dominant cause of cardiac arrest, which is cardiac disease, and far overrepresenting acute traumas such as motor vehicle accidents, near drowning, or gun violence; third, "CPR succeeded more frequently on television than in the real world as reflected in the medical literature. On all three shows combined, 75% of the patients were alive immediately after their cardiac arrests, and 67% appeared to survive in the long term" [[34]: p. 1581], whereas in reality, rates of long-term survival after cardiac arrest as reported in the medical literature vary from 2% to 30% for arrests outside a hospital, and from 6.5% to 15% for arrests that take place inside a hospital [34]; and, finally, on television, the outcome of CPR was typically portrayed as either full recovery or death—in only one case of all episodes watched was a patient's outcome more complicated. But, as discussed earlier in the context of nonmaleficence, "controversy surrounds the use of CPR precisely because the procedure can lead to prolonged suffering, severe neurologic damage, or an undignified death" [[34]: p. 1581].

The differences between the perception given on television and the reality of resuscitation are grossly disproportionate. Addressing these differences in some way might help moderate expectations and give patients and families a better sense of what equitable and fair outcomes might look like. A more just distribution of knowledge about the reality of resuscitation might temper expectations of patients and their families regarding the outcomes of actual resuscitation practices.

Other issues of fairness and equity also are impactful in the provision of resuscitation. It is often difficult to clearly separate the interests of the patient from the physician's individual and collective interests, despite having codes of conduct. Berkman et al. say, "Physician organizations (local, state, and national) have codified their standards. However, complying with these standards, it should be understood, may not always fulfill the moral norms as the codes have "often appeared to protect the profession's interests more than to offer a broad and impartial moral viewpoint or to address issues of importance to patients and society" [30].

This review of the four principles of ethics for clinical resuscitation practices in the medical community is meant to demonstrate some of the complexity which the practitioners (physicians, medical practitioners, and the general public) must handle as they negotiate codes as ancient as the Bible and as modern as defibrillator technology in the context of real-life care. The question which we will take up shortly is, how will these codes and the dilemmas that stem from their real-life application apply when using the metaphor of resuscitation to describe not individuals but entire species?

But next, let us turn to the second of the two metaphors I've chosen to investigate based on its frequent appearance in references to de-extinction projects, resurrection.

6.3 The Ethics of Resurrection

"Nothing can save us that is possible: we who must die demand a miracle."

—W. H. Auden, *For the Time Being: A Christmas Oratorio* [44]

Resurrection is a word in American culture packed with religious and philosophical undertones (a particularly potent symbol for evangelical Christians), but also used regularly in popular culture (for instance, in the third installment in the Matrix trilogy, appropriately named *Resurrection* given the techno-theological foundations of that movie series). Resurrection is a common mythological theme in many cultural traditions around the world. Resurrection even appears in scientific literature surrounding species de-extinction, as can be seen from no fewer than a thousand peer-reviewed article titles on the topic available through a simple library search for "resurrection of extinct species" (as of this publication). What, then, are the meanings of the term "resurrection" in the culture of the Western Judeo-Christian tradition and mythologically, and what is its relation to morality and, ultimately, ethics?

"Resurrection" is, by the standard OED definition, "the rising of the dead as a tenet of religious belief; spec. (in Christian thought) that which is to occur at the Last Judgement." [45]. But while there is perhaps a stronger emphasis on the concept of the dead in this definition as of the living, theologians writing on the topic are more likely to address the significance of the resurrection as a deific act of creation, outside of time, space, and perhaps even reason.

6.3.1 Themes of the Resurrection

Simon Perry writes in his 2012 book *Resurrecting Interpretation: Technology, Hermeneutics, and the Parable of Lazarus and the Rich Man,* Chapter 4: Resurrection as the Chiasmus of History that, "At the 'center of time' stands the climactic Christ-event. The resurrection of Jesus is not a resurrection *from* creation but sets off a chain reaction that is the resurrection *of* creation and thereby involves every part of the created order" [[46]: p. 115]. Not merely a "re-creation" of an ordinary human, the resurrection symbolizes a reordering of the Universe such that all humans can be delivered into God's good grace. And this ethic is not merely Christian. In her contribution to the Oxford Dictionary of Theological Ethics (2007), titled "Creation and Ethics," Lisa Cahill states that "to invoke creation is not to bracket religious identity while pursuing common human moral values. Rather, creation is a symbolic point of unity among religious traditions. It can underwrite religious commitment to uphold continuities and commonalities in human moral experiences and obligations across traditions" [[47]: p. 7] The inevitable moral quality of the resurrection is universal and, truly, divine.

Why might scientists adopt such a word? In part, the popular usage is a function of the general "disenchantment of the world" of Max Weber [48]. Another reason may be that in accepting this theologically inspired metaphor, advocates of de-extinction technology (people such as Stewart Brand and the ironically named George Church) are attempting to subsume its deep meaning that brings it an element of social capital [49]. In fact, Brand has said since 1968 (in relation to his first project, the Whole Earth Catalog) that "We are as gods, we might as well get good at it" [50], making that claim again in a debate arguing for the advancement of de-extinction projects, such as his own *Revive and Restore* project (in that instance, Brand and his debate partner George Church clearly lost) [51]. Certainly, the effort for the protection and conservation of living species is important as scientists recognize that we appear to be fast entering the early stages of what may be the Earth's sixth mass extinction [52], not coincidentally shortly after recognizing the Anthropocene Era. Having tools such as de-extinction at our fingertips to address such rapid evolutionary changes seems wise, but nonetheless the ethical debate continues to rage [7, 8, 10, 33, 53–62].

Before looking more specifically at the differences between these spiritual and scientific ideas, first we must ask, is the alleged breach between religious morality and scientific knowledge truly so great?

6.3.2 Science and Religion

The camps of scientific and religious thinking and talking have been recognized as representing a split between faith and reason in the light of new knowledge that is continually being produced from the scientific institutions of Western cultures. For those "concerned with the basic conceptual and methodological problems of religious language, and here the most significant influence has undoubtedly been science... Science seems to yield indubitable knowledge on which all men can agree. Its apparent objectivity contrasts with the subjectivity of religion" [63]. Perhaps the greatest nadir moment for the two institutions was the Scopes trial of 1925. Clearly there is at least the perception of a large gap ever since between evidence-based knowledge and faith-based knowledge. However, our conception regarding the size of the chasm between the two institutions is, according to John Evans, overestimated [64].

Evans examines the relationship between science and religion by examining the degree to which the "systemic knowledge" that informs belief systems are compatible. The weak version of "systemic knowledge" suggests that some facts and beliefs are incompatible between science and religion, whereas most are not. Says Evans:

> The strong version of the systemic knowledge relationship... requires that a person could not believe scientific claims based on scientific methodology about why plants move and also believe in a nontestable claim like the existence of God. You could also not believe two on-the-ground fact claims—one religious and one scientific—even if the religious one is never examined by science. For example, a Catholic [with a strong systemic knowledge belief system] could not believe in scientific claims about global warming and in the Resurrection, even if there is no scientist who ever makes a claim about the resurrection (e.g., there is no scientific discipline of anti-Resurrection studies). [*author's note: a scientist with a strong systemic knowledge belief system might argue that the study of biology is intrinsically an anti-Resurrection discipline.*] [[64]: p. 10]

It is revealing that those with a weak systemic knowledge framework—which does not make any claim to the strength of belief itself, but rather the strength of the relationship between the factual versus the non-factual claims that occur within the systems—can rise to prominent positions within each institution. There are many leading theologians who clearly accept the institution of science. Although the rumors of Pope Francis having a master's degree in chemistry have been debunked (he did graduate with the equivalent of an American associate degree as a chemical technician according to Snopes), he has been a staunch advocate of the science regarding climate change [65]. The institution of politics even poked its head into the fray when U.S. politician Rick Santorum commented that the Pope should "leave science to the scientists" [[65]: np]. Clearly,

Santorum either has a strong systemic knowledge belief profile, or he was embracing such a belief publicly to appeal to his conservative voter base. (It would be interesting to do a study on the degree to which politics influences the maintenance of such strong systemic knowledge ideologies).

Likewise, religiously affiliated scientific leaders have occupied positions of great prominence throughout the history of science. Sir Isaac Newton was known to be a staunch Protestant [66] and who can forget Einstein's famous phrase "God does not play dice with the Universe"? While often misunderstood to mean that Einstein was referring to a personal deity and or affirming a belief in destiny, which he wasn't, he still invokes an element of the divine in a 1954 letter explaining this quote: "If something is in me which can be called religious then it is the unbounded admiration for the structure of the world so far as our science can reveal it" [67]. According to Crosby, Einstein "saw the two as mutually dependent in that religion provided the faith that inspired scientific inquiry and science provided the rationality to analyze and understand religious traditions" [[68]: p. 186].

Francis Collins, the leader of the United States' Human Genome Project in the 1990s (the first organization to fully map the human genome) is a confirmed fundamentalist, guitar-playing, evangelical Christian. He has written extensively about how he conjoins the relationship between faith and science in his 2007 book *The Language of God: A Scientist Presents Evidence for Belief* [69]. In a 2022 interview, Collins talks explicitly about his integration of scientific and religious perspectives: "I never have encountered a conflict between what I know as a Christian and what I know as a scientist. I have to be careful about which kind of question is being asked. If it's about, you know, how does the Golgi apparatus of the cell work?—that's a scientific question. If it's a question about, why am I here?—well, science isn't helping me very much with that. But I can ask both those questions on any given day, and I'm not having a wall around them" [70].

In fact, Evans says that "most of the larger religious traditions in the U.S. teach that there is not a conflict between their religious belief system and the methodological naturalist version of a scientific belief system... Anything about the natural world that is potentially *demonstrable* will have a naturalistic explanation (like the cause of hurricanes), but non-demonstrable claims that no scientist cares about—like the Resurrection—may have a supernatural explanation. But, since the resurrection is not a violation of methodological naturalism, there is no conflict between religion and science" [[64]: p. 87].

While pockets of strong systemic knowledge still exist and could potentially grow, this weak version of systemic knowledge has become the default perspective on the part of the U.S. public, and to a greater extent in Europe. This in part explains the lack of concern with the terminology used in de-extinction narratives that invokes the chief and central miracle of Christianity, the Resurrection. This does not necessarily indicate a demythologization of religion, but it does signal a way in which the language is being used differently. Barbour [71] explains that "difficulties in religious language have been described by many authors in recent years. In Germany, Rudolf Bultmann has said that modern man can no longer speak of a God who acts in nature and history and has proposed a 'demythologized' version of the gospel... In the United States, three theologians who found themselves unable to accept theistic assertions were presented in the popular press as the 'Death of God' movement. These men are symptomatic of a widespread questioning of classical formulations" [[63]: np].

Howsoever the term *resurrection* is being adopted—spiritually or in a more secular sense—it has the power to alert us to several ethical themes: creation, yes, but also restoration, revival, and rebirth. Let us take the latter three themes and see what moral and ethical dimensions are involved therein.

6.3.3 Resurrection as Restoration, Revival, and Rebirth

From an ethical perspective, the term *restoration* evokes a sensibility of the return to a just state after an injustice is recognized and ameliorated. Perhaps no finer definition of restorative justice exists than that of Archbishop Desmond Tutu, who, reflecting on South Africa's Truth and Reconciliation Commission, wrote: "[in] restorative justice, the central concern is not retribution or punishment... in the spirit of *ubuntu* [humanity], the central concern is the healing of breaches, the redressing of imbalances, the restoration of broken relationships, a seeking to rehabilitate both the victim and the perpetrator, who should be given the opportunity to be reintegrated into the community he has injured by his offence" [[72]: p. 305].

The emphasis here on healing through reintegration and renormalization in a society which had suffered over 40 years of apartheid is a striking example of a practical approach to restoring justice attempted in a divided and broken society. This remarkable restorative process has not healed all wounds but has led to a more just society. The path to justice is a theme consistent with the meaning of the Resurrection, as this commenter on a blog titled *Do Justice* notes:

> One way that Jesus' resurrection impacts me is through understanding that the resurrection means the restoration and victory of true justice. Jesus, who is fully God and fully human, a prophet, teacher, and messiah was crucified, the only living being without any sin... [Yet] Jesus, the Son of God, was condemned as a sinner. Everyone who knew Jesus understood that this was a total injustice but no one did anything to stop this from occurring or even comment that this was unjust because they were afraid of the repercussions. [73]

Howard Zehr poses three questions about current restorative practices: "Are we really delivering justice for victims? Are we taking harms that have larger social, economic, political dimensions and treating them as individual wrongs? Are communities healthy enough to be doing the job we are asking them to do?" [72]. These pragmatic questions focus not only on the victim, but more pointedly emphasize the community and the institutions that, if not addressed, might lead back to the same injustices that occurred in the first place. Restorative justice involves more than just redemption for an individual, it promises communal restitution, perhaps creating a resurrection, a renewed life of the culture and society in which the restoration occurs.

This kind of cultural resurrection is also known as a *revival*, sometimes referred to as a Great Awakening (and another metaphor frequently used in the de-extinction literature). According to scholars of revivalism [74], there have been either three or four Great Awakenings in U.S. history and some argue the U.S. is experiencing a current Great Awakening of evangelicals who have embraced national politics and dug their heels in to some extent (certainly more than the Pope) against integrating natural and religious knowledge systems. This is distinct from the third Great Awakening (1890–1920) about which Mcloughlin writes:

> While Christianity had learned to cope with the Copernican astronomy and the geological challenge to biblical chronology, the crisis that arose in orthodoxy after 1890 was unlike anything Western culture had ever faced... Human beings, like birds and beasts, were pawns driven by forces beyond their control. Or so the positivists, psychologists, and naturalists of the more skeptical sort maintained. The prophets of the Third Great Awakening had to undertake an enormous rescue operation to sustain the culture. They had to redefine and relocate God, provide means of access to him, and sacralize a new world view. ([74]: p. 152)

Part of this redefinition involves the Christian concept of *stewardship*. To what extent is it the responsibility of humans to provide for non-human animals? And is this conveyed in some way within scripture?

According to Hermit, "ecological appropriation of Biblical texts throws ample light on the multilateral relation between God, humanity and [the] non-human world and can be further used to restore relations between the three. Appropriation of biblical texts facilitates applying texts outside of their original historical context to give insight to new issues. In this case Biblical texts can be analysed from ecological points of view..." [[75]: p. 101]. Hermit goes on to say that these views need to recognize the inherent stewardship humans have over the non-human world given their blessed position under the covenant. "Environmental stewardship is based on the concept that the Earth Community is an invaluable gift from God for which we are the caretakers" [[75]: p. 101]. This responsibility, combined with new technologies, would then place in our hands the dual goals of preservation and, perhaps, rebirth of particularly those species that we may have been responsible for driving to extinction in the first place. And the list is long.

6.3.4 Rebirth and Authenticity

Resurrection has primarily been described here as a rebirth and a transformation within the context of Christian theology. But what of other mythological connections to the notion of rebirth and transformation? It is a common theme, worldwide and eternal. In Greek mythology, Persephone is unwillingly taken by Hades to the Underworld and, despite Demeter and Zeus' protest, though returned to the living world for half the year, must return to the Underworld the other half of the year (due to her pomegranate fetish, yet another apple in the Garden). In Slavic mythology, "Jarylo, the tenth son of the lightning god Parun, was kidnapped on the day of his birth and raised by the god of the dead. Spring marked his return to the world" [[76]: p. 1]. These, of course, are seasonal themes, recognizing and marking the natural patterns and rhythms of annual time.

Resurrection themes also cover a more individualized rebirth than the seasonal. These range enormously from the divine to the macabre. Stories of vampires, ghouls, and zombies are, in a very literal sense, resurrection stories.

Gods and humans are not the only ones to be resurrected; animals are also subject to rebirth. The most famous of these is likely to be a bird known as the phoenix, legends about which can be found across Egypt, Europe, and Asia. Garfield-White says the phoenix "is a creature associated with the sun and fire. It is often said that when the bird dies, it bursts into flame, and a new phoenix is born from the ashes of the old.... Because the phoenix appears in so many myths and cultures, there are many different versions of the legend" [[76]: p. 3].

The rebirth story is even more ubiquitous if extended to the "hero's journey." Joseph Campbell outlined the hero's journey in this way: "The hero's death and resurrection is a model for the casting off of the old life and moving into the new" [[77] p. 71]. What is common to the hero's journey across mythologies is that the outcome for the hero is, first, the development of an authentic self, and second, the return with the "elixir," an object or body of knowledge that restores the hero's culture and allows it to renew, transform, and remain vibrant. According to Efthimiou and Franco, "heroism experts interviewed concur that the heroic involves a deep acknowledgment of the other, whether that is a physical person or an abstract ideal. Interviews indicate that the heroic is a call to realize a radical evolutionary state that is defined by reciprocity and the presence of an authentic self that has an inextricable relationship with everything else. This denotes a heightened form of intelligence deeply rooted in evolutionary and transformative processes, similar to the ideas expressed in transpersonal views of psychology" [[78] p. 40]. These "evolutionary and transformative processes" are resonant with the narrative of de-extinction and those performing it.

6.4 The Ethics of De-Extinction

"A new species would bless me as its creator and source; many happy and excellent natures would owe their being to me."

—Shelley, *Frankenstein* [[79] p. 101]

I first noticed that the specific metaphors of resuscitation and resurrection that I have just discussed were important to de-extinction narratives when chancing upon an academic article by Cottrell, Jensen, and Peck titled "Resuscitation and resurrection: The ethics of cloning cheetahs, mammoths, and Neanderthals" [7]. The colon after the two metaphors—prefacing but separating these concepts from both ethics and from the animals to which they were referring—struck me as oddly, and perhaps unintentionally, meaningful. So, I followed up on the investigation of the ethical foundations of each metaphor, as represented above, and next I will examine the relationship of these metaphors to current ethical discussions of de-extinction.

The ethical principles, issues, and dilemmas of the metaphors used to describe de-extinction can be compared to formal analyses of the ethics, technology and the rationales behind de-extinction. To summarize, the metaphor of resuscitation carries with it four ethical principles: autonomy, beneficence, nonmaleficence, and distributive justice. The metaphor of resurrection carries with it the ethical principles connected to the topics of creation, restorative justice, and rebirth/authenticity.

6.4.1 De-Extinction and Resuscitation

On the part of resuscitation, how do the four principles outlined by Beauchamp and Childress [26] in their discussion of resuscitation ethics align with the ethics of de-extinction?

The principle of autonomy does not offer a promising start for meaningful comparison. It is clearly the wrong analogy. Wooly mammoths were not signing DNRs and there was no MOLST form tacked to the Neanderthal's cave wall. Autonomy for an individual– or in this case, species—means they have rights-related decision-making ability, but this is always preemptive to death/extinction and even if it weren't, would presumably require the cognitive ability for the creature/species of their future demise. Agency is a requirement for autonomy in many philosophies; without the decision-making ability, autonomy is a moot proposition.

But need agency be defined merely as the will of an individual? And what counts as an individual? A person? Does your pet cat have agency? Some philosophies presume so. Where then is the line drawn? Must it be *one* entity, or could it include many? In some institutions, personhood has been assigned variably to entities which are likely to be consciously unaware (fetuses, for example) and in others to multiple persons (such as corporations). Can an entire population of organisms be another of these categories which might qualify for autonomous moral status, despite extinction? If Dawkins can propose that "the gene is the basic unit of selfishness" [[80] p. 39), can we not reverse the process and talk about a selfish species? Campbell and Whittle [[10] p. 119] ask, "Could we not identify a species with its gene pool, and in this way substantiate the claim that a species 'wants' to remain extant and perpetuate itself into the future?" (They answer their rhetorical question with a resounding "no").

It's a stretch, but this kind of thought experiment might end with the notion that if a species were not extinct, its survival would be its prime directive, so to speak. In that instance, it might be imputed that if there were agency it would be used to preserve said agency, to maintain autonomy and the continuity of agency. And if that is the case, returning a species to existence—de-extinction—would be ethically justified. It's a stretch and, simply because of the number of premises involved, perhaps this is a specious argument (pun intended). One might also object to the very definition of species, or even the reality of the concept of species (I will take up these arguments later in the section on authenticity). But this is only the first of the principles correlating resuscitation ethics with de-extinction ethics.

The principle of autonomy may also be linked to the utilitarian perspective, recognizing that decision-making is a balancing act which involves judgment calls between the benefits and costs resulting from the actor's decision [[81] p. S50]. This consequentialist perspective is a teleological analysis that examines courses of action and who they will benefit or cost, attempting to maximize the efficacy of the outcome for the greatest number of actors (usually people). In these cases, the means, in other words, justifies the ends.

In fact, this benefit/cost analysis maps nicely over the following two principles of resuscitation ethics defined by Mentzelopoulos et al. above [25]. *Beneficence* was Hippocrates' principle of "do good," which on its face is an applied benefits-production analysis. *Nonmaleficence* was Hippocrates' principle of "at least do no harm," an applied cost-limiting analysis. So, what are the goods that might arise from de-extinction and what are its potential costs?

The utilitarian ethical analysis of benefit and cost is probably the most common analysis found in both popular and scholarly accounts of de-extinction [7, 8, 10, 53, 62, 64, 81]. It is at the center of debates about the topic as well [51, 82].

6.4.1.1 Benefits to gain (beneficence)

The benefits discussed in the literature are many. Browning [53] provides an excellent summary of these (and equally fine analysis of each). She lists five clear benefits of de-extinction: ecological benefits, aesthetic benefits, financial benefits, restorative benefits, and scientific benefits. Ecological benefits involve the ways in which de-extinction might retrieve a species that could provide aid to the recovery of specific environments. Much has been made of the idea that wooly mammoths, in large enough numbers and transplanted to the Siberian tundra, might create a steppe-like environment that could improve carbon-retention which would be impactful on the extent of climate change. To me, this sounds like an end searching desperately for a means; i.e., it is an inauthentic, reverse-engineered reason for proceeding with the de-extinction of mammoths. But perhaps this benefit *is* real and, in addition to other reasons, it serves as yet another to proceed with deep de-extinction (the resuscitation of long-extinct species).

Aesthetic benefits (or the argument from "cool", which seems to me a far more authentic case than the former) are more difficult to assess as moral reasons to engage in de-extinction *ceteris parabus*. The relationship between beauty and goodness is a question of the ages, going back before Socrates' causes and Plato's forms [[8] p. 175]. Slater and Clatterbuck [54] weigh in on the aesthetics of de-extinct species with this clever analogy to the *Mona Lisa*:

> Suppose that the Mona Lisa is destroyed in a fire but that we are able to reconstruct an exact, molecule-for-molecule replica of the painting. Does the Mona Lisa persist across this temporal gap? Have we still lost something valuable that has not been restored by the existence of the replica? Plausibly, the answer to the first question is "no", in part because the answer to the second question is "yes". Part of what makes the Mona Lisa valuable is its historic properties, that it was painted by actual hand of Leonardo da Vinci at a particular time and place. The replica can perhaps restore the loss of the painting's aesthetic value (as something pleasurable to look at) or can serve as the same kind for the purposes of art theorists, but it cannot restore its historical value. Therefore, relative to this kind of interest, it does not count as the same painting. (p. 19)

By this account, a newly created "de-extinct" species should not be viewed as a precise resuscitation of the original species, but rather as a slick simulacrum of such. Is the beautiful good? Perhaps, but it may not be true. To paraphrase the iconoclastic musician Frank Zappa, "Information is not knowledge. Knowledge is not wisdom. Wisdom is not truth. Truth is not beauty. Beauty is not love. Love is not de-extinction. De-extinction is THE BEST!" [83].

In this examination of the benefits of de-extinction, financial benefits might be considered, although that analysis might carry questionable ethical weight, given that the prospect of gene-editing could change the course of natural history on our planet. Antonelli and Perrigo [55] question the motives of late capitalism arguing that biologists must push back against financial interests. "If we as biologists are not vocal about our values, we risk being poorly prepared to combat major societal challenges — including the ongoing sixth mass extinction on the planet. This is particularly crucial in a time when strong financial and political interests undermine the values of biodiversity" [[55] p. 581].

Restorative benefits, briefly, mean "a responsibility we have to resurrect those species for which we were the cause of their extinction" [[53] p. 798]. I will cover restorative justice in the next section on the ethics of the resurrection metaphor, so I refer the reader to that area of the essay for a discussion on this topic.

Scientific benefits include the knowledge gained from basic research and the development of scientific techniques that may lead to new technologies that can benefit humankind. The anthropocentric bias here is clear, but this has been a long-held value in the scientific community, at least, and in the frame of the enlightenment especially. Humanism is technologism at its heart. And so, technological progress becomes paramount. And this technology represents a pinnacle of human engineering–one that is evolutionarily unprecedented and whose outcome is extraordinarily uncertain and unpredictable. This seems to me more warning than benefit and invokes a sense of humanistic hubris that will be addressed again in the section on resurrection ethics about "playing God."

6.4.1.2 Costs to avoid (de-extinction nonmaleficence)

Considering the moral benefits of human science or aesthetic value evokes a sense of anthropocentric bias. Nonmaleficence would indicate an examination of the potential risks of harm inherent in the current practice of de-extinction. Other ethical perspectives take into account the moral qualities of the species involved, or the larger ecological systems in which they live. Interestingly, non-anthropocentric perspectives are often framed negatively, or in terms that the costs a particular technology might have on the natural environment. As mentioned above, the risk of harm to ecosystems is not perfectly known in this early science of de-extinction, but some insight can be gained from the scientific literature on invasive species and genetically modified organisms, or GMOs. In fact, Greely believes that de-extinction should not be considered a separate classification of ethical analysis from GMOs, given that they are essentially equivalent—de-extinct species will, de facto, be genetically modified organisms) [56]. The details of this literature are outside of the scope of this essay but suffice to say that the scientific community agrees generally [56, 57] that there is too little data to reliably predict the outcomes on ecosystems of GMO species (CRISPR critters?) contributing their laboratory-created genetics into the gene pools of natural species, creating the potential for all varieties of chimera. This could be a very serious cost to be weighed against some questionable benefits.

Another cost could be to the individual animals themselves. Animal welfare is a main theme in many ethical studies and de-extinction is no exception [7, 8, 56–58, 81, 83]. Since humans would be responsible for bringing animals into existence, they should then hold responsibility for their welfare in life and minimize the suffering of the animal. This seems simple enough, but complications abound.

The methods of de-extinction provide one example. Cloning is complicated and not always a successful engineering enterprise. In 2003, to create the very first clone of an extinct species—the *Pyrenean ibex*, a kind of European mountain goat also called a *bucardo*—fifty-seven somatic cell nuclear transfers (SCNT) were attempted, of which seven of them implanted as embryos and only one came to term. The surviving clone lived for approximately 6 minutes, being unable to breathe effectively as lung development underwent a genetic deformity— atelectasis, a collapsed lung, and the development of a third lobe, disconnected from the bronchial tubes [84, 85]. These kinds of experiments test the limits of suffering but are presumably required to advance the science and create viable organisms.

To prevent suffering of animals when they do reach term and can survive independently in numbers large enough to interbreed and naturally resume their evolutionary course (the goal of most de-extinction projects), a significant resource investment would be required. In the case of large animals, such as the wooly mammoth, Lynn Rothschild (evolutionary biologist and astrobiologist) estimates that, according to the Performing Animal Welfare Society, a healthy adult elephant would require $70,000 annually to sustain it [51]. To sustain a herd of a hundred mammoth (a soft minimum for interbreeding) would cost nearly a million dollars a year. While not insurmountable, this is a significant economic cost.

Other costs include those potentially to human welfare. De-extinct species entering an ecosystem (sometimes radically different from the ones in which they originally adapted) add unknown influences that could create the potential for new disease vectors and alter the food chain creating conditions of famine, effects that would not only be problematic for humans but for the entire ecosystem. Such problems, however, might be ameliorated with consistent management and oversight. In fact, some scientists argue that de-extinction could help ecosystems by re-instantiating and reinvigorating the biodiversity of nature. On the face of it, this seems sensible, as reintroducing especially recently extinct species can only serve to hold off biodiversity declines. But in practice, de-extinction could lead to far more lackadaisical conservation efforts stemming from such a promise that "we can just bring them back." As Campagna et al. put it, "The promise of de-extinction 'feels good' because it counteracts the malodorous legacy associated with losing a species. In our view, 'de-extinction' is the latest example of a long play of language that has given conservation efforts a tragically false sense of accomplishment and has worsened the conservation crisis" [81]. In their fascinating article that gives more attention to how we talk about de-extinction than to the scientific and biological processes involved, they suggest that creating better metaphors for the need for conservation is halted by the enchanting language used in de-extinction narratives. "Our lack of attention to considering language as evidence for how we refer to, understand, signify, communicate, imagine, and evaluate a conservation problem leaves gaps in our thinking that get filled with audacious metaphors, daring rhetoric, and other enchantments— what Wittgenstein calls "bewitchments" of language. . ." ([81] p. S49). Perhaps resuscitation and resurrection are two of these "audacious metaphors." The cost of succumbing to the alluring language of de-extinction could affect the degree to which people take conservation seriously. "If people take a 'Jurassic Park' future seriously, they will do even less to stem the building sixth great mass extinction," says population scientists Paul and Anne Ehrlich [86].

It is important to recognize the contribution of language to human transcendence. It was the power of human language that was responsible for the transcendence of humans from tribalism to techno-culture. This is not necessarily to equate language with "the good," yet without language humans would likely still be stuck in the Garden. A universal code that continually improves in terms of its viral-like capacity to spread among humans, language provides the 'cipher machine function' giving us the ability to translate meaning between many systems, math (proofs and logic), physics (atomic forces), chemistry (molecular manipulation), biology (DNA manipulation), even those of other animals in some cases [87–89], fundamentally altering our world in the process. It is only through our application of language that we have decoded the illusory secrets of life and nature. Perhaps *not* engaging in projects of scientific progress puts that kind of transcendence at risk?

6.4.1.3 Distributive justice and de-extinction

The benefit-to-cost measurement, while transactional and pragmatic, is only one ethical perspective on de-extinction. As with resuscitation ethics, de-extinction can be evaluated in terms of distributive justice as well. Similar to benefit/cost analysis, but with greater attention to *relative* allocation, distributive justice involves an analysis of resources to see how and if required resources are being utilized justly and equitably. Such resources might include money, time, equipment, attention, people, and raw materials.

In 2021, the de-extinction for-profit firm *Colossal Biosciences* announced that they had procured $15 million from investors (including private equity firms and self-help guru Tony Robbins, hotel magnate Paris Hilton, and Thomas Tull, a tech investor who produced the movie *Jurassic World*) toward their initial project for the de-extinction of the wooly mammoth [90]. Co-founded by George Church (Harvard molecular biologist) and Ben Lamm (entrepreneur in media apps and video game software), *Colossal* has garnered significant global media attention and by 2022 had accumulated another $60 million from investment. By 2023, it had captured $225 million in funding, adding movie star Chris Hemsworth and In-Q-Tel (the CIA's venture capital arm) among others [91] to their list of supporters. The company now claims it is valued at $1 billion. Stewart Brand's *Revive and Restore* project, while not exclusively invested in de-extinction, does feature de-extinction as a major priority in its non-profit work to conserve and restore ecological biodiversity [92]. Other projects have included the apparently now defunct Australian Museum scientific initiative known as the "Lazarus Project" and another called "The Ark" [59]. Of the "Lazarus Project", Lean writes: "The Lazarus Project is an evocative title for a de-extinction project. A group of scientists, working under the moniker of the biblical resurrectee, is working to 'resurrect' the extinct Gastric Brooding Frog. This fascinating Australian frog gestates its own tadpoles within its stomach, 'giving birth' from its mouth to small frogs. . . The product of this ambitious process will undoubtedly be very similar to a Gastric Brooding Frog but is it of the same species?" [[9] p. 571–572]. This question of identity and authenticity will be addressed in the next section.

With respect to distributive justice, ethicists wonder how these initiatives might be affecting more traditional conservation methods and draw attention away from new efforts to influence more directly and substantially increasing rates of extinction and ecological damage such as climate change [8, 10, 57, 60, 81]. Some of these other areas include facilitated adaptation, gene therapy, rewilding, and wildlife management strategies, as well as efforts that go beyond resuscitative efforts linked directly to species survival/renewal, such as de-exploitation and de-growth. Certainly, we do not want to put all our eggs in one basket, but limited resources might make massive projects devoting huge amounts of resources more ethically suspect than others.

6.4.2 De-Extinction and Resurrection

Moving on to the second and perhaps more audacious of our metaphors, resurrection, three ethical points of view have come to the fore. Skrimshire perfectly outlines the mythic qualities of the application this metaphor to de-extinction when he writes:

> What is the relationship between this materially fleshy hope in the restoration of creation (a re-materialised, plenitudous Ark) and the eschatological debates about the resurrection of the body? Is there something of the early Christian's belief in the resurrection of the body, and the veneration of material flesh (of Christ in the Eucharist, and holy martyrs in relics) in the redemptive offering of the dead flesh of extinct species? A key feature of de-extinction projects is its mythic narrativisation: a heroic quest to retrieve the very origins and stuff of life that offer redemption to all species. The romanticized story of the hunt for the Mammoth (for the purposes of DNA cloning) by Russian scientists across icy tundra is an example of this very search for the holy grail of preserved flesh that bears the code to new life. [[59] p. 12–13]

In this passage, Skrimshire neatly summarizes the three areas of attention shown here to "resurrection ethics." Resurrection involves creation ethics, restorative ethics, and the ethics surrounding authenticity and revival. How do these concepts map onto the prospects of de-extinction?

6.4.2.1 Creation, history, and de-extinction

In the literature on resurrection, much space has been given to the understanding of the Resurrection as an act of pure creation, a phenomenon, as noted earlier, that is perhaps outside of our understanding of space/time. However, O'Donovan suggests that the eternal metaphor poses a dilemma for creation doctrine:

> The patristic church correctly discerned that to think of the world as infinitely extended in time carried the heretical suggestion that history contained all possibilities immanently within itself. (This suggestion was no less implied in the cyclical conceptions which were current in the classical world: the cyclical concept implies that even sameness and regularity can be generated out of temporal movement, and it was no mistake that Nietzsche, in carrying historicism to its limits, found himself drawn back to this idea.) To defend the finitude of history as the object of God's creative decree, the early Fathers spoke of creation as absolute origin, *ex nihilo*. The possibilities of history were defined by the terms of the gift of being which God conferred upon the world. To put the same point in the language of another theological era, creation is a 'covenant'. [[93] p. 62]

The relevant covenant here is the Noahic Covenant, the sign of which is the rainbow and which applies not only to humanity but to all living creatures, an "everlasting covenant between God and every living creature of all flesh that is on the earth" (Gen. 9:11). Noah and the generations of his progeny were then required to procreate and not murder other humans, because mankind was made in the image of God.

I am not a Biblical scholar, but from my reading this covenant does not appear to intrinsically prohibit procreative, or in the case of de-extinction, re-creative, efforts—in fact, given that humans were made in his image, the Noahic Covenant seems to encourage rather than discourage "playing God." Yet, playing God is almost universally now seen as one of the deepest prohibitions and a sign of profound anthropocentric hubris.

One of the most often heard ethical reservations that has followed behind scientific progress since the origins of the scientific revolution has been that of "playing God." Before and since Shelley's *Frankenstein* [79] have the moral implications been scrutinized of altering the universe around us (and especially ourselves) using human cultural adaptations. The technologies devised by humans are (to us) ingenious, but time and time again the consequences of these adaptations have not been entirely positive, for ourselves or the natural environment around us. Thus, the scrutiny on de-extinction as an emerging technology designed specifically and explicitly to bypass a fundamental law of nature: natural selection.

At this point, "playing God" is a phrase that invokes an almost hackneyed quality, so often is it used as to be meaningless. "This expression has... become, as publicist and science journalist Philip Ball remarks, a lazy journalistic cliché and an alarmist slogan. However, it is a cliché that may still enable writers or NGOs to attract the attention of a large audience" [[61] p. 260]. Appearing in the titles of hundreds of journal articles and thousands of periodicals, the phrase is used up, shriveled and hollow, especially in an era that could be construed as "disenchanted" (despite the Great Awakenings). Alternatively, it could be seen as surviving as a particularly viral meme in the media and scientific cultures, making it a clearly adaptive idea. I prefer the latter interpretation, recognizing that the systemic knowledge of science and religion can co-exist in an ethical discussion of de-extinction. Thus, we can harness both the teleological ends-means arguments and the deontological moral virtue arguments.

According to Greely, the "playing God" critique comes more from bioethics than environmental ethics (although geo-engineering may be an exception) [56]. The connection to our own bodies and their biological foundations is perhaps one reason for this favoritism. Greely suspects that de-extinction issues would be taken on a more case-by-case basis in terms of their effects on the natural environment, whereas synthetic biology generally, and its applications within de-extinction projects particularly, triggers a more holistic sense of ethical concern [56]. It also may have to do with the connection of de-extinction specifically and genetic modification at the molecular level particularly to the paradox of creation.

If we are to follow a principle like the Noahic Covenant and be agents of creation in God's image, then it follows that we should be willing to go ahead with de-extinction practices as a form of creation for nature's (and God's) sake. However, Cohen points out a flaw with the "creation for creation's sake" argument:

> The philosophical argument against performing de-extinction for nature's sake is the following. If nature has intrinsic value, this

> value must entail independence (autonomy) from human purpose, activity and interest. Humans cannot restore natural intrinsic value, since all restoration is, *qua* human creation, *artifactual*—the precise antonym of the natural. An artifact's value is instrumental to the human purpose in creating it; it hence, *ipso facto* loses all intrinsic value. Relatedly, Bernard Williams spoke of a "paradox" when "we have to use our power to preserve a sense of what is not in our power." One cannot manipulate nature in order to preserve its autonomy; yet only as autonomous is it intrinsically valuable. If correct, this would undermine the ecological value of de-extinction. [[8] p. 167]

In particular, the paradox of using our creative power to re-create the species we have forced to extinction as a direct result of hubristic human behavior (the dodo bird, thylacine, and white rhino come immediately to mind, and perhaps the mammoth) is at the root of the ethical dilemma of playing God.

Campagna points out that "according to Yuval Noah Harari, the scientific revolution gave birth to humanist religions, in which humans replaced gods. In that transition, humans acquired a unique and sacred essence that is the source of all meaning and authority in the universe, and everything that happens is judged as good or bad depending on its impact on people. The problem is that behaving like gods on the scale that we are doing is a moral mistake that threatens all life" [[81] p. S52].

In 2022, when the *Colossal* project founder Ben Lamm was brought the question of playing God, he replied: "I don't look at it as playing God.... We need to ensure that the ecosystems that developed without us continue to develop with us, it's all about human integration. So, we're just trying to advance technologies that can help conservation, and the by-product is building technologies that can also help humanity" [94].

6.4.2.2 Restorative justice and de-extinction

The restoration inherent in the Resurrection was an intervention intent on undoing the moral offense of the crucifixion (for all time). The resurrection metaphor as applied to the de-extinction of species carries the power of this same logic: but in this case, humanity is both Pilate and Jesus, extinct-or and re-creator.

What does restorative justice look like regarding de-extinction from the point of view of Zehr's three questions: "Are we really delivering justice for victims? Are we taking harms that have larger social, economic, political dimensions and treating them as individual wrongs? Are communities healthy enough to be doing the job we are asking them to do?" [Zehr in [72] p. 305]

To the first question, "Are we really delivering justice for victims?", it must be asked: are extinct species "victims"? This would probably have to be taken on a case-by-case basis. For species such as the dodo, whose demise was recorded by the Dutch settling Mauritius, it might be an open and shut case (there is "smoking gun" evidence that humans caused the extinction of the species). For others, like the wooly mammoth or Neanderthals, there is some circumstantial evidence that *Homo sapiens* might have been involved in their extinction, but it is hardly an open and shut case. For yet others, there is no evidence whatsoever. Then there is the question of indirect effects leading to extinction, such as climate change or habitat destruction.

Once victimhood is established, the question of "delivering justice" must be addressed. Is de-extinction using the current technology of CRISPR gene edits and SCNT cloning truly reviving a species? First, is the resultant organism a member of the same species that went extinct? (I will cover this further below in the section on *authenticity*). The extent of stewardship over a species recovery would have to be more involved than simply reproducing individuals in a lab using synthetic biology, which again introduces risks to ecosystems and the welfare of the resurrected individuals themselves. It would take a significant amount of time to establish that the species as a whole (i.e., populations that can continue intergenerational reproduction successfully) is truly de-extinct.

The second question, "Are we taking harms that have larger social, economic, political dimensions and treating them as individual wrongs?" [72]. This question is essentially sociological and is asking if institutionally there are problems that led to the injustice in the first place. There is no easy answer to this question; books could be devoted to this topic, involving the larger, institutional reasons that the world appears to be experiencing its sixth mass extinction and what role (if any) humans have played in it.

Zehr's third question, "Are communities healthy enough to be doing the job we are asking them to do?" [72]. In examining the just restoration of a species, this question seems to be asking, is it ethical to resurrect a species into an environment (the species' "community") that is so significantly different than the one it was originally adapted to that the species' survival would be questionable? Given the potential for continued climate change, this seems a very relevant question.

These are difficult questions to answer clearly for individuals and doubly so for entire species, and the ethical dilemmas that they bring up are complicated. O'Donovan writes that "the Resurrection... demands respect for 'the natural structures of life in the world, while looking forward to their transformation.' It fulfills the demands of 'natural justice' with a 'divine justice that was more complete and more satisfying than any eye had yet seen or ear heard of'" [O'Donovan in [93] p. 74–75]. On this account it seems like a good idea, but if conditions cannot be secured for "true justice," perhaps de-extinction might be rethought on these grounds.

6.4.2.3 Authenticity, re-birth, and de-extinction

The question of authenticity comes down to Campbell and Whittle's summation here: "Is the resurrection of an extinct species genuinely possible, or not? Will organisms produced by de-extinction technology be authentic new members of the species that died out, or just convincing fakes?" [[10] p. 49]. After all, other arguments could be seen as moot if de-extinction is, in fact, not de-extinction by virtue of inauthenticity. Yet, after an exhaustive analysis they arrive at the conclusion that the arguments for and against de-extinction are equivalent:

> The result is, as we have seen, a mixed bag. Of the arguments for de-extinction, most are weak, but some are forceful. The same goes for arguments against de-extinction: most are weak; some are forceful. Where does this leave us? In a complicated place! This should be no surprise. The technologies of de-extinction are not

> intrinsically good or bad. They are simply tools—extraordinarily powerful tools—that, like any tools, can be used either wisely or stupidly. De-extinction is neither good nor bad in and of itself. Whether it is good or bad depends instead on the details of the particular case in question, and on how the technology is applied. If it is used wisely and carefully, for the betterment of the environment, and for the betterment of the lives of future people, then it is good. If it is used foolishly or carelessly, then it is bad. [[10] p. 116]

In any case, is authenticity even necessary to evaluate de-extinction ethically? Lean doesn't think so. "While there have been a range of arguments which aim to establish that identity is important in de-extinction I have found them all lacking. The conceptual argument is particularly weak and based on a naïve understanding of the science of de-extinction" [[9] p. 586].

And there is the metaphysical problem, one that is highlighted in our postmodern moment: What's real anymore? Baudrillard's [95] analysis of simulacra comes to mind. Is a de-extinct member of an extinct species the ultimate simulacra? It occurs to me that this could be the basis of a fantastic science fiction story, where in the future humans continually go extinct, prompting pre-programmed AI to continually resurrect and reestablish humanity, an eternal cycle not unlike the phoenix. Or perhaps it has already been written.

It should be noted that not all resurrections are "successful," even in mythology. Many resurrection tales are of the cautionary type, where the rebirth is unsuccessful or partial, as in the case of Perspephone, or of zombies and vampires. This aligns well with de-extinction, as in the case of cloning, where it takes many tries for a somatic cell nuclear transfer to implant, or gene-editing which can create chimeras. And there is a profound potential that use of gene-editing will result in the creation of so-called "frankenspecies" [62].

Can we save ourselves from ourselves? Can we return with the elixir of life and revive the culture? Is de-extinction the hero or the villain in the story? Will it renew humanity or further lead to its demise? Is the value of our own species linked in some meaningful way to the value we hold in the protection, conservation, and potential de-extinction of other species? Or is the value in life akin to the value held in the exclusively natural selection of evolution and extinction? These are just a few of the questions raised from an inquiry into the ethical contemplation of species resurrection.

6.5 Conclusion

Not since Copernicus' heliocentric model has such an audacious project been speculated upon by the institutions of scientific inquiry. But, as Harari notes, the pace of change has an amplifying effect upon consequences [Harari in [81] p. S52]. Whereas it was offensive to Rome, it would take several centuries before that discovery truly altered the world through cartography and shipping. In this case, from the invention of CRISPR to its application to revive deeply extinct species, it will almost certainly be fewer than two decades. All this without any rational imperative to do so. "These projects are, for the most part, not time sensitive. The targeted animals will not become more extinct the longer we wait" [[53] p. 801]. Nonetheless, de-extinction science moves forward in a seemingly inevitable way.

In the context of resuscitation, the argument from autonomy saw many holes, chiefly coming from the fact that extinct species can hardly be considered autonomous actors, and that there is no duty to the dead. The beneficence/nonmaleficence argument provided many ways that de-extinction might benefit humanity and ecosystems, but also show that there may be significant costs to pay in terms of ecosystem risk—the degree of potential maleficence is truly an unknown. And the distributive justice argument simply leaves us with the question of whether de-extinction draws resources away from conservation efforts or whether it poses as a literal life preserver to be thrown to a vast number of endangered species.

The resuscitation metaphor evokes conceptions of the materialistic realm of medicine and the objective human body. The value of life in resuscitation is quite simple: preserve it when possible. But degrees of complication arise when the future welfare of the patient and community is uncertain, in terms of possible side effects of treatment, unknown wishes of the patient, and issues of equity in the community. Translated to de-extinction this raises concerns over possible side effects of gene-editing and cloning on a mass scale, unknown outcomes on species becoming de-extinct, and which species should undergo de-extinction.

In the context of resurrection, the issues summarized were the hermeneutics of creation (playing God), the potential for restorative justice, and the identity issues related to authenticity. In examining the case of traditional Christian resurrection narratives, God is the creator and man is (part of) His creation, made in His image and redeemable after death. Taken less biblically, resurrection mythology could be taken as a nature narrative, whereby who is to say that humans could not be a creator and re-creator? The potential for restoration, revival, and rebirth are themes of the Christian resurrection, and these are also found in other resurrection narratives around the world, part of Campbell's monomyth [77].

Applied to de-extinction, the use of technologies that are unique, new, powerful, and designed specifically to alter the very process of evolution and the natural selection of life on Earth, from a theological point of view is the height of human hubris and playing the God role. But at the same time, proponents argue that it is the ultimate stewardship of the Earth, a way to deliver Noah's covenant and the key to the future. It promises not merely metaphor but the *material reality* of resurrection. Some argue that restorative justice is necessary for the lack of stewardship humans have shown species whose extinction we have unjustifiably caused, and that de-extinction is a means to redeem such sins. Others simply ask, what do we owe dead species? And they might add that the species being "resurrected" are inauthentic simulacrum of extinct species, not the 'real thing.' The retort is, what does it matter? Was Lazarus the same when he returned from the dead? Was Jesus?

Resurrection narratives exist in the idealistic realm of ancient histories and divine meaning. The value of life in the resurrection metaphor is the eternal. It represents the cycle of death and rebirth, decay and renewal, an infinite resuscitation. But it also signals that in being reborn, we are changed, improved, and closer to God. Each iteration is a Hegelian move toward the absolute spirit, and life, fragile though it is, will be continually sustained and improved. To the scientists who embrace this metaphor, that is likely to be the subtextual message in their use of it: in referring to de-extinction as resurrection, they are saying 'this time we can do it right; with this science we can restore the world; with this technology we can truly embrace the value of life.' And it is with that hope that de-extinction projects may go forward, for as with every human enterprise, go forward they shall.

References

1. Card, O. S. 1995. *Alvin Journeyman: The Tales of Alvin Maker IV*. 1 ed. New York: Tor Books.
2. National Geographic. 2023. The Frozen Zoo. Overheard. https://www.nationalgeographic.com/podcasts/overheard/article/episode-8-frozen-zoo. Accessed October 15.
3. Carlson, K. 2023. Austin Entrepreneur's Startup Lands $60 Million as It Tries to Bring Back the Woolly Mammoth. *Austin American-Statesman*. https://www.statesman.com/story/business/2022/03/09/wooly-mammoth-colossal-60-million-ben-lamm-george-church/94424 30002/. Accessed October 15.
4. Saul, J., and S. McBride. 2023. Dodo Bird DNA Has Startup Valued at $1 Billion to Bring Back Extinct Species. Bloomberg.com.
5. Guenot, M. 2023. Billion-Dollar Startup Plans to Bring the Dodo Back from the Dead, and Claims It Can Also Revive the Woolly Mammoth. *Business Insider*. https://www.businessinsider.in/science/news/billion-dollar-startup-plans-to-bring-the-dodo-back-from-the-dead-and-claims-it-can-also-revive-the-woolly-mammoth/articleshow/97532025.cms. Accessed October 15.
6. Human Cloning Policies. 2023. *Center for Genetics and Society*. https://www.geneticsandsociety.org/internal-content/human-cloning-policies. Accessed October 15.
7. Cottrell, S., J. L. Jensen, and S. L. Peck. 2014. Resuscitation and Resurrection: The Ethics of Cloning Cheetahs, Mammoths, and Neanderthals. *Life Sciences, Society and Policy* 10: 3. https://doi.org/10.1186/2195-7819-10-3.
8. Cohen, S. 2014. The Ethics of De-Extinction. *NanoEthics* 8: 165–178. https://doi.org/10.1007/s11569-014-0201-2.
9. Lean, C. H. 2020. Why Wake the Dead? Identity and De-Extinction. *Journal of Agricultural and Environmental Ethics* 33: 571–589. https://doi.org/10.1007/s10806-020-09839-8.

10. Campbell, D. I., and P. M. Whittle. 2017. *Resurrecting Extinct Species: Ethics and Authenticity*. Springer.
11. Siipi, H., and L. Finkelman. 2017. The Extinction and De-Extinction of Species. *Philosophy & Technology* 30: 427–441. https://doi.org/10.1007/s13347-016-0244-0.
12. Black, R. 2013. The Promise and Pitfalls of Resurrection Ecology. *Science*. March 12.
13. Neuman, S. 2021. Scientists Say They Could Bring Back Woolly Mammoths. But Maybe They Shouldn't. *NPR*, September 15, sec. Science.
14. *Reviving Extinct Species: The Possibilities of CRISPR Technology*. 2023.
15. Biello, D. 2023. Efforts to Resuscitate Extinct Species May Spawn a New Era of the Hybrid. *Scientific American*. https:// www.scientificamerican.com/article/lost-species-revived-from-dna-and-restored-to-nature/. Accessed October 15.
16. The Quagga Project. 2023. *The Quagga Project*. https://www.quaggaproject.org/. Accessed October 15.
17. Fox, A. 2023. Elizabeth Ann Is the First Cloned Black-Footed Ferret | Smart News | Smithsonian Magazine. https://www.smithsonianmag.com/smart-news/elizabeth-ann-first-cloned-black-footed-ferret-180977065/. Accessed October 15.
18. Picheta, R. 2021. Meet Elizabeth Ann the Ferret: The First Endangered American animal to Be Cloned. *CNN*. February 19.
19. Davis, M. 2022. Panel Discusses History of Black-Footed Ferret Rescue. *Powell Tribune*. October 4.
20. The Nobel Prize in Chemistry 2020. 2024. *NobelPrize.org*. https://www.nobelprize.org/prizes/chemistry/2020/summary/. Accessed January 12.
21. Burn, S. J., ed. 2012. *Conversations with David Foster Wallace*. 1 ed. Jackson (Miss.): University Press of Mississippi.
22. Rubulotta, F., and G. Rubulotta. 2013. Cardiopulmonary Resuscitation and Ethics. *Revista Brasileira de Terapia Intensiva* 25: 265. https://doi.org/10.5935/0103-507X.20130046.
23. Martin, P., J. Courtade, and P. Martin-Reviewed by Jason Courtade. 2020. CPR Facts and Stats – What Is the CPR Success Rate? *ProCPR*. June 25.
24. Cheema, M. A., W. Ullah, H. M. A. Abdullah, S. Haq, A. Ahmad, and A. Balaratna. 2019. Duration of in-hospital cardiopulmonary resuscitation and its effect on survival. *Indian Heart Journal* 71: 314–319. https://doi.org/10.1016/j.ihj.2019.09.002.

25. Mentzelopoulos, S. D., A.-M. Slowther, Z. F., C. Sandroni, T. Xanthos, C. Callaway, G. D. Perkins, et al. 2018. Ethical Challenges in Resuscitation. *Intensive Care Medicine* 44: 703–716. https://doi.org/10.1007/s00134-018-5202-0.
26. Beauchamp, T., and J. Childress. 2013. *Principles of Biomedical Ethics*. 7th ed. Oxford University Press.
27. Marco, C. A. 2005. Ethical Issues of Resuscitation: An American Perspective. *Postgraduate Medical Journal* 81: 608–612. https://doi.org/10.1136/pgmj.2004.030429.
28. Mentzelopoulos, S. D., K. Couper, P. V. de Voorde, P. Druwé, M. Blom, G. D. Perkins, I. Lulic, et al. 2021. European Resuscitation Council Guidelines 2021: Ethics of Resuscitation and End of Life Decisions. *Resuscitation* 161. Elsevier: 408–432. https://doi.org/10.1016/j.resuscitation.2021.02.017.
29. Burns, J. P., and R. D. Truog. 2016. The DNR Order after 40 Years. *The New England Journal of Medicine* 375: 504–506. https://doi.org/10.1056/NEJMp1605597.
30. Varkey, B. 2020. Principles of Clinical Ethics and Their Application to Practice. *Medical Principles and Practice* 30: 17–28. https://doi.org/10.1159/000509119.
31. O'Neill, O. 1984. Paternalism and Partial Autonomy. *Journal of Medical Ethics* 10: 173–178.
32. Petty, K., N. DeGarmo, R. Aitchison, and et al. Ethical Considerations in Resuscitation. *Disease-a-Month*: 217–220.
33. Cahill, L. S. 2010. The Art of Dying: At the End of Life, How Can People of Faith Prepare for a Good Death? *Sojourners Magazine*, June.
34. Diem, S. J., J. D. Lantos, and J. A. Tulsky. 1996. Cardiopulmonary Resuscitation on Television — Miracles and Misinformation. *New England Journal of Medicine* 334. Massachusetts Medical Society: 1578–1582. https://doi.org/10.1056/NEJM199606133342406.
35. Downar, J., T. Luk, R. W. Sibbald, T. Santini, J. Mikhael, H. Berman, and L. Hawryluck. 2011. Why Do Patients Agree to a "Do Not Resuscitate" or "Full Code" Order? Perspectives of Medical Inpatients. *Journal of General Internal Medicine* 26: 582–587. https://doi.org/10.1007/s11606-010-1616-2.
36. The History of the Hippocratic Oath. 2015. *Northeastern*. June 3.
37. Perkins, G. D., C. W. Callaway, K. Haywood, R. W. Neumar, G. Lilja, M. J. Rowland, K. N. Sawyer, M. B. Skrifvars, and J. P. Nolan. 2021.

Brain Injury after Cardiac Arrest. *The Lancet* 398: 1269–1278. https://doi.org/10.1016/S0140-6736(21)00953-3.

38. Ethical Issues of Resuscitation. 2024. https://www.acep.org/patient-care/policy-statements/ethical-issues-of-resuscitation. Accessed January 12.
39. American Heart Association CPR & First Aid. 2024. *cpr.heart.org*. https://cpr.heart.org/en/. Accessed January 12.
40. Vearrier, L. 2024. Do-Not-Resuscitate Tattoos: Are They Valid? *ACEP Now*. https://www.acepnow.com/article/do-not-resuscitate-tattoos- are-they-valid/. Accessed January 12.
41. Training Requirements in OSHA Standards.
42. Free Safety Posters—National Safety Council. 2024. https://www.nsc.org/workplace/resources/free-posters. Accessed January 12.
43. CPR
Steps |Perform CPR. 2024. *Red Cross*. https://www.redcross.org/take-a-class/cpr/performing-cpr/cpr-steps. Accessed January 12.
44. *For the Time Being*. 2013.
45. Resurrection—Quick search results |Oxford English Dictionary. 2024. https://www.oed.com/search/dictionary/?scope=Entries&q=resurrection&tl=true. Accessed January 12.
46. Perry, S. 2012. *Resurrecting Interpretation*. Eugene, Or: Pickwick Publications.
47. Cahill, L. S. 2007. Creation and Ethics. In *The Oxford Handbook of Theological Ethics*, ed. Gilbert Meilaender and William Werpehowski, 0. Oxford University Press. https://doi.org/10.1093/oxfordhb/9780199227228.003.0002.
48. Weber, M. Science as a Vocation. In *New York: Free Press*.
49. Bourdieu, P. 1986. The Forms of Capital. In *Handbook of Theory and Research for the Sociology of Education*, 241–258. New York: Greenwood Press.
50. Brockman, J. 2024. We Are as Gods and Have to Get Good at It | Edge.org. https://www.edge.org/conversation/stewart_brand-we-are-as-gods-and-have-to-get-good-at-it. Accessed January 12.
51. *LIVE DEBATE – Don't Bring Extinct Creatures Back to Life*. 2019.
52. Rull, V. 2022. Biodiversity Crisis or Sixth Mass Extinction? *EMBO Reports* 23. John Wiley & Sons, Ltd: e54193. https://doi.org/10.15252/embr.202154193.
53. Browning, H. 2018. Won't Somebody Please Think of the Mammoths? De-Extinction and Animal Welfare. *Journal of Agricultural and Envi-*

ronmental Ethics 31: 785–803. https://doi.org/10.1007/s10806-018-9755-2.

54. Slater, M. H., and H. Clatterbuck. 2018. A Pragmatic Approach to the Possibility of De-Extinction. *Biology & Philosophy* 33: 4. https://doi.org/10.1007/s10539-018-9615-5.
55. Antonelli, A., and A. Perrigo. 2018. The Science and Ethics of Extinction. *Nature Ecology & Evolution* 2. Nature Publishing Group: 581–581. https://doi.org/10.1038/s41559-018-0500-z.
56. Greely, H. T. 2017. Is De-Extinction Special? *Hastings Center Report* 47: S30–S36. https://doi.org/10.1002/hast.749.
57. Robert, A., C. Thévenin, K. Princé, F. Sarrazin, and J. Clavel. 2017. De-Extinction and Evolution. *Functional Ecology* 31: 1021–1031. https://doi.org/10.1111/1365-2435.12723.
58. Jørgensen, D. 2013. Reintroduction and De-Extinction. *BioScience* 63: 719–720. https://doi.org/10.1525/bio.2013.63.9.6.
59. Skirmshire, S. 2016. Rewriting Mortality: A Theological Critique of Geoengineering and De-Extinction. In *Theological and Ethical Perspectives on Climate Engineering: Calming the Storm*, 103–126. Lanham, MD: Lexington Books.
60. Jennings, B. 2017. The Moral Imagination of De-Extinction. *The Hastings Center Report* 47 Suppl 2: S54–S59. https://doi.org/10.1002/hast.753.
61. van den Belt, H. 2009. Playing God in Frankenstein's Footsteps: Synthetic Biology and the Meaning of Life. *Nanoethics* 3: 257–268. https://doi.org/10.1007/s11569-009-0079-6.
62. Welchman, J. 2021. Does Justice Require De-Extinction of the Heath Hen? In *Animals in Our Midst: The Challenges of Co-Existing with Animals in the Anthropocene*. Switzerland: Springer Nature.
63. Barbour, Ian G. 1974. *Myths, Models and Paradigms*. New edition. New York, NY: HarperCollins College Div.
64. Evans, J. H. 2018. *Morals Not Knowledge: Recasting the Contemporary U.S. Conflict between Religion and Science*.
65. Mikkelson, D. 2015. FALSE: Pope Francis Has a Master's Degree in Chemistry. *Snopes*. June 4.
66. Jones, S. 2011. A Brief Survey of Sir Isaac Newton's Views on Religion. In *Converging Paths to Truth: The Summerhays Lectures on Science and Religion*. Salt Lake City, Utah: Desert Books.
67. Dickerson, K. 2024. One of Einstein's Most Famous Quotes Is Often Completely Misinterpreted. *Business Insider*. https://www.businessin-

sider.com/god-does-not-play-dice-quote-meaning-2015-11. Accessed January 13.

68. Crosby, D. 1994. Einstein on Religion. Pittsburg State University: 186.
69. Collins, F. S. 2007. *The Language of God: A Scientist Presents Evidence for Belief.* First Paperback Edition. New York, NY: Free Press.
70. Next chapter: Ex-NIH Head Francis Collins Works to Bridge the Country's Divides. 2022. *Morning Edition.* NPR.
71. Barbour, I. G. 1976. *Myths, Models, and Paradigms: A Comparative Study in Science and Religion.* New York: Harper & Row.
72. Broughton, G. 2009. Restorative Justice: Opportunities for Christian Engagement. *International Journal of Public Theology* 3: 299–318. https://doi.org/10.1163/156973209X438265.
73. Pablo, K. S. 2022. Jesus' Resurrection and Social Justice. *Do Justice.* April 29.
74. McLoughlin, W. G. 1980. *Revivals, Awakening and Reform.* Chicago History of American Religion. Chicago, IL: University of Chicago Press.
75. Hermit, J. T. Environmental Stewardship: Emerging Trends in Christian Ecological Ethics. *International Journal of English and Literature (IJEL)* 5: 97–104.
76. Garfield-White, G. 2021. Resurrection in Mythology. *Mythos.* April 2.
77. Campbell, J. 1995. *Reflections on the Art of Living: A Joseph Campbell Companion.* New York: Harper Perennial.
78. Efthimiou, O., and Z. Franco. 2018. Heroic Intelligence: The Hero's Journey as an Evolutionary Blueprint. *Journal of Genius and Eminence* 2. https://doi.org/10.18536/jge.2017.02.2.2.04.
79. Shelley, M. 1985. *Frankenstein, or, The Modern Prometheus.* New York: Oxford University Press.
80. Dawkins, R. 1989. *The Selfish Gene.* Oxford: Oxford University Press.
81. Campagna, C., D. Guevara, and B. Le Boeuf. 2017. De-Scenting Extinction: The Promise of De-Extinction May Hasten Continuing Extinctions. *Hastings Center Report* 47: S48–S53. https://doi.org/10.1002/hast.752.
82. TEDxDeExtinction | TED. 2024. https://www.ted.com/tedx/events/7650. Accessed October 15.
83. Zappa, F. 1979. *Watermelon in Easter Hay.* Vol. Joe's Garage, Act III. Zappa Records.
84. Folch, J., M. J. Cocero, P. Chesné, J. L. Alabart, V. Domínguez, Y. Cognié, A. Roche, et al. 2009. First Birth of an Animal from an Extinct Subspecies

(Capra pyrenaica pyrenaica) by Cloning. *Theriogenology* 71: 1026–1034. https://doi.org/10.1016/j.theriogenology.2008.11.005.

85. Piña-Aguilar, R. E., J. Lopez-Saucedo, R. Sheffield, L. I. Ruiz-Galaz, J. de J. Barroso-Padilla, and A. Gutiérrez-Gutiérrez. 2009. Revival of Extinct Species Using Nuclear Transfer: Hope for the Mammoth, True for the Pyrenean Ibex, But Is It Time for "Conservation Cloning"? *Cloning and Stem Cells* 11: 341–346. https://doi.org/10.1089/clo.2009.0026.
86. The Case Against De-Extinction: It's a Fascinating but Dumb Idea. 2024. *Yale E360*. https://e360.yale.edu/features/the_case_ against_de-extinction_its_a_fascinating_but_dumb_idea. Accessed January 20.
87. Berne, R. W. 2023. When the Horses Whisper: The Wisdom of Wise and Sentient Beings. *(No Title)*. https://cir.nii.ac.jp/crid/1130000797907728512. Accessed March 14.
88. Grandin, T., and C. Johnson. 2006. *Animals in Translation: Using the Mysteries of Autism to Decode Animal Behavior*. 1 ed. Orlando: Harcourt.
89. Shah, S. 2023. The Animals Are Talking. What Does It Mean? *The New York Times*, September 20, sec. Magazine.
90. Luce, I. D. 2021. Colossal Co-Founder Ben Lamm Explains Why Resurrecting the Woolly Mammoth Will Save the Planet. September 21.
91. Pasternack, A. 2023. How (and Why) a Billion-Dollar Startup Plans to Bring Back the Dodo. *Fast Company*. January 31.
92. Revive & Restore |Genetic Rescue to Enhance Biodiversity. 2024. https://reviverestore.org/. Accessed January 20.
93. O'Donovan, O. 1994. *Resurrection and Moral Order: An Outline for Evangelical Ethics*. Revised edition. Leicester: Wm. B. Eerdmans Publishing Co.
94. Founder Of Mammoth "De-Extinction" Company Responds To Accusations Of "Playing God." 2022. *LADbible*. February 12.
95. Baudrillard, J. 1994. *Simulacra and Simulation*. Translated by Sheila Faria Glaser. 33601st ed. Ann Arbor, Mich: University of Michigan Press.

Chapter 7

Animal-Centered Technology and Sustainable Development: An Earth Odyssey

Clara Mancini
Animal-Computer Interaction Lab,
School of Computing and Communications, The Open University,
Milton Keynes, United Kingdom
clara.mancini@open.ac.uk

7.1 Introduction

The Dawn of Man, the opening sequence of Kubrik [1]'s film *2001: A Space Odyssey*, famously provides an iconic and poignant imagining of the moment when humans discover technology:

> *In a primordial landscape, great apes survive by foraging alongside other herbivores, by dodging the attacks of carnivores and by competing with rival troops for access to water. One fateful dawn, the apes awaken to the sight of a monolith, its geometric dark form stark against the rugged desertic landscape, a rising sun and crescent moon perfectly aligned above it. The alien intelligence that*

Animals, Ethics, and Engineering: Intersections and Implications
Edited by Rosalyn W. Berne

ISBN 978-981-5129-56-4 (Hardcover), 978-1-003-65231-1 (eBook)
www.jennystanford.com

> *the object embodies changes the way the apes look at the world. So, when next rummaging for food amongst scattered bones, one of them suddenly realises that a femur can be used as a tool to kill the animals alongside whom he had been foraging and who will thus become his prey. Soon after, all members of the ape's troop brandish the same kind of weapon, which allows them to defeat a rival troop by killing one of its members and scaring away the others, and thus seize control of the local water source. From then on, the dawning man is set onto a path of technological development that will lead him, millions of years later, to colonise space.*

The juxtaposition of the sequence's closing shot (a bone thrown in the air by one of the apes) with the next sequence's opening shot (the elongated shape of a military satellite orbiting in space) conveys the transformative function that technological innovation has had in human evolution, as well as its inherent association with violence. More specifically, though, *The Dawn of Man*'s portrayal of humans' primitive discovery highlights some key implications of technological innovation: technology frees its possessors from limiting physical and environmental conditions, thus giving them access to previously unattainable resources and standing; by the same token, technology creates separation and power asymmetry, between those who possess it and those who do not, advantaging the former and disadvantaging the latter, who can thus be outcompeted or turned into a resource to be exploited; the impact of technology, both on those who possess it and on those who do not, is amplified by the possessor's ability to transmit knowledge and by technology's consequent propagation. These considerations beg a fundamental question that remains unexplored in the film: how should our species address the responsibility that arises from our ability to develop technology that allows us to overcome biological and environmental limitations, and exert seemingly unlimited power over other beings?

This chapter makes the case that the field of Animal–Computer Interaction (ACI), which focuses on researching and designing technology *for* and *with* animals, could provide the animal-centered perspective needed to address this question. To this end, it begins by discussing the impacts that technology can have on animals (Section 7.3), before introducing ACI's animal-centered perspective on technological development (Section 7.4) and defining it in relation to the concepts of animal dignity and justice (Section 7.5). The chapter then outlines conceptual frameworks and methodological approaches (Section 7.6), and discusses the ethics of animal-centered research and design (Section 7.7), before elaborating on the wider political implications and how researchers and designers could increase their work's influence by demonstrating its importance for the pursuit of a global sustainable future (Section 7.8).

7.2 Technological Development and Its Impact on Animals

For millennia, we humans have developed and used technology to control and shape our environment, to protect ourselves from some of its elements whilst taking advantage of others, and to promote interactions with it that would enable us to do both. For example, we have built ever more elaborate dwellings to provide shelter for us and our activities; we have devised ever more efficient methods for farming plants and animals to produce food and clothing; and we have developed ever more pervasive communication and transportation systems across which information and artefacts have been travelling at growing speed. Whilst technology has enabled our species to colonize the planet and harvest its riches, the power it has given us has also increasingly distanced us from the natural world, by significantly reducing, or altogether eliminating, the need for us to negotiate with the nonhuman beings with whom we share the planet, and whom we have thus been able to displace and exploit with little consequence. However, there is now growing awareness that, by making it possible for us to exert unbridled domination over the natural world, technology has also allowed us to destroy the very ecosystems on which all living beings, including humans, depend [2].

Although some technologies (e.g., for carbon capture, for renewable energy production, for environmental monitoring) are widely regarded as instrumental to mitigate the damage that human expansion has caused to the planet, and to support environmental restoration and ecosystems' conservation, others (e.g., for genetic engineering, for nuclear development) are perceived with ambivalence as potential existential threats to humanity (as they enable, e.g., the production and deployment of catastrophically deadly pathogens or weapons). In 1968, *2001: A Space Odyssey*, already seemed to warn us about the dangers posed by technological progress, represented by super-intelligent computer HAL9000's decision to kill the crew of its spaceship for fear of being terminated and to protect the space mission that had been assigned to it. This warning resonates today more than ever, as increasingly pervasive and powerful computing technologies, such as the Internet of Things (IoT) and Artificial Intelligence (AI), are both celebrated as essential tools to achieve global sustainable development (e.g., [84]) and cautioned against as dangerous creations that will soon escape human control and enslave or, worse, annihilate our species and possibly all sentient life (e.g., [3, 4]).

To address concerns over the impact of these emerging technologies, some governance bodies have issued ethics guidelines for responsible technological innovation (e.g., [5]), which have been broadly subscribed to by major corporations in the sector (e.g., [6, 7]). Academic researchers are also increasingly investigating ethical issues and possible responsible approaches regarding the development and deployment of these technologies (e.g., [8, 9]). Within the mainstream discourse, almost invariably, proposals for responsible technological innovation have been concerned with protecting and benefitting humans, even though trillions of other animals are impacted by human technology, whether intentionally or unintentionally, directly or indirectly. In this regard, researchers have recently begun to raise concerns (e.g., [10–12]) and, for example, Singer and Tse [11] have advocated greater consideration for the impact of AI technologies, which might be specifically designed to interact with animals (e.g., robotic milking systems), which might not be designed to interact with animals but might do so accidentally (e.g., self-driving cars), or which might not interact directly with animals but might influence human behavior towards them (e.g., algorithms selecting social media content).

Focusing on the practical impacts of AI systems, Coghlan and Parker [13] have proposed a framework to classify the different ways in which AI technologies may harm animals, accounting for intentional, unintentional, direct and indirect harms. In the authors' analysis, AI systems may be specifically designed (e.g., AI-enabled drones developed by poachers to track and kill endangered animals) or deliberately misused (e.g., drones used in conservation being hacked by ill-intentioned individuals) to cause *intentional harm* to animals in ways that are socially *condemned*; or they may be designed to support activities that cause intentional but socially *acceptable* harm to animals (e.g., precision farming technologies enabling further farming intensification; novel brain-computer interfaces being developed through invasive animal tests; AI-enabled drug discoveries leading to more animals being used in experiments). AI systems may also have *unintentional* harmful impacts on animals, whether *direct* (e.g., self-driving cars not being designed to recognize small animals, or malfunctioning and thus failing to recognize them) or *indirect* (e.g., proliferation of AI systems damaging animals' habitats; AI systems use in husbandry practices resulting in animal neglect; algorithmic bias reinforcing anthropocentric prejudice). In addition, the authors point out that harms may ensue when AI systems, which could benefit animals (e.g., for veterinary medicine), are not developed or deployed (e.g., due to lack of funding); they refer to these as *forgone benefits*.[1]

Although Coghlan and Parker [13]'s concerns relate specifically to AI systems, their considerations are relevant to computing technologies more broadly, whose development has by default been informed by anthropocentric interests and has all too often failed to account for the interests of nonhuman stakeholders. But what is the alternative?

7.3 An Animal-Centered Perspective on Technological Development

Animal–Computer Interaction (ACI) [15, 16] is a growing field of research and practice that takes an animal-centered perspective on the study and design of interactive systems, aiming to develop frameworks and methods that enable animals to participate in

---

[1] This paragraph and the one above are reused in adapted form from [14].

the design process as legitimate stakeholders and contributors [15, 17–20]. ACI has been strongly influenced by the field of Interaction Design (ID) [21], which focuses on the study and design of interactive systems, informed by disciplines such as psychology, ergonomics, engineering, informatics, social sciences and product design. But it extends the boundaries and core values of ID [21–23] to nonhuman animals, whether the interaction is active and intentional [17], active and unintentional [24], passive and intentional [25] or passive and unintentional [15], dyadic and direct [18] or distributed and indirect [26]. Consistent with the tenets of ID, if it is to be used for a specific purpose, as with operant devices, interactive technology is expected to afford good usability for animal users (e.g., being easy to learn how to use, helping users to perform a task efficiently—[27, 28]); if it is to be worn, as with biotelemetry devices, interactive technology is expected to provide good wearability for animal wearers (e.g., being imperceptible and unobtrusive, or at least acceptable to the wearer—[29, 30]). In any case, interactive technology is expected to provide, mediate or lead to good experience for animal stakeholders (e.g., being motivating and stimulating for a user, and not interfering negatively with the daily experience of a wearer).[2]

Technology designed to conform to animals' characteristics and to meet their needs can be highly beneficial for them. For example, assistance dogs are trained to perform a wide range of tasks on behalf of their assisted humans but, to do so, must interact with an environment that is typically too complex and whose features are inconsistent with their sensory, cognitive, physical and behavioral characteristics. This renders their work unnecessarily challenging, limiting their ability to perform and negatively impacting their welfare [32]. In response to this problem, ACI researchers have designed a variety of ambient and wearable interfaces [27, 28, 33] informed by dogs' biological characteristics to mediate and simplify their interaction with the environment, thus enabling them to work in ways that are both more effective and less demanding. Conversely, animals who live in captive environments, large carnivores being a prime example [34], typically endure conditions which lack the

[2] This paragraph is reused in adapted form from [31].

complexity of the natural settings in which their species has evolved, and the sensory, cognitive and physical stimulation that those settings would afford, the lack of which results in suboptimal welfare typically manifest in the form of stereotypic behaviors. To address this problem, ACI researchers have experimented with a range of interactive applications [35–37] designed to engage resident animals' capacities and enrich their daily experience despite the environmental limitations imposed by spatial constraints.

Furthermore, human behaviors that negatively affect animals' welfare and survival, for example through inappropriate management practices or through the indiscriminate prioritization of human interests, may derive from a lack of knowledge or appreciation of animals' capacities, conditions and needs, and of the threats that human activity poses for them. In this regard, tools developed to detect and interpret behavioral signs that might otherwise escape human attention can help researchers to recognize when an animal is experiencing different affective states, such as pain [38]. Moreover, applications developed to monitor the presence and behavior of vulnerable species in non-intrusive ways can be valuable for supporting conservation efforts [39], by providing insight into the pressures placed on animals by human activity and by supporting strategies that can help reduce human-wildlife conflict [40]. ACI researchers have also found that witnessing animals' interaction with technology, for example in zoos, can influence visitors' perception of their capacities, fostering greater appreciation of their intelligence and increased empathy towards their emotional experience [41].

However, over the years, ACI researchers have also highlighted the risks that computing technology poses for animals, even when developed with the best intentions. For example, researchers have been mindful that (as mentioned above—[13]) tracking technologies developed to support conservation efforts could be sabotaged and exploited to animals' detriment [42]. Similarly, they have pointed out that data collected to monitor the welfare of farmed animals could be used by the farming industry to intensify the animals' exploitation [43]. ACI researchers have also expressed concern that the wearables often used for tracking purposes can cause distress or, worse, physically injure animals or disrupt activities that are essential for their survival, or for the survival of other individuals or groups within their ecosystem [44]. Similarly, concerns have been raised regarding the use of technologies designed to care for companion animals, which may result in them being neglected as their guardians delegate care-taking responsibility to computing devices [45]; or regarding the use of quantifying technologies, which may lead to an incorrect interpretation of and response to animals' welfare needs, as well as to a weakening of the human-animal bond [46]. More generally, ACI researchers have been alert to the interspecies differences, communication barriers [47] and power dynamics [48] which characterize human-animal relations and which are likely to result in technologies that serve the interests of humans rather than those of animals. These risks are compounded by the fact that, as some have pointed out, human-animal relations are shaped through a multiformity of technology-mediated interactions [49], which are impacted by complex contextual forces [37] in ways that are not always obvious.

These considerations raise a key question for ACI researchers: what do we understand as animal-centered technology and how could we pursue animal-centered design?

7.4 Defining Animal-Centeredness

Animal welfare frameworks have often informed researchers' understandings of the notion of animal-centeredness. For example, some [50, 51] have proposed that ACI work should be consistent with the Five Freedoms [52], which frame welfare in terms of freedom from thirst, hunger and malnutrition; from discomfort; from pain, injury and disease; from fear and distress; and freedom to express normal behavior. Others [41] have based their animal-centered design framework on the Five Domains Model [53], which defines welfare in terms of positive and negative influences related to animals' nutrition, physical environment, and health; behavioral interactions with the environment, with other animals and with humans; and the mental states that result from all these influences. Whilst animal welfare frameworks have informed user-centered design in ACI, aspects of welfarist approaches have been deemed incompatible [54] with the field's advocated non-speciesist [15] and multispeciesist [55] perspective. Indeed, van der Linden [43] has argued that notions of animal-centered design ought to go beyond affording good usability for animal users. Modelled on the concept of human-centered design, as underpinned by the value of human dignity [56], the author proposes a value-based notion of animal-centeredness grounded in animal dignity as defined by the Swiss Animal Welfare Act of 2005 [57]). On this basis, the author distinguishes technologies that are *acknowledging* of animals but still prioritize human interests; technologies that are *focused* on animals and that support human-animal relationships by respecting animal welfare; and technologies that are *centered* on animals and that empower them by respecting their dignity and reducing human power.

Although animal dignity is an important consideration when defining animal-centeredness, the difficulty with drawing from animal welfare legislation is that existing legal frameworks invariably prioritize human interests, which means that the values of both welfare and dignity are not treated as absolutes but rather are defined in relation to human goals, and ultimately subordinated to them. As a case in point, in the Swiss legislation, "*strains*" imposed on the animal that can be justified by "*overriding interests*" do not constitute a "*disregard for the animal's dignity*" (Art.3-a), where said interests are not determined by the animals themselves but ultimately by those who manage them. Similarly, according to the British Animal Welfare Act (2006, [58]), it is illegal to cause an animal to suffer "*unnecessarily*", where the necessity of causing an animal to suffer is invariably determined by humans and ultimately subordinated to human interests. In other words, quite at odds with the idea of animal dignity, these fundamentally anthropocentric frameworks fail to properly recognize and protect animals' *agency* and *autonomy*, as well as their fundamental interest not to be subordinated to human interests (e.g., being used in scientific experiments against their will, being imprisoned and killed in farming practices).

In this regard, Mancini et al. [59] propose a definition of animal-centered design grounded in the political philosophy of Nussbaum

[60, 61] and her *capability approach* to animal justice. In contrast to influential contractarian approaches [62, 63], according to which justice is a negotiated contract among subjects capable of rational reasoning and self-representation, Nussbaum argues that a theory of justice should account for asymmetries of power, capacity and moral rationality, and extend justice to anyone for whom it is relevant. While animals may not be able to negotiate with humans, they are nevertheless subjects of justice for whom justice principles must be negotiated. For the author, all those who share the condition (and vulnerability—[64]) of *animality*, and who have an ability to feel and a perspective on the world, strive to pursue basic *capabilities* that are important for their flourishing and that they are entitled to have the opportunity to pursue as *autonomous* agents endowed with intrinsic *dignity* [60]. Basic capabilities include *life*, *health* and *bodily integrity*; using one's *senses*, *imagination* and *thought*; making meaningful *choices* and *plans*; forming relevant *affiliations*; *playing*; having *control* over one's environment [61]. In Mancini et al.'s [59] work, these capabilities inform a set of questions against which ACI researchers and practitioners can assess their projects (see Table 7.1).

Within the capability approach, animal maltreatment is not an affront to human dignity (as in contractarianism), but rather an affront to *animal dignity*; the emphasis is not on humans' obligation to treat animals with compassion (akin to welfarist approaches), but rather on animals' *entitlement* to a just treatment. Rather than being measured based on the balance between pleasure and pain (as in utilitarian approaches—[65]), justice is measured based on the *opportunities* animals have to engage in activities they value [66], regardless of whether these might cause pain. For Nussbaum, depriving individual animals of the opportunity to pursue basic capabilities cannot be justified in the name of collective interests; and, although this applies especially for sentient animals, moral consideration is also warranted for animals who might not be sentient, because restricting their autonomy may frustrate their capabilities regardless. Likewise, (differently from utilitarian and welfarist views—[67]) at least for animals who experience life as extended over time, death is harmful even if painless, because it interrupts life activities and capabilities in which the animals are invested. In other

Table 7.1 Questions to help assess the animal-centeredness of ACI work against animal capabilities identified by Nussbaum [60, 61]

Assessing the Animal-Centeredness of ACI Projects	
Animal capabilities [Nussbaum 2006]	**Examples of questions for researchers**
Life	*Does the work directly or indirectly impact animals' entitlement to life?*
Bodily health	*Does it support or hamper animals' health in the short or long term?*
Bodily integrity	*Does it respect or injure animals' bodily integrity directly or indirectly?*
Senses, imagination, Thought	*Does it enhance or worsen animals' sensory & cognitive ability at any point?*
Emotions	*Does it provide nurturing or warping emotions in any way?*
Practical reason	*Does it enable or frustrate animals' capacity to frame goals and plans?*
Affiliation	*Does it allow animals to have rewarding intra- and inter-species affiliations?*
Other species	*Does it allow animals to benefit from multispecies interdependencies?*
Play	*Does it directly or indirectly support or suppress animals' play behavior?*
Control over Environment	*Does it restrain or enable animals to control and manage their environment?*

Source: Redrawn from [59].

words, within Nusbaum's capability approach to animal justice, the recognition of animals' intrinsic dignity informs an animal-centered perspective that is not conditional upon human interests and that emphasizes animals' autonomy as well as well-being. At the same time, the approach acknowledges the need to mediate among the capabilities of different species and does not demand animal equality or the abolition of all instrumental relations with animals, so long as animals' basic capabilities are fulfilled at least to an adequate threshold, above which inequalities are not unjust.

As pointed out by Mancini et al. [59], these characteristics make the capability approach to animal justice particularly useful when defining animal-centeredness in ACI. But how should researchers approach animal-centered design in practice?

7.5 Designing *for* and *with* Animals

Mancini [54] has proposed that a user-centered approach to the design of interactive systems for animals has implications regarding both the design features of the systems in question, which need to be informed by the characteristics and requirements of the animals for whom they are intended, and the methods employed during the research and design process, which need to enable animal users to express their requirements through appropriate forms of participation compatible with their characteristics. In turn, this has implications for the "*ethical perspective adopted by researchers and designers, which needs to foster the conditions for appropriate forms of engagement with users and enable their autonomous involvement in the design process as legitimate stakeholders and design contributors*" [p. 3]. In this regard, a range of ACI work has focused on the different facets of user-centered and participant-centered design, drawing on a multidisciplinary background to deliver novel ACI design frameworks and methods.

For example, drawing on User-Centered Design (UCD—[23]) and Biosemiotics (the study of signification across the natural world—[68]), researchers have identified animal-centered interaction design principles [16, 36] and developed interface design frameworks [51, 69, 70] to inform the development of applications that afford good usability and experience for animal users. Drawing on UCD and Ethology (the study of animal behavior in naturalistic settings—[71]), researchers have developed wearable design frameworks [29, 30] to inform the development of devices that minimally interfere with the bodily functions, daily activities and lived experiences of animal wearers. Drawing on the abovementioned Five Domains of Animal Welfare model [53], researchers have developed structured approaches [41] to define and iteratively refine animal-centered objectives for developing technologies that enhance animals' life.

Iteration is a key aspect of the interaction design process in general [21] and even more so when the aim is developing technologies for animals [72]. While it is essential that the perspective of animal stakeholders informs design outcomes, human-animal interspecies differences and communication barriers make identifying animal-centered requirements, developing designs that adequately meet those requirements and evaluating the resulting prototypes particularly challenging. This almost inevitably makes multiple iterations of the design cycle necessary. To give animals representation throughout the design process over subsequent iterations, ACI researchers have adopted various approaches, which involve the direct participation of animal stakeholders (e.g., usability tests with prospective animal users—[27, 73, 74]) and that of human proxies (e.g., interviews or workshops with animal guardians, caretakers and experts—[15, 41, 75]). Approaches developed for enabling animals to engage with research set-ups, and thus enable designers to capture animals' expressed needs and preferences, may be more or less structured, depending on the type of application being developed and on the stage of the design process.

On the one hand, naturalistic observations of animals' interactions with objects in their habitual surroundings may be employed to inform novel design ideas based on animals' spontaneous behaviors and choices [18, 19]. On the other hand, instrumented experimental protocols in controlled settings may be employed to evaluate specific aspects of prototypes by measuring animals' behavioral and physiological responses [74, 76]. Animal participants may be trained to interact with prototypes leveraging associative learning principles through positive reinforcement (e.g., when desired behaviors are rewarded with food treats). Training tends to be used when the aim is developing tools that animals might be expected to use for specific tasks which are not part of their evolved behavioral repertoire but which can increase their ability to exert control over and express agency within their environment (e.g., a dog may need to learn how to interact with a new interface to open a door but, once learned, he can use his new skill to move across spaces - [33, 77]). Conversely, when the aim is designing enriching, aesthetic, and ludic experiences for animals (e.g., in zoos—[36]; shelters—[78]; homes—[18]), the interaction needs to be intrinsically rewarding, which makes the use of training methods to elicit design requirements and evaluate prototypes less appropriate. In these cases, researchers tend to use more open-ended approaches, for example, by making prototypes available for animal participants to freely engage with and by deriving design insights from their interactions (or lack thereof) [18, 19].

While some have questioned whether, and under what conditions, animals might be considered participants [47] or even users [69], Mancini and Lehtonen [77] have stressed the importance of recognizing animals' participation regardless of what form their engagement might take. According to the authors, "*anthropomorphic notions of participation, which assume that part-taking agents should have shared understandings of, and shared goals for, design activities*" lead to "*anthropocentric discourses and practices*" which "*exclude, devalue, dismiss, delegitimize, or render invisible the participatory contribution*" of those who do not possess capacities that only some humans possess "*to enter symbolic, deferred, abstract conversations, away from the specificities of the 'here and now'*" (p. 914). Instead, argue the authors, designing with animals should be conceived as "*a process of incremental orientation towards an optimal final outcome that may never be reached, but that can be approximated*" so that "*attending to the process by carefully crafting spaces and procedures that foster the emergence of participatory engagement is arguably more important than any interim design outcome*" (p. 914).

In other words, the issue is not whether humans and animals share understandings and goals, but rather whether research set-ups, and the interactions they afford, allow partaking animals to make sense, express volition and exercise choice in relation to the activities they are involved in, so they can respond "*through compliance, diversion or disruption*" (p. 914), thus shaping the design process. To this end, the authors have identified features that characterize multispecies participatory research set-ups:

- *biological salience* (offering participants reinforcers that are meaningful to them given their biological characteristics and that motivate them to engage)
- *signal reliability* (affording interactions based on signals that participants can trust and that facilitate stable associations and sensemaking)
- *engagement options* (allowing participants opportunities to express preferences among different design options, as well as different ways of engaging with or disengaging from the research process)

- *contingencies variations* (whereby motivators, signals and available options are modulated to help assess the extent to which a participant wants to engage and what preferences they have).

Here the question is: what ethical frameworks might inform research set-ups that present those characteristics and foster participation by enabling animals' *semiotic*, *volitional* and *choice-full* engagement and, thus, allowing them to negotiate design outcomes (p. 914)?

7.6 The Ethics of Animal-Centered Research and Design with Animals

Ethical concerns regarding the involvement of animals in research activities have been at the forefront in ACI since early on [15, 48, 54, 79], with an emphasis on protecting animals' welfare (e.g., by treating each individual participant with respect and care regardless of species) and autonomy (e.g., by allowing participants to withdraw from research activities at any time). The need to obtain informed consent for the animals' involvement (e.g., from guardians, carers, legal authorities) has also been highlighted [31, 54]. Some early proposals of ethical guidelines [79] were informed by existing frameworks regulating the use of animals in research, underpinned by Russel and Burch [80]'s widely accepted principles of humane research, known as the Replacement, Reduction and Refinement (3Rs) framework. The framework was developed to address the ethical dilemma that arises from the use of animals in research which may harm them and to which they are unable to consent, but which is considered necessary to achieve a greater good. Specifically, *Replacement* requires that, whenever possible, animals be replaced with alternative non-animal methods, or that a sentient species be replaced with a less sentient one; *Reduction* requires that the number of animals involved be reduced to the minimum necessary to achieve statistical power; *Refinement* requires that research procedures and practices be refined to minimize their impact on the animals involved.

Whilst recognizing the importance of continually refining research protocols, ACI researchers [31, 54] have highlighted the limitations of the 3Rs and of the regulatory frameworks it underpins. For example, under such frameworks, animals may be used in research that has no relevance for them as individuals or even for their species, if the potential benefits to society are deemed to warrant the cost to the animals; where considered, the possibility for animals to withdraw from research procedures is subordinated to the aims of the research, with no explicit provision for animals to consent to their involvement; some species are granted greater consideration than others, not on the grounds of their greater capacity to suffer, but based on public concerns; it is possible to reduce the number of animals affected by reusing individuals for multiple procedures; and the killing of animals at the end of procedures is permissible provided it is performed humanely. In other words, these frameworks reflect a research-centered approach and, in this regard, an assumption that animals are "*substitutable components of an experimental apparatus*" whereby "[they] *and their subjective experience (whatever this might be) are regarded as the abstract-able, add-able, subtract-able, interchangeable quantities of an experimental equation*" ([54], p. 225).

Thus, although all ACI research must comply with existing regulations, researchers have pointed out that these are insufficient to support animal-centered research and design. To address this gap, building on Mancini [54]'s previous work on ACI ethics, Mancini and Nannoni [31] have proposed an Animal-Centered Research framework (ACRf) and associated scoring template (see Table 7.2) that researchers and institutional ethical review bodies could use to assess the ethical standard of a project against four fundamental principles:

- *Relevance* requires that animals be involved in research only if this is directly relevant and beneficial to them, ideally as individuals but at least as a species (e.g., using mice as a test species to develop a technology for assistance dogs would not be appropriate, but working with dogs who do not necessarily do assistance work might be acceptable

Table 7.2 ACRf scoring template (the term 'partakes' refers to all those who take part in the research)

Ethics Standard	**Compliance with principles of animal-centered research**			
	Relevance to partakers	*Impartiality* toward partakers	*Welfare* of partakers	*Consent* of partakers
5. Very high	procedure is directly relevant and highly beneficial for partakers	individuals receive highest consideration regardless of their capacities	procedure enhances partakers' welfare	partakers are enabled to choose whether and how to engage with procedure
4. High	procedure is relevant but benefits may not be direct or immediate	individuals receive high consideration but not as much as others with more capacities would	procedure does not impact negatively on partakers' welfare	partakers are mostly able to choose whether and how to engage with procedure
3. Moderate	procedure has some relevance but benefits are only indirect and only in future	individuals receive some consideration but notably less than more capable ones would	procedure has minor impact on partakers' welfare	partakers have limited ability to choose whether and how to engage with procedure
2. Low	procedure has little relevance and benefits are only indirect and only in future	individuals receive significantly less consideration than more capable ones would	procedure has significant negative impact on partakers' welfare	partakers are mostly not allowed to dissent or withdraw from procedure
1. Very low	procedure has no relevance whatsoever and no benefits for partakers even indirectly or in future	individuals receive very little or no consideration compared to more capable ones	procedure has severe negative impact on partakers' welfare	partakers are not allowed to dissent or withdraw from procedure in any way

Source: Redrawn from [31].

provided their involvement in the research activities was compatible with or, better still, enhanced their welfare)

- *Impartiality* requires that all those involved are protected on the grounds of their role as participants and not of their taxonomic classification and any characteristics on which such classification might be based (e.g., species, sentience), or societal values and consequent practices (e.g., food vs. companion animals). In brief, researchers should acknowledge and respect the characteristics of individual participants as equally deserving of consideration and care according to their welfare needs
- *Welfare* requires that researchers always prioritize the well-being of their research participants and that the animals' best interests prevail over the interests of science and society. To this end, researchers should respect the animals' needs and wants, and protect their biological integrity (i.e. their physical and psychological health) and autonomy (i.e. their ability to express and pursue their wants), by refraining from disrupting their daily life patterns and routines, by affording them freedom of expression and control over the research environment and the research process, and by preserving their life
- *Consent* requires that individual animals' informed consent always be garnered in two complementary and equally necessary forms: *mediated* consent from legal guardians and welfare experts (or ethical review bodies) who have the animals' best interests at heart and who are in a position to understand the implications of the animals' involvement in the research; and *contingent* consent from the animals themselves, who are the only ones who can assess the contingencies before them and how these affect them in the 'here and now', as expressed by their willingness to engage and by their chosen modalities of engagement.

However, while normative approaches such as the ACRf can help researchers plan animal-centered design activities, they may not be enough to support them when, during the research, they face unexpected, ethically charged situations, which require them to make impromptu decisions. In this regard, Ruge and Mancini [70] have proposed and ethics toolkit for helping researchers to make ethically sound situated decisions as their work progresses. The toolkit consists of several templates, each asking a series of questions in relation to different ethical values, helping researchers involved in a research project to articulate their own ethical baselines as well as the ethical baseline of the project as a whole. The function of the templates is to enable researchers to develop a series of ethical guiding statements to which they can subsequently refer when making decisions during research activities (e.g., regarding the way they are, could be or should be interacting with the animals involved). The toolkit, the authors suggest, can foster a practice of active reflection within research teams, helping researchers to maintain their commitment to a project's ethos in the face of unexpected ethical challenges.

Plausibly, without an animal-centered ethical approach to working with animals, developing animal-centered technology for them is unlikely. But what are the wider implications for animals' situation beyond specific ACI projects?

7.7 The Politics of Animal-Centered Research and Design

Mancini et al. [59] have highlighted the need for ACI researchers to consider the forces that may shape ACI projects and how these may affect animals (e.g., as in the case of technology that is designed to monitor farm animals' welfare but may end up being used to intensify their exploitation—[43]). The authors have pointed out the inherent tension that exists in ACI, between researchers' endeavor to address the immediate needs of animals in specific situations through technological interventions and these interventions' wider implications for the animals' situation beyond specific ACI projects. In this regard, they suggest that thinking politically about ACI may help question the standards according to which the animal-centeredness of ACI research and design is measured.

Grounded in a tradition of political thinking within interaction design, Dombrowski et al. [81] have proposed a *social justice-oriented interaction design* approach to help researchers and designers work ethically and responsibly, by "*attending to power relations*" and "*foregrounding politics surrounding oppression*", with a view to addressing systemic social justice problems through design processes and outcomes. The authors' approach comprises six design strategies (*Designing for Transformation, Designing for Recognition, Designing for Reciprocity, Designing for Enablement, Designing for Distribution, Designing for Accountability*), which designers could adopt to design for social justice, and three design commitments (*Commitment to Conflict, Commitment to Reflexivity, Commitment to Ethics*), which would enable designers to pursue the proposed strategies during the design process. Mancini et al. [59] reinterpret Dombrowski et al. [81]'s strategies and commitments to support animal-centered design, grounding them in the values of multispecies justice [60, 61]. Additionally, the authors propose three new design strategies to support designers' political engagement in ACI:

- *Designing for Disruption* of incumbent systems, norms and practices that go against animals' basic capabilities and that designers would like to address or change
- *Designing for Reconfiguration* by redefining design-related frameworks whose current anthropocentric definition legitimizes the dismissal of animals' role in the design process or justifies design outcomes that discriminate against them
- *Designing for Pollination* by seeking opportunities to enable animal-centered values to penetrate socio-economic systems in which animals exist but which are not animal-centered, to help such systems evolve towards animal justice

The authors propose also a new *Commitment to Expanded Empathy*: while acknowledging that capturing the perspective of another being is impossible, they argue that there is value in the very act of trying, and that empathic design methods might help designers gain a deeper and more holistic understanding of animals, greater respect and concern towards them, and increased motivation to act for them.

Notwithstanding calls for considering the wider implications of ACI projects, work in the field has tended to concentrate on realizing the potential benefits of technology for animals; at the same time, ethical concerns regarding its potential negative impacts on animals have tended to focus on applications that specifically target them. However, Mancini [14] has argued that, at a time when technologies of increasing power and reach are encroaching on both managed and free-living animals in unprecedented ways, there is a need for ACI to extend its remit by: focusing on the risks of harm that technology poses for animals, as well as its potential to benefit them; and paying attention to the impacts of technology that does not necessarily target animals, as well as technology that specifically targets them. In this regard, a key question raised by the author is how to engage stakeholders, such as governance bodies, to create opportunities for debate and conditions for influencing policies that could eventually reflect ACI's ethos and maximize ACI's benefits for animals and society more broadly. To this end, the author makes two complementary proposals.

Firstly, to engage governance bodies, researchers could reinterpret human-centric ethical frameworks produced by those very bodies, to reflect a multispecies perspective against which to foreground impacts of technologies on animals and raise related animal-centered ethical concerns. This could be helpful for three reasons: ACI researchers' work that addresses the impacts of computing technologies on animals is arguably more likely to engage governance bodies, if it is framed in terms of those bodies' existing policies; examining technology's impacts against the frameworks that underpin those policies could make ACI researchers' work more relatable for a range of other stakeholders (e.g., non-governmental organizations, media, general public) interested in the public discourse on emerging technologies and familiar with related policies; examining impacts on animals through the lens of human-centric frameworks could highlight implicit (e.g., socially ingrained) anthropocentric bias inherent in computing systems and inform debate on the importance of considering animals' interests in the development of policies that regulate technological innovation. As an illustrative example, Mancini [14] refers to the *Ethics Guidelines for Trustworthy AI* produced by the European Commission's High-Level Expert Group on Artificial Intelligence (AI HLEG 2019), and reinterprets the four ethical principles underpinning the guidelines—*respect for human autonomy*; *prevention of harm*; *fairness*; and *explicability*—from a multispecies perspective. The author notes how the guidelines' recommendation to "*...pay particular attention to situations involving more vulnerable groups such as children, persons with disabilities and others that have historically been disadvantaged or are at risk of exclusion, and to situations which are characterised by asymmetries of power or information...*" (p. 2) is relevant for animals just as it is for humans. That is, animals are among those "others that have historically been disadvantaged", given their subordinated position within human legal and socio-economic systems [59]; "or are at risk of exclusion" from receiving AI's potential benefits and from being protected from potential harms, given the "asymmetries of power and information" characterizing human-animal relations [48]. By examining the case of precision farming systems and of anthropocentric smart cities through the lens of the Guidelines' four multispecies principles, the author identifies numerous ethical considerations that arise in those contexts and discusses how such considerations might be foregrounded for the eventual attention of governance bodies regulating the development of AI systems.

Secondly, Mancini [14] argues that, to extend the influence of ACI onto any regulated settings where animals and computing technologies might come into contact, researchers would need to show the relevance of their work to the societal priorities set by those governance bodies. Specifically, researchers would need to show how the animal-centered perspective championed by ACI is instrumental to the pursuit of those priorities. In this regard, the author refers to the United Nations (UN)' Sustainable Development Goals (SDGs—[82]), adopted in 2015 "*as a universal call to action to end poverty, protect the planet, and ensure that by 2030 all people enjoy peace and prosperity*", balancing "*social, economic and environmental sustainability*" by leveraging "*creativity, knowhow, technology...from all of society*" [82]. While, says the author, the relevance of emerging technologies to the SDGs' agenda is widely acknowledged, the overall perspective on their potential is decidedly anthropocentric, so the challenge for ACI researchers is to articulate how their work can contribute towards realizing a global sustainable future and how, to this end, ACI's animal-centered perspective is key to achieving a balance between "social, economic and environmental sustainability". In particular, considering as illustrative examples four of the SDGs (*good health and well-being, sustainable cities and communities, life below water* and *life on land*), the author highlights different ways in which ACI work could support the UN's agenda, for example, by:

- designing, deploying and evaluating technologies for monitoring, analyzing and interpreting animal-related phenomena in different settings (e.g., presence, behavior, well-being and needs of species, groups, individuals) ensuring that the design of monitoring tools and protocols is informed by animal-centered frameworks
- designing, deploying and evaluating technologies that afford animals effective interactions with their environment (e.g., enabling meaningful choices, autonomous access to relevant resources, management of social relations) maximizing, not only utility and usability, but also compatibility with animals' *capabilities* and minimizing human-animal conflicts
- designing, deploying and evaluating animal-centered technologies that engage different human communities in the monitoring of and providing for animal well-being (e.g., raising awareness of human activities' impact on some animals, highlighting the human–animal well-being nexus, coordinating collective animal care activities), thus fostering attitudes of understanding, respect and safe interspecies relations.

Ultimately, the author encourages ACI researchers and designers to take responsibility for both the potential benefits and risks that increasingly powerful and pervasive technologies present for animals, by leveraging governance bodies' frameworks to foreground impacts of technologies on animals; and by demonstrating how an animal-centered perspective on technological innovation could support global societal priorities.[3]

7.8 Conclusion

As I write, humanity is continuing to expand and technology is continuing to advance at an unprecedented pace, both phenomena emblematically represented by the rapid growth of global urbanization [83]. Smart technology is expected to play a pivotal role in ensuring that urbanization is sustainable, enabling cities to meet the *environmental, economic, social* and *cultural* needs of its inhabitants with *equity* and *inclusivity* [84, 85]. But many [86] argue that, beyond (anthropocentric) sustainability, the design of smart cities should provide for multispecies cohabitation. As human settlements continue to swallow up the land that many other animals call home, displaced creatures have nowhere to go but urban spaces, so cities are playing an increasingly important ecological role. However, Mancini et al. [87] have noted how, in imaginings of future smart sustainable cities (e.g., vincent.callebaut.org), which blend elegant structures and lush vegetation to represent anthropocentric utopias, animals are typically segregated in indoor production systems or altogether absent. Such "sterilized" visions of smart sustainable cities seemingly imply a worldview in which animals are either resources to be controlled and exploited (e.g., 'food animals'), or competitors to be expunged and eliminated (e.g., "pest animals"). In contrast, the authors envisage the *just smart city*, grounded in Nussbaum [60, 61]'s capability approach. The just smart city extends sustainability values [88] to animal dwellers, recognizing their entitlement to a dignified and flourishing existence according to their species' needs and affording them, with *equity* and *inclusivity*, the opportunity to pursue biological goals relevant to their basic capabilities at least to an adequate extent. This would include *environmental goals* (such as building, inhabiting, and traversing structures and spaces), *economic goals* (such as acquiring and managing resources), *social goals* (such as forming and maintaining

[3] This paragraph and the four above it are adapted from Mancini 2023b [14].

social groups), and *cultural goals* (such as gaining and transmitting knowledge and skills). The authors propose a framework for designing just smart cities, which takes a relational approach to multispecies coexistence [89], and in which smart technology decenters the human by mediating interspecies relations and redistributing control among city dwellers. Over time, this would inform the design of urban spaces that different species can *make sense of* and *interact with effectively* as they *pursue their biological goals* to fulfil their basic capabilities. Ultimately, the authors argue that, as ecological microcosms, cities exemplify our ability to relate to and coexist with nature; therefore, designing just smart cities could provide a roadmap to achieving ecological justice and sustainable development for the whole planet.

At the end of *2001: A Space Odyssey*, after the monolith appears at Dr Bowman's deathbed, Kubrick leaves us with the vision of a giant fetus, floating in space, bathed in cosmic light, eyes wide open, seemingly representing the birth of a new, superior humanity, and the promise of a different future, which concludes the space odyssey narrated by the film. However, cinematic narrations aside, our reality is that no alien intelligence is likely to transform us into superior beings any time soon. Ours, the one we need to embark on with great urgency, is not a space odyssey; it is an earth odyssey. It is up to us to use the intellectual and emotional capacities, which millions of years of evolution alongside our fellow earthlings has endowed us with, to address the existential challenges that face our planet, so that it can sustain all earthly life into the future. While technology can certainly play an important role in the pursuit of a sustainable future, this will arguably depend on whether we will continue to use it to exploit and outcompete other creatures, thus further alienating ourselves from the natural world and from the multispecies relationships that sustain us, or whether we will be able to leverage it to reduce our separation from nature and live in more harmonious relation with the beings with whom we share our common home. For the latter to happen, though, we need to start thinking 'outside the (human) box' and stop prioritizing human interests to everyone else's detriment. To this end, ACI's animal-centered perspective could, not only yield design outcomes and processes that improve animals' lives and human-animal relations, but arguably also produce ways of thinking and practices that are more attentive and sensitive to and more empathic towards other beings, helping us to meet the responsibility that arises from our ability to develop technology and from the power that technology gives us. In brief, ACI provides us with an *opportunity* to help reconfigure multispecies relations and our place in the world for a more inclusive, equitable and sustainable world.

References

1. Kubrik, S. 1968. 2001: A Space Odyssey (1968) - IMDb.
2. AnuMeena Care Foundation. 2023. The Dark Side of Technology: Harm to Ecosystem and Humanity |LinkedIn. *LinkedIn Pulse*, September 13.
3. Conn, A. 2015. Benefits & Risks of Artificial Intelligence. *Future of Life Institute*. https://futureoflife.org/ai/benefits-risks-of-artificial-intelligence/ (accessed 23.02.2025).
4. Yudkowsky, E. 2023. The Open Letter on AI Doesn't Go Far Enough. *TIME*. March 29. https://time.com/6266923/ai-eliezer-yudkowsky-open-letter-not-enough/ (accessed 23.02.2025).
5. AI HLEG. 2019. Ethics Guidelines for Trustworthy AI. European Commission's High-Level Expert Group on Artificial Intelligence. https://digital-strategy.ec.europa.eu/en/library/ethics-guidelines-trustworthy-ai (accessed 23.02.2025).
6. Google. 2023. Google AI Principles. *Google AI*. https://ai.google/responsibility/principles/ (accessed 23.02.2025).
7. Microsoft. 2023. Empowering responsible AI practices | Microsoft AI. https://www.microsoft.com/en-us/ai/responsible-ai (accessed 23.02.2025).
8. Domínguez, A., E. Klein, C. Raab, and J. Stewart. 2020. Ethical and Responsible IoT: The Edinburgh Initiative. *European Journal of Law and Technology* 11. School of Law, Queen's University Belfast: 1–37.
9. Reyero Lobo, P., E. Daga, H. Alani, and M. Fernandez. 2023. Semantic Web Technologies and Bias in Artificial Intelligence: A Systematic Literature Review. *Semantic Web* 14: 745–770. https://doi.org/10.3233/sw-223041.
10. Bendel, O. 2018. Towards Animal-Friendly Machines. *Paladyn, Journal of Behavioral Robotics* 9: 204–213.

11. Singer, P., and Y. F. Tse. 2023. AI Ethics: The Case for Including Animals. *AI and Ethics* 3: 539–551. https://doi.org/10.1007/s43681-022-00187-z.
12. Ziesche, S. 2021. AI Ethics and Value Alignment for Nonhuman Animals. *Philosophies* 6: 31. https://doi.org/10.3390/philosophies6020031.
13. Coghlan, S., and C. Parker. 2023. Harm to Nonhuman Animals from AI: a Systematic Account and Framework. *Philosophy & Technology* 36: 25. https://doi.org/10.1007/s13347-023-00627-6.
14. Mancini, C. 2024. Responsible ACI: Expanding the Influence of Animal-Computer Interaction. In *Proceedings of the Tenth International Conference on Animal-Computer Interaction*, 1–9. ACI '23. New York, NY, USA: Association for Computing Machinery. https://doi.org/10.1145/3637882.3637895.
15. Mancini, C. 2011. Animal-Computer Interaction: A Manifesto. *Interactions* 18: 69–73. https://doi.org/10.1145/1978822.1978836.
16. Mancini, C., S. Lawson, and O. Juhlin. 2017. Animal-Computer Interaction: The Emergence of a Discipline. *International Journal of Human-Computer Studies* 98: 129–134. https://doi.org/10.1016/j.ijhcs.2016.10.003.
17. Robinson, C. L., C. Mancini, J. van der Linden, C. Guest, and R. Harris. 2014. Canine-Centered Interface Design: Supporting the Work of Diabetes Alert Dogs. In *Proceedings of the SIGCHI Conference on Human Factors in Computing Systems*, 3757–3766. CHI '14. New York, NY, USA: Association for Computing Machinery. https://doi.org/10.1145/2556288.2557396.
18. Westerlaken, M., and S. Gualeni. 2016. Becoming with: Towards the Inclusion of Animals as Participants in Design Processes. 1–10. https://doi.org/10.1145/2995257.2995392.
19. Hirskyj-Douglas, I., J. C. Read, and B. Cassidy. 2017. A Dog Centred Approach to the Analysis of Dogs' Interactions with Media on TV Screens. *International Journal of Human-Computer Studies* 98: 208–220. https://doi.org/10.1016/j.ijhcs.2016.05.007.
20. Webber, S., M. Carter, W. Smith, and F. Vetere. 2020. Co-Designing with Orangutans: Enhancing the Design of Enrichment for Animals. In *Proceedings of the 2020 ACM Designing Interactive Systems Conference*, 1713–1725. DIS '20. New York, NY, USA: Association for Computing Machinery. https://doi.org/10.1145/3357236.3395559.
21. Sharp, H., J. Preece, and Y. Rogers. 2019. *Interaction Design: Beyond Human-Computer Interaction*. John Wiley & Sons.

22. Norman, D. 1986. Cognitive Engineering. In *User Centered System Design: New Perspectives on Human-Computer Interaction*, 31–61. Lawrence Erlbaum Associates. https://doi.org/10.1201/b15703-3.
23. Norman, D., and S. W. Draper. 1986. *User Centered System Design*. Lawrence Erlbaum Associates. https://doi.org/10.1201/b15703-3.
24. Mancini, C., R. Harris, B. Aengenheister, and C. Guest. 2015. Re-Centering Multispecies Practices: A Canine Interface for Cancer Detection Dogs. In *Proceedings of the 33rd Annual ACM Conference on Human Factors in Computing Systems*, 2673–2682. CHI '15. New York, NY, USA: Association for Computing Machinery. https://doi.org/10.1145/2702123.2702562.
25. Cheok, A., R. Tan, R. Peiris, O. Fernando, J. Soon, I. Wijesena, and Y.-P. Sen. 2011. Metazoa Ludens: Mixed-Reality Interaction and Play for Small Pets and Humans. *Systems, Man and Cybernetics, Part A: Systems and Humans, IEEE Transactions on* 41: 876–891. https://doi.org/10.1109/TSMCA.2011.2108998.
26. Aspling, F., and O. Juhlin. 2017. Theorizing Animal-Computer Interaction as Machinations. *International Journal of Human-Computer Studies* 98: 135–149. https://doi.org/10.1016/j.ijhcs.2016.05.005.
27. Zeagler, C., S. Gilliland, L. Freil, T. Starner, and M. Jackson. 2014. Going to the Dogs: Towards an Interactive Touchscreen Interface for Working Dogs. In *Proceedings of the 27th Annual ACM Symposium on User Interface Software and Technology*, 497–507. UIST '14. New York, NY, USA: Association for Computing Machinery. https://doi.org/10.1145/2642918.2647364.
28. Robinson, C., C. Mancini, J. van der Linden, C. Guest, L. Swanson, H. Marsden, J. Valencia, and B. Aengenheister. 2015. Designing an Emergency Communication System for Human and Assistance Dog Partnerships. In *Proceedings of the 2015 ACM International Joint Conference on Pervasive and Ubiquitous Computing*, 337–347. UbiComp '15. New York, NY, USA: Association for Computing Machinery. https://doi.org/10.1145/2750858.2805849.
29. Valentin, G., J. Alcaidinho, and M. M. Jackson. 2015. The Challenges of Wearable Computing for Working Dogs. In *Adjunct Proceedings of the 2015 ACM International Joint Conference on Pervasive and Ubiquitous Computing and Proceedings of the 2015 ACM International Symposium on Wearable Computers*, 1279–1284. UbiComp/ISWC'15 Adjunct. New York, NY, USA: Association for Computing Machinery. https://doi.org/10.1145/2800835.2807925.
30. Paci, P., C. Mancini, and B. A. Price. 2019. Wearer-Centered Design for Animal Biotelemetry: Implementation and Wearability Test of a Prototype.

In *Proceedings of the 2019 ACM International Symposium on Wearable Computers,* 177–185. ISWC '19. New York, NY, USA: Association for Computing Machinery. https://doi.org/10.1145/3341163.3347750.

31. Mancini, C., and E. Nannoni. 2022. Relevance, Impartiality, Welfare and Consent: Principles of an Animal-Centered Research Ethics. *Frontiers in Animal Science* 3. Frontiers. https://doi.org/10.3389/fanim.2022.800186.
32. Mancini, C., S. Li, G. O'Connor, J. Valencia, D. Edwards, and H. McCain. 2016. Towards Multispecies Interaction Environments: Extending Accessibility to Canine Users. In *Proceedings of the Third International Conference on Animal-Computer Interaction,* 1–10. ACI '16. New York, NY, USA: Association for Computing Machinery. https://doi.org/10.1145/2995257.2995395.
33. Jackson, M. M., G. Valentin, L. Freil, L. Burkeen, C. Zeagler, S. Gilliland, B. Currier, and T. Starner. 2015. FIDO—Facilitating Interactions for Dogs with Occupations: Wearable Communication Interfaces for Working Dogs. *Personal and Ubiquitous Computing* 19: 155–173. https://doi.org/10.1007/s00779-014-0817-9.
34. Veasey, J. S. 2020. Can Zoos Ever Be Big Enough for Large Wild Animals? A Review Using an Expert Panel Assessment of the Psychological Priorities of the Amur Tiger (*Panthera tigris altaica*) as a Model Species. *Animals* 10. Multidisciplinary Digital Publishing Institute: 1536. https://doi.org/10.3390/ani10091536.
35. Grant, A., V. Kankaanpää, and I. Hirskyj-Douglas. 2023. Humble Beginnings: Developing Touch- and Proximity-Input-Based Interfaces for Zoo-Housed Giraffes' Audio Enrichment. *Proceedings of the ACM on Human-Computer Interaction* 7: 434:175–434:197. https://doi.org/10.1145/3626470.
36. French, F., C. Mancini, and H. Sharp. 2020. More Than Human Aesthetics: Interactive Enrichment for Elephants. *Proceedings of the 2020 ACM Designing Interactive Systems Conference.* Eindhoven Netherlands: ACM: 1661–1672. https://doi.org/10.1145/3357236.3395445.
37. Webber, S., M. Carter, W. Smith, and F. Vetere. 2017. Interactive Technology and Human–Animal Encounters at the Zoo. *International Journal of Human-Computer Studies* 98: 150–168. https://doi.org/10.1016/j.ijhcs.2016.05.003.
38. Broomé, S., M. Feighelstein, A. Zamansky, G. C. Lencioni, P. H. Andersen, F. Pessanha, M. Mahmoud, H. Kjellström, and A. A. Salah. 2023. Going Deeper than Tracking: A Survey of Computer-Vision Based Recognition

of Animal Pain and Emotions. *International Journal of Computer Vision* 131: 572–590. https://doi.org/10.1007/s11263-022-01716-3.

39. Nieusma, D. 2021. Into the Wild: Into the Wild: Opportunities and Challenges for Animal-Computer Interaction in Wildlife Conservation. 1–7. https://doi.org/10.1145/3493842.3493904.
40. Grinko, M., T. Ertl, K. Aal, and V. Wulf. 2021. Transitions by Methodology in Human-Wildlife Conflict: Reflections on Tech-Based Reorganization of Social Practices. In *Computing within Limits*. LIMITS. https://doi.org/10.21428/bf6fb269.7c2f2cd2.
41. Webber, S., W. Smith, M. Carter, and F. Vetere. 2023. Watching Animal-Computer Interaction: Effects on Perceptions of Animal Intellect. In *Proceedings of the Ninth International Conference on Animal-Computer Interaction*, 1–14. ACI '22. New York, NY, USA: Association for Computing Machinery. https://doi.org/10.1145/3565995.3566035.
42. Paci, P., C. Mancini, and B. Nuseibeh. 2022. The Case for Animal Privacy in the Design of Technologically Supported Environments. *Frontiers in Veterinary Science* 8: 784794. doi: 10.3389/fvets.2021.784794.
43. van der Linden, D. 2021. Interspecies Information Systems. *Requirements Engineering* 26: 535–556. https://doi.org/10.1007/s00766-021-00355-3.
44. Paci, P., C. Mancini, and B. A. Price. 2016. Towards a Wearer-Centred Framework for Animal Biotelemetry. In: *Proceedings of Measuring Behaviour* 2016 (Spink, A. J., ed.), 25–27 May 2016, Dublin.
45. Grillaert, K., and S. Camenzind. 2016. Unleashed Enthusiasm: Ethical Reflections on Harms, Benefits, and Animal-Centered Aims of ACI. In. https://doi.org/10.1145/2995257.2995382.
46. Lawson, S., B. Kirman, C. Linehan, T. Feltwell, and L. Hopkins. 2015. Problematising Upstream Technology through Speculative Design: The Case of Quantified Cats and Dogs. In *Proceedings of the 33rd Annual ACM Conference on Human Factors in Computing Systems*, 2663–2672. CHI '15. New York, NY, USA: Association for Computing Machinery. https://doi.org/10.1145/2702123.2702260.
47. Lawson, S., B. Kirman, and C. Linehan. 2016. Power, Participation, and the Dog Internet. *Interactions* 23: 37–41. https://doi.org/10.1145/2942442.
48. French, F., I. Hirskyj-Douglas, H. Väätäjä, P. Pons, S. Karl, Y. Chisik, E. Nannoni, et al. 2021. Ethics and Power Dynamics in Playful Technology for Animals: Using Speculative Design to Provoke Reflection. In *Proceedings of the 24th International Academic Mindtrek Conference*,

91–101. Academic Mindtrek '21. New York, NY, USA: Association for Computing Machinery. https://doi.org/10.1145/3464327.3464366.

49. Aspling, F. 2020. Unleashing Animal-Computer Interaction: A Theoretical Investigation of the "I" in ACI. Doctoral dissertation. Stockholm: Department of Computer and Systems Sciences, Stockholm University, 978-91-7911-082-6 (ISBN) 978-91-7911-083-3 (ISBN)
50. Väätäjä, H. 2014. Animal Welfare as a Design Goal in Technology Mediated Human-Animal Interaction. In *Proceedings of the 2014 Workshops on Advances in Computer Entertainment Conference*, 1–8. ACE '14 Workshops. New York, NY, USA: Association for Computing Machinery. https://doi.org/10.1145/2693787.2693797.
51. Freil, L., C. Byrne, G. Valentin, C. Zeagler, D. Roberts, T. Starner, and M. Jackson. 2017. Canine-Centered Computing. *Foundations and Trends® in Human-Computer Interaction* 10: 87–164. https://doi.org/10.1561/1100000064.
52. Brambell, F. W. C. 1965. *Report of the Technical Committee to Enquire into the Welfare of Animals Kept Under Intensive Livestock Husbandry Systems*. London: H.M.S.O.
53. Mellor, D. J., N. J. Beausoleil, K. E. Littlewood, A. N. McLean, P. D. McGreevy, B. Jones, and C. Wilkins. 2020. The 2020 Five Domains Model: Including Human-Animal Interactions in Assessments of Animal Welfare. *Animals: an open access journal from MDPI* 10: 1870. https://doi.org/10.3390/ani10101870.
54. Mancini, C. 2017. Towards an Animal-Centred Ethics for Animal–Computer Interaction. *International Journal of Human-Computer Studies* 98: 221–233. https://doi.org/10.1016/j.ijhcs.2016.04.008.
55. Westerlaken, M. 2020. Imagining Multispecies Worlds. Malmö Universitet. DOI: 10.24834/isbn.9789178771059.
56. Buchanan, R. 2001. Human Dignity and Human Rights: Thoughts on the Principles of Human-Centered Design. *Design Issues* 17: 35–39. https://doi.org/10.1162/074793601750357178.
57. Animal Welfare Act of 2005. 2005. December 16. Swiss Animal Welfare Act. https://www.fedlex.admin.ch/eli/cc/2008/414/en (accessed 23.02.2025).
58. Participation. 2024. (British) Animal Welfare Act 2006. Text. Statute Law Database. https://www.legislation.gov.uk/ukpga/2006/45. Accessed May 4.
59. Mancini, C., O. Hirsch-Matsioulas, and D. Metcalfe. 2023. Politicising Animal-Computer Interaction: an Approach to Political Engagement

with Animal-Centred Design. In *Proceedings of the Ninth International Conference on Animal-Computer Interaction*, 1–11. ACI '22. New York, NY, USA: Association for Computing Machinery. https://doi.org/10.1145/3565995.3566034.

60. Nussbaum, M. C. 2006. *Frontiers of Justice: Disability, Nationality, Species Membership*. Harvard University Press. https://doi.org/10.2307/j.ctv1c7zftw.
61. Nussbaum, M. C. 2023. *Justice for Animals: Our Collective Responsibility*. New York (N.Y.) London: Simon & Schuster.
62. Rawls, J. 2001. *Justice as Fairness: A Restatement*. Harvard University Press. https://doi.org/10.2307/j.ctv31xf5v0.
63. Rawls, J. 1972. *A Theory of Justice*. Cambridge, MA: Harvard University Press.
64. Deckha, M. 2015. Vulnerability, Equality, and Animals. *Canadian Journal of Women and the Law* 27: 47–70. https://doi.org/10.3138/cjwl.27.1.47.
65. Singer, P. 1980. Animals and the Value of Life. In *Matters of Life and Death: New Introductory Essays in Moral Philosophy*, ed. T. Regan. New York: McGraw-Hill.
66. Sen, A. 2010. *The Idea of Justice*. Penguin Books, London, ISBN: 978-0141-03785-1.
67. Dawkins, M. S. 2012. *Why Animals Matter: Animal Consciousness, Animal Welfare, and Human Well-Being*. New York: Oxford University Press.
68. Favareau, D. 2024. *Essential Readings in Biosemiotics*. Springer Dordrecht. https://link.springer.com/book/10.1007/978-1-4020-9650-1. Accessed May 4.
69. Resner, B. I. 2001. Rover@Home: Computer Mediated Remote Interaction between Humans and Dogs. Thesis, Massachusetts Institute of Technology.
70. Mancini, C. 2023. A Biosemiotics Perspective on Dogs' Interaction with Interfaces: An Analytical and Design Framework. *Interaction Studies* 24. John Benjamins: 201–224. https://doi.org/10.1075/is.22027.man.
71. Paci, P., C. Mancini, and B. A. Price. 2017. The Role of Ethological Observation for Measuring Animal Reactions to Biotelemetry Devices. In *Proceedings of the Fourth International Conference on Animal-Computer Interaction*, 1–12. ACI '17. New York, NY, USA: Association for Computing Machinery. https://doi.org/10.1145/3152130.3152144.
72. Hirskyj-Douglas, I. 2023. Systematic Iterative Design of Interactive Devices for Animals: Guidance and Reflections. *Interaction Studies*.

Social Behaviour and Communication in Biological and Artificial Systems 24: 225–256. https://doi.org/10.1075/is.22045.hir.

73. Ruge, L., and C. Mancini. 2020. A Method for Evaluating Animal Usability (MEAU). In *Proceedings of the Sixth International Conference on Animal-Computer Interaction*, 1–12. ACI '19. New York, NY, USA: Association for Computing Machinery. https://doi.org/10.1145/3371049.3371060.
74. Byrne, C., L. Freil, T. Starner, and M. M. Jackson. 2017. A Method to Evaluate Haptic Interfaces for Working Dogs. *International Journal of Human-Computer Studies* 98: 196–207. https://doi.org/10.1016/j.ijhcs.2016.04.004.
75. Nelson, J. K., and P. C. Shih. 2017. *CompanionViz*: Mediated Platform for Gauging Canine Health and Enhancing Human–Pet Interactions. *International Journal of Human-Computer Studies* 98: 169–178. https://doi.org/10.1016/j.ijhcs.2016.04.002.
76. Foster, M., R. Brugarolas, K. Walker, S. Mealin, Z. Cleghern, S. Yuschak, J. C. Clark, et al. 2020. Preliminary Evaluation of a Wearable Sensor System for Heart Rate Assessment in Guide Dog Puppies. *IEEE Sensors Journal* 20: 9449–9459. https://doi.org/10.1109/JSEN.2020.2986159.
77. Mancini, C., and J. Lehtonen. 2018. The Emerging Nature of Participation in Multispecies Interaction Design. In *Proceedings of the 2018 Designing Interactive Systems Conference*, 907–918. DIS '18. New York, NY, USA: Association for Computing Machinery. https://doi.org/10.1145/3196709.3196785.
78. Pons, P., A. Carrion-Plaza, and J. Jaen. 2019. Remote Interspecies Interactions: Improving Humans and Animals' Wellbeing through Mobile Playful Spaces. *Pervasive and Mobile Computing* 52: 113–130. https://doi.org/10.1016/j.pmcj.2018.12.003.
79. Väätäjä, H. K., and E. K. Pesonen. 2013. Ethical Issues and Guidelines When Conducting HCI Studies with Animals. In *CHI '13 Extended Abstracts on Human Factors in Computing Systems*, 2159–2168. CHI EA '13. New York, NY, USA: Association for Computing Machinery. https://doi.org/10.1145/2468356.2468736.
80. Russell, W. M. S., and R. Burch. 1959. *The Principles of Humane Experimental Technique*. London, UK: Metheun.
81. Dombrowski, L., E. Harmon, and S. Fox. 2016. Social Justice-Oriented Interaction Design: Outlining Key Design Strategies and Commitments. DIS'16: *Proceedings of the 2016 International Conference on Designing Interactive Systems*. ACM Press, New York, 656–671. https://doi.org/10.1145/2901790.2901861.

82. UNDP United Nations Development Programme. 2024. Sustainable Development Goals. *UNDP*. https://www.undp.org/sustainable-development-goals. Accessed May 4.

83. United Nations. 2014. *2014 Revision of the World Urbanization Prospects*. Report.

84. Angelidou, M., A. Psaltoglou, N. Komninos, C. Kakderi, P. Tsarchopoulos, and A. Panori. 2017. Enhancing Sustainable Urban Development through Smart City Applications. *Journal of Science and Technology Policy Management* 9: 146–169. https://doi.org/10.1108/JSTPM-05-2017-0016.

85. Bibri, S. E., and J. Krogstie. 2017. Smart Sustainable Cities of the Future: An Extensive Interdisciplinary Literature Review. *Sustainable Cities and Society* 31: 183–212. https://doi.org/10.1016/j.scs.2017.02.016.

86. Heitlinger, S., M. Foth, and R. Clarke, ed. 2024. *Designing More-than-Human Smart Cities: Beyond Sustainability, Towards Cohabitation*. Oxford University Press.

87. Mancini, C., D. Metcalfe, and O. Hirsch-Matsioulas. 2024. Justice by Design: The Case for Equitable and Inclusive Smart Cities for Animal Dwellers. In *Designing More-than-Human Smart Cities: Beyond Sustainability, Towards Cohabitation*, ed. S. Heitlinger, M. Foth, and R. Clarke, 0. Oxford University Press. https://doi.org/10.1093/9780191980060.003.0011.

88. ITU. 2014. *Agreed Definition of a Smart Sustainable City*. Geneva: International Telecommunications Union (ITU), https://globalcitieshub.org/en/international-telecommunication-union-itu/ (accessed 23.02.2025).

89. Graham, M. 2014. Aboriginal Notions of Relationality and Positionalism: A Reply to Weber. *Global Discourse* 4: 17–22. https://doi.org/10.1080/23269995.2014.895931.

Chapter 8

Ethical and Societal Issues Occasioned by Xenotransplantation*

Bernard E. Rollin
Department of Philosophy, Colorado State University
Fort Collins, Colorado, USA (Deceased)

Summary

The dream of transferring bodily organs from animals to humans goes back to antiquity, as articulated in the myth of Daedalus and Icarus. But the genuine possibility of xenotransplantation is

*This chapter is from an article in *Animals*, 2020, 10, 1695; doi:10.3390/ani10091695 www.mdpi.com/journal/animalsSeptember 2020. Published: 19 September as an open access article distributed under the terms and conditions of the Creative Commons Attribution (CC BY) license (http://creativecommons.org/licenses/by/4.0/). This chapter is included in this volume because of the direct connection of biomedical engineering to xenotransplantation. Since the 2020 publication date of the article reprinted here, in 2022 the University of Maryland Medical Center in Baltimore, Maryland, performed a pig heart transplantation on David Bennett, Sr., the first human to receive a transplanted pig heart. He survived two months beyond the procedure. On September 20, 2023, another pig heart transplantation was performed at the same institution on recipient Lawrence Faucette, a 58-year-old man with end-stage heart disease. He lived for six weeks following surgery.

Animals, Ethics, and Engineering: Intersections and Implications
Edited by Rosalyn W. Berne

ISBN 978-981-5129-56-4 (Hardcover), 978-1-003-65231-1 (eBook)
www.jennystanford.com

largely a creature of the last half-century. That possibility raises a plethora of genuine ethical issues, as well as spurious ones occasioned by societal ignorance of science. Abstract: There are three sorts of issues associated with genetic engineering and, by implication, with xenotransplantation. These are dangers associated with the technology, animal welfare issues, and the claim that genetic engineering represents a technology that humans should not embark upon. Using the hearts of pigs for humans in need of transplants has been a major issue in xenotransplantation. There are dangers associated with such use, such as immunological rejection of the organ, endogenous viruses infecting the recipients, and issues of privacy. In addition, the issue of fair distribution of organs arises. Animal welfare issues also arise, most notably the living conditions of the donor animals, issues notably present in confinement agriculture. A major issue emerges from animals' being kept under conditions that fail to meet the needs dictated by the animals' biological and psychological natures. Xenotransplantation animals will be kept under deprived laboratory conditions that similarly fail to meet the animals' natures. This is a significant concern for society in general. There are also issues of "bad ethics" arising from scientists' disavowal of ethical concerns in science. This in turn, coupled with societal ignorance of science, creates a climate for proliferation of religious and other non-rational concerns, such as the claim that xenotransplantation violates God's will. These spurious concerns can only be ameliorated when public understanding of science improves, and scientific understanding of ethics increases.

8.1 Introduction

In 1995, I published what I believe to be the first book ever written on the social and ethical issues associated with the genetic engineering of animals, entitled The Frankenstein Thing: Ethical and Social Issues in the Genetic Engineering of Animals [1]. I used the iconic image of the Frankenstein monster as the symbol for society's view of biotechnology. I disambiguated three different aspects of the myth that I called "There are certain things humans were not meant to do", "Rampaging monsters", and "The plight of the creature". Aspect one referred to the widespread belief, primarily for religious reasons, that certain areas should not be studied. Aspect two referred to dangers attendant on human manipulation of species. Aspect three referred to the suffering of the created being. All of these are strikingly present in the issue of using animal organs for xenotransplantation.

While aspect one was of great social prominence, I did not see it as a genuine ethical issue, though much of society surely did. I in fact enunciated what I called "a Gresham's law for ethics". Recall that Gresham's law affirms that "bad money drives good money out of circulation". For example, after World War I in Germany it took a wheelbarrow full of deutschemarks to buy a loaf of bread, so devalued was the currency. In such a situation, no one paid off major debts in gold, but rather in worthless currency. Similarly, I argued that "bad ethics drives good ethics out of circulation", since such aspect one claims as "genetic engineering violates God's will" are far "sexier" than genuine ethical issues. After Dolly the cloned sheep was announced, for example, a Time/Warner survey conducted the next week revealed that 75% of the US public believed that the creation of Dolly violated God's will.

Aspect two referred to various kinds of danger or harm associated with genetic engineering; for example, environmental dangers or the potential for new pathogens. While certainly counting as ethical issues, they were equally matters of prudence. Aspect three, the effect of genetic engineering on the welfare of the animals engineered, was a truly ethical issue, as it almost invariably pitted the well-being of the animals against human well-being. And, as we know from the rise of industrial agriculture and the history of animal research, human benefit usually wins.

The template I have just outlined works very well for all aspects of biotechnology including agriculture, animal research, creation of animal models of human genetic disease, and the topic at hand, xenotransplantation. In fact, xenotransplantation provides a textbook example of all of the above categories. Approximately 3500 to 4000 people seek heart transplants per year in the US. As a result, alternatives to human donors have been sought. One putatively plausible alternative has been to use animals, specifically pigs, although historically baboons and other animals have been utilized. There are no cases reported of humans surviving with pig hearts.

Consider the use of pig hearts for people needing heart transplants. (This is overwhelmingly the primary use of xenotransplantation that has been discussed.) There exist very real dangers in using animal organs for transplantation, specifically on microbiological, physiological, and immunological levels. There are also animal welfare issues associated with such uses, albeit subtle ones. Finally, there are extremely clear examples of bad ethics supplanting and eclipsing good ethics in discussions of xenotransplantation.

8.2 Dangers Associated with Xenotransplantation

Pigs have been the animals of choice for organ donation, specifically heart, but also kidneys and islet cells of the pancreas. There are numerous reasons for this. Pig organs, particularly the heart, are comparable in size to that of humans. In addition, pigs, as domesticated animals, can easily be raised in relatively controlled confinements, which is not the case with nonhuman primates. In addition, because of genetic relatedness with nonhuman primates, there is a greater risk of disease spread in the case of primates. Additionally, pigs have relatively large litters—up to a dozen piglets at a time—and have a relatively short gestation period; three months, three weeks, and three days. For these reasons, certain governments, for example that of the UK, have strongly discouraged and even banned nonhuman primates as candidates for transplantation. See the Nuffield Council on Bioethics's comprehensive report on xenotransplantation [2]. (The logic implicit in the use of pig hearts for human transplantation also holds in the case of other transplantable organs such as kidneys and livers, and islet cells for elimination of diabetes.)

As of mid-April 2019, an Israeli scientist has announced achieving three-dimensional printing of the human heart [3]. It is as yet too early to tell if this is a practicable alternative to xenotransplantation. Since it is created from the patient's own cells, the problem of immunological rejection may be circumvented.

The most significant issue with using animals for a source of transplanted organs (xenotransplantation) for humans is immunological rejection of the organ, with the human immune system recognizing the foreign organ as "not-self" and correlatively rejecting it. In what is known as "hyper-acute rejection", the body begins to reject the organ virtually as soon as it is implanted [4].

In the case of genetically engineered pigs, however, a small amount of human genetic material can be injected into the developing pig embryo, so that the resulting piglet is not recognized as foreign. In 2016, National Institutes of Health (NIH) researchers announced that a pig heart had been kept alive in a baboon for three years. The use of immune-system-suppressing drugs has also reduced the probability of rejection.

An additional major danger with xenotransplantation are the endogenous retroviruses carried by pigs, which are capable of making humans very sick although not harming the pigs [4]. These viruses are called porcine endogenous retroviruses (PERVs). One salient example of a PERV was the virus likely responsible for an epidemic of swine flu in 2009, which took about a quarter of a million human lives. Another example is the Nipah virus, which caused an epidemic of encephalitis in Singapore and Malaysia. One can speculate that some of these viruses could conceivably not be recognized until such time as they created disastrous epidemics. Antibiotic-resistant bacteria are also capable of being transplanted into humans through the use of swine organs. In addition to infecting a grant recipient, some of these diseases may spread "horizontally" across a community, serving as a danger to public health. All of this clearly illustrates the great extent to which xenotransplants represent grave danger.

In addition, concerns for safety for the transplant recipient, as well as for public health, put a severe limitation on the privacy of such recipients, as they need to be monitored for pathogens dangerous to them and others, for what may possibly be very extended lengths of time, which in turn affects the privacy not only of the recipient, but also of his or her family, friends, work associates and others with whom he or she is in contact. Thus, there arises an ethical tension between the good of the recipient and the good of society. Correlatively, it is extremely unlikely that the identity of those receiving xenotransplants can be kept confidential, so it is quite possible that such people will be seen as significantly different, or even as "freaks", particularly if the recipient is a child and will likely be subject to taunting with locutions like "pig heart" regularly thrown at them, which in turn can cause significant psychological damage. Another ethical issue associated with xenotransplantation is the question of how such organs would be distributed. What is a fair way to assure equitable distribution of both risks and benefits [5]?

In sum, there are a variety of dangers associated with xenotransplantation, both for recipients and for the public at large. As yet, however, none of these projected dangers have come to pass, in part because there have been no successful xenotransplants. Governments have been extremely cautious about allowing unrestricted research into xenotransplantation to proceed, presumably because of societal hesitation regarding all aspects of biotechnology.

8.3 Animal Welfare Issues Associated with Xenotransplantation

Some people are instinctively horrified by the idea of raising pigs to put their organs into human bodies. A reality check should diminish that reaction. The number of animals raised for xenotransplantation will probably number in the thousands. In contrast, the number of pigs killed in a year in the world for food is over a billion [6]! If we can cavalierly kill that many pigs in pursuit of bacon, ham, and sausages, we should surely not be horrified at killing them in order to save human lives.

The ethical issue of concern to the majority of society is not the killing of animals, as the above number attests. There is little reason to believe that animals fear death or anticipate that it is coming. Lacking human linguistic capacity, animals cannot conceive of death, described by Heidegger as "awareness of the possibility of the impossibility of [their] being". While animals certainly communicate, they do not possess the ability to transcend the moment and imagine the future. In all fairness, there are those who do see death as a welfare issue, since opportunities for future pleasures are denied to the animal.

Rather, what troubles society in general is the question of how the animals live, and the unnatural conditions under which they are housed and maintained. Attesting to this point has been the elimination of such severe systems as confinement veal production in wooden crates where the animals are tethered so as to keep the flesh soft, and fed a diet designed to produce borderline anemia so that consumers can eat "white veal". Gestation crates—essentially 60 by 2.50 by 30 cages that are so small that they do not allow sows to turn around or even stand up—have been banned in Scandinavia for decades, and the EU committed beginning in 2013 to phase them out. The UK and New Zealand have already done so. Smithfield, the largest pork producer in the US, began eliminating gestation crates in 2008, and finished the process in 2019.

Were pigs for food raised the way they were traditionally kept, outdoors on pasture with barns to retreat to in severe weather, the severe confinement necessitated by keeping swine used for organ transplants would be blatantly unacceptable. Traditional rearing of all farm animals was accomplished by placing the animals under conditions meeting the needs following from their biological and psychological natures—what Aristotle called their telos [7], the "pigness" of the pig, the "cowness" of the cow. In fact, raising animals under these natural conditions was an essential component of the "good husbandry" that was the key to agricultural success for 10,000 years, until after the Industrial Revolution.

"Husbandry" is derived from the Old Norse phrase "hus bond", meaning "bonded to the household". Husbandry has been termed "the ancient contract with animals", where, as in any fair contract, both parties benefit from the relationship. Virtually until the 20th century, husbandry was the key concept in animal agriculture.

The essence of husbandry was care. Humans put animals into the most optimal environment congenial to the animals' not only surviving but thriving, the environment for which they had evolved and been selected. The better the animals did, the better farmers did. Sanctioned by their own self-interest, humans provided farm animals with sustenance, shelter, protection from predation, such medical attention as was available, help in birthing, food during famine, water during drought, safe surroundings, and comfortable appointments. The Noah story is emblematic of this "ancient contract". In the Noah story, we learn that even as God preserves humans, humans preserve animals. In the 23rd Psalm, the Psalmist points out that God is to humans as shepherd are to sheep [8].

The singular beauty of husbandry is that it was at once an ethical and prudential doctrine. It was prudential in that failure to observe husbandry inexorably led to ruination of the person keeping animals. Not feeding, not watering, not protecting from predators, not respecting the animals' physical, biological, physiological, and psychological needs and natures, meant your animals did not survive and thrive, and thus neither did you. Thus, no formally articulated animal ethic was needed. Animal husbandry in essence became the basis for what was the newly civilized society and the leisure time necessitated by the development of culture, as Thomas Hobbes pointed out in Leviathan. Thus, husbandry was about putting square pegs into square holes, round pegs into round holes, and creating as little friction as possible while doing so. The industrialization of confinement agriculture, as it developed in the 20th century, enabled the forcing of square pegs into round holes, round pegs into square holes, i.e., animals are placed into confinement systems for which they are not naturally suited, by virtue of the development of what I have called "technological sanders", such as antibiotics, vaccines, air-handling systems, that severed the traditional connection between productivity and welfare. The use of antibiotics for disease prevention and growth promotion has become a major societal issue, given the risk of developing antibiotic-resistant pathogens [9].

Swine raised for organ transplantation will certainly not be kept under traditional husbandry conditions. Rather, they will be kept much in the manner of laboratory animals, under confined, sterile conditions that minimize the risk of pathogen proliferation and keep the animals sufficiently healthy to provide a (relatively) safe source for transplantation. Although such conditions are far better than agricultural conditions in terms of animal health, they are equally deficient in accommodating the animals' biological and psychological natures. Thus, we may confidently affirm that while such animals will be healthier, and thus have increased welfare in a single dimension, their overall welfare, i.e., the respect for the needs emerging from their telos, will be no better [10].

There are numerous accounts of "animal welfare" circulating in contemporary thought, ranging from an animal experiencing more pleasure than pain (the utilitarian view) to my own account based in the thinking of Aristotle. Aristotle believed that things in the natural world would best be explained teleologically, i.e., by reference to their basic function or nature, on the model of biological organisms. This is what Aristotle [7] called their telos (i.e., their nature). Post-Newtonian science is mechanistic, with physics serving as the model for good explanations. When I began to work on animal ethics, it dawned on me that telos was the most reasonable vehicle for describing our moral obligations to animals. Telos accommodates an animal's biological and psychological needs and nature—the "pigness of a pig", the "dogness of a dog". As such, it is extremely congenial to common sense. Implicit reference to telos is what militates against the acceptability of small cages for pigs or veal calves, or any cages for zoo animals. Furthermore, the notion of telos fits beautifully with Western political theories according rights to humans based on the protection of their human nature, and accords beautifully with the Platonic dictum that ethics should proceed from pre-existing ethics. (i.e., that animal ethics would be best derived from our established human ethics).

When I discuss animal welfare with the scientific community, I am usually told that it is solely a matter of "sound science". While science is certainly relevant, questions of animal welfare are at least partly "ought questions", questions of ethical obligation. The concept of animal welfare is in part an ethical concept to which science brings relevant data. When we ask about an animal's welfare, or about a person's welfare, we are asking about what we owe the animal and to what extent.

In the early 1980s, the US agricultural community issued a document known as the CAST Report, which affirmed that "what we owe animals and to what extent is simply what it takes to get them to create profit" [11]. This implies that animals are well off if they only have food, water, and shelter. At about the same time, the British Farm Animal Welfare Council (FAWC) declared that animal welfare includes the animal's physical and mental state and implies both "fitness and a sense of well-being" [12]. This was in turn defined in terms of the famous five freedoms: freedom from hunger and thirst, freedom from discomfort, freedom from pain injury or disease, freedom to express normal behavior, and freedom from fear and distress. (The FAWC definition is closely connected to what I call telos.)

Clearly the two definitions contain very different notions of our moral obligations to animals (and there is an indefinite number of other definitions). Which is correct cannot be decided by gathering facts or doing experiments—indeed, which ethical framework one adopts will in fact determine the shape of science studying animal welfare. Thus, sound science does not determine your concept of welfare; rather your concept of welfare determines what counts as sound science! Certainly, society leans in the direction of the FAWC definition. In my books on animal ethics [10, 13], I have argued that respect for animal nature (telos) is the fundamental basis for the moral status of animals, even more so than the ability to feel pain. (Some animals caught in steel-jawed traps will chew their legs off to escape!).

One can surmise that a telos-based/FAWC view of welfare will continue to dominate societal pressure for farm animal welfare. However, in the case of swine organs being used to save lives, the situation is not so clear. In the case of dogs and primates used in science, society has moved relentlessly in the direction of environments for these animals that at least partially accommodate their social, biological, and psychological natures. Undoubtedly, conditions for keeping swine destined for organ donation will continue to improve, although within the constraints of gnotobiotics, i.e., control of pathogenic microorganisms. There is no reason that such swine cannot enjoy significantly enriched confinement as societal awareness of farm animal welfare continues to grow. In the end, as I have said many times before in many contexts, they are having their lives taken by humans; the least we can do for them is to make sure that what life they do have is a good and pleasant one. This should include adequate space, an enriched environment, preservation of social groups, good handling by stockpeople, adequate veterinary care, proper diet, climate control. However, the attempt to keep these animals free of pathogens significantly limits their environment and thus their ability to actualize their telos, i.e., their psychological and biological natures. It is unlikely, given these concerns, that animals could be kept outdoors in a pastoral manner, as controlling the extensive environment microbiologically is near impossible.

8.4 Issues of "Bad Ethics"

Many articles on xenotransplantation stress the importance of societal involvement in the formation of policy regarding xenotransplantation, for example [14]. The ethical issues we have discussed so far are genuinely ethical issues or issues of danger. Thus, it seems to be obligatory to also pay attention to what we have referred to as "bad ethics". We shall now demonstrate that integrally involved in bad ethics is appalling social lack of understanding of science, as well as the scientific community's denial of the relevance of ethics to science.

As I have described before, there is a perfect storm combining these two factors that militate in favor of the proliferation of bad ethics [15]. Using the United States as my source, we can get a clear sense of how ignorant the general public is about science. In a 2004 paper, neuroscientist and physician Dr. Keith Black affirmed that research shows the following: "One half of the American public does not know the earth goes around the sun once a year. A 1996 National Assessment of Educational Progress survey found that 43 percent of high school seniors did not meet the basic standard for scientific knowledge" [16]. A 2015 Pew Charitable Trusts report on scientific literacy indicates that 40% of the US public believes that humans coexisted with the dinosaurs [17]. And a 2015 survey showed that 80% of the public wanted mandatory labeling of food indicating the presence of DNA [18]!

A 2014 Gallup poll cited in a National Center for Science Education report stated that only a very small percentage (19%) of the US public has accepted the naturalistic position that "man has developed over millions of years from less advanced forms of life. God had no part in this process" [19]. Correlatively, a 2017 Gallup poll indicated that "the percentage of Americans who actively want creation not evolution to be taught in schools may seem relatively modest at 30%, but this number is significantly larger than the percentage who actively want evolution to be taught over creation" [20].

Things are no better from the science perspective. Science has done little to address societal ignorance. Some years ago, I, a philosophy professor with a reasonable background in science, began to plan to teach a full-year introductory honors biology course along with a biologist with interest in philosophy. With all the major changes taking place in biology, particularly in the area of biotechnology, as well as the ever-increasing societal fermentation regarding the use of animals in research, we were concerned that biology students were ill-prepared to engage these concerns. We received funding from the National Science Foundation (NSF) and began looking for suitable textbooks. We were appalled at what we found. A recurrent theme invariably found in the "throat-clearing" introductions to these textbooks (generally making brief mention of Linnaeus, Darwin, etc.) was the claim that science was "value-free" and, more importantly, that science has nothing to do with ethics. Such a position represented a legacy of the philosophical position known as "logical positivism" that dominated science for much of the 20th century.

If one asks most working scientists what separates science from religion, speculative metaphysics, or shamanistic world views, they would unhesitatingly reply that it is an emphasis on validating all claims through sense experience, observation, or experimental manipulation. This component of scientific ideology can be traced directly back to Newton who proclaimed that he did not "feign hypotheses" (hypotheses non fingo) but operated directly from experience [21].

The insistence on experience as the bedrock for science continues from Newton to the 20th century, where it reaches its most philosophical articulation in the reductive movement known as Logical Positivism, a movement that was designed to excise the unverifiable from science and, in some of its forms, to axiomatize science so that its derivation from observations was transparent. Examples of positivist targets were Bergson's (and other biologists') talk of life force (élan vital) as separating the living from the non-living, or the embryologist Driesch's postulation of "entelechies" to explain regeneration in starfish.

Although logical positivism took many subtly different and variegated forms, the message, as received by working scientists, and passed on to students (including myself), was that proper science ought not allow unverifiable statements, including ethical judgments.

What does all this have to do with ethics? Quite a bit, it turns out. The philosopher Ludwig Wittgenstein, who greatly influenced the logical positivists, once remarked that, if you take an inventory of all the facts in the universe, you will not find it a fact that killing is wrong [22]. In other words, ethics is not part of the furniture of the scientific universe. You cannot, in principle, test the proposition that "killing is wrong". It can neither be verified nor falsified. So, in Wittgenstein's view, empirically and scientifically, ethical judgments are meaningless. From this, it was concluded that ethics is outside of the scope of science, as are all judgments regarding values, rather than facts. The slogan that I had in fact learned in my science courses in the 1960s, and which has persisted to the present, and is still being taught in too many places, is that "science is value-free" in general, and "ethics-free" in particular.

It is therefore not surprising that when scientists were drawn into social discussions of ethical issues, they were every bit as emotional as their untutored opponents. It is because their ideology dictates that these issues are nothing but emotional, that the notion of rational ethics is an oxymoron, and that he who generates the most effective emotional response "wins". So, for example, in the 1970s and 1980s debate on the morality of animal research, scientists either totally ignored the issue, or countered criticisms with emotional appeals to the health of children. For example, in one film entitled "Will I Be All Right, Doctor?" (the question asked by a frightened child of a pediatrician), made by defenders of unrestricted research, the response was as follows: "Yes, if they leave us alone to do what we want with animals." So appallingly and unabashedly emotional and mawkish was the film that, when it was premiered at the American Association for Laboratory Animal Science (AALAS) meetings at a subsection of laboratory animal veterinarians, a putatively sympathetic audience, the only comment forthcoming from the audience came from a veterinarian, who affirmed that he was "ashamed to be associated with a film that is pitched lower than the worst anti-vivisectionist clap-trap!".

Positivist thinkers felt compelled to explain why intelligent people continued to make moral judgments and continued to argue about them. They explained the former by saying that when people make assertions such as "killing is wrong", which seem to be statements about reality, they are in fact describing nothing. Rather, they are "emoting", expressing their own revulsion at killing. "Killing is wrong" really expresses "Killing, yuk!" rather than describing some state of affairs. And when we seem to debate about killing, we are not really arguing ethics (which one cannot do any more than you and I can debate whether we like or do not like pepperoni), but rather disputing each other's facts. So, a debate over the alleged morality of capital punishment is my expressing revulsion at capital punishment while you express approval, and any debate we can engender is over such factual questions as whether or not capital punishment serves as a deterrent against murder.

Perhaps the most extreme example of the denial of ethics occurred when the director of the NIH (the major US government funding agency for biomedical science) made a personal visit to Michigan State University in the early 1990s. Let us bear in mind that there are two presuppositional requirements for being director of the NIH. One is being an outstanding scientist, which this gentleman was. But equally important is a sense of diplomacy allowing one to negotiate very treacherous waters. The students became aware of his visit, and invited him to address the premedical club. A number of students asked him about the ethical issues occasioned by biotechnology. He was apparently unguarded in his remarks, not realizing that a student reporter for the school paper was present. He opined that "though scientific advances like genetic engineering are always controversial, science should never be hampered by ethical considerations" [23]. When I quote that remark to my students, even freshmen, and ask them who said that in the 20th century, they invariably say "Hitler", and are extremely shocked when I tell them that it was the director of the NIH!

The position of the scientific community on ethics had the predictable effect of encouraging emotionally based religious and non-rational positions articulated by society under the rubric of ethics, as we indicated earlier with regard to the cloning of Dolly the sheep. Thus, science's denial of ethics when combined with society's unclear understanding of ethics creates "a perfect storm" blocking a reasonable articulation of the ethical issues of xenotransplantation, and opening the doorway for bad ethics.

Scholars have been carefully tracking European opinions on numerous changes in scientific, social, and cultural life by way of a vehicle known as the Eurobarometer. These surveys have been conducted under the aegis of a sociologist named George Gaskell based at the London School of Economics [24]. Gaskell has a significant interest in biotechnology and in attitudes thereto. In their writings, Gaskell and his associates have checked on attitudes towards six modalities of biotechnology—genetic testing for genetic diseases, cloning human cells and tissues, genetically modified enzymes for producing environmentally friendly soaps, genetically modified crops, genetically modified foods, and xenotransplantation.

Surprisingly, of all these instances of biotechnology, the one receiving the least support was xenotransplantation, by quite a significant majority. This is true each time polls were conducted. For example, in a 1999 article in Science, George Gaskell and his associates showed that a higher percentage of the European public opposed xenotransplantation than any other biotechnological modality. Prima facie, this appears to be incomprehensible. After all, xenotransplantation is most directly relevant to saving lives. Gaskell has also repeatedly made the point that moral concerns are more important to the public in evaluating biotechnology than considerations of risk and safety [25]. Obviously, xenotransplantation is seen as presenting weighty moral concerns.

Many people articulated what we have called "spurious moral concerns" (bad ethics), i.e., reasons for rejecting transplanting organs from animals, including concerns about the "unnatural" dimension of such a procedure. Others saw xenotransplantation as "playing God" or "crossing species barriers". Such claims are easily defeated. If transplanting organs from animals to humans is unnatural, so too are building flying machines, damming the Nile, exterminating pathogenic insects, and the myriad activities in which we engage to make nature more congenial to human life and flourishing. In the same manner, we can put paid to claims that "crossing species barriers violates God's will". One can plausibly claim that if God did not wish humans to blur the lines between species, as we have in fact done repeatedly in plant agriculture, why would He have given us the ability to do so? (In fact, many theologians who support biotechnology make precisely this claim.) The key point is that theologically-based opinions do not in and of themselves constitute rational ethical concerns.

I personally believe that much of the opposition to xenotransplantation stems from the appalling degree of social illiteracy regarding transplant science. Many people in society, for example, still retain the ancient belief that the heart is where personality resides and is the locus and seat of emotions, thought, and will, a belief entertained by both the ancient Greeks and the ancient Egyptians. This idea of course persists in the belief that love is founded in the heart, and in such phrases as "he has a kind heart". (Hence the myth of Cupid firing arrows into the heart.) This naïve view in turn manifests itself in the belief that if one acquires a heart from a pig, one will have the emotions and feelings of a pig, and thus be rendered considerably less than human. Such a naïve view can and does actually impede the development of therapies that fill the perennial need for heart transplants. Essentially, this belief suggests that if one acquires a heart from a pig, one will be transformed into a being possessing the mind and feelings of a pig. Once again, we see the extent to which ignorance of science creates bad ethics, even when what is at stake is preserving life!

To recapitulate, transplanting organs from animals to people is largely condemned because of people's extremely primitive views of biology and the mind. Once again, we find that bad biology ramifies in bad ethics, which in turn provides a basis for rejecting life-saving biotechnology. Thus, we see that creating a scientifically literate public is essential for societal acceptance of biotechnology. Let us recall that no less an ethical luminary than the Pope condemned the cloning of Dolly the sheep, based upon the belief that "all animals are entitled to a normal birth". It seems that the Pontiff was unaware of the fact that Dolly had in fact experienced a normal birth! This kind of thinking, or rather lack thereof, serves no one well.

8.5 Conclusions

As we indicated at the beginning of our discussion, and as the rise of xenotransplantation makes plain, biotechnology presents the potential for changing human life in ways that are unprecedented. And, as Gaskell's research indicates, moral considerations regarding biotechnology are putatively of greater concern to society than even issues of safety. At the same time, societal ignorance of both science and ethics militate strongly against rational solutions to ethical issues, and even elevate non-issues to the highest rank of ethical concerns, for example, "violating God's will" or operating "against nature". Such a mistake, as our discussion indicates, can impede using biotechnology to save human life and alleviate suffering.

This in turn indicates the desperate need for better education of both the public and the scientific community in ethics, and of society in general in science, given the abysmal social ignorance of science we illustrated at the beginning of our discussion.

While recognition of this issue is comparatively easy, its resolution is far more difficult. I have both taught and advocated for the integration of ethical issues into science education as presuppositional to rational solutions of ethical issues arising out of scientific developments. I have also seen that such education makes for better scientists who have a sense of social responsibility. I have also argued for the teaching of science, and in particular biological science, to include discussion and elucidation of ethical issues. This has proven to be far more difficult. Impediments to such an approach include the fact that it is historically unprecedented and that most scientists have been brainwashed to believe that science is "value-free in general, and ethics-free in particular", as demonstrated in our discussion. Further impediments arise out of the fact that most people teaching science have themselves not been educated in ethical issues arising out of science, or even how to begin to approach such issues. For these reasons, it is difficult to know where and how to begin with inculcation of an ethics component into science education. The same problem is evident in the research community's cavalier attitude towards the ethical issues attendant upon animal use in research, an attitude that distances the scientific community from the general public, as societal concern with the treatment of animals increases exponentially. If, as leaders of the scientific community have repeatedly affirmed, scientific progress is absolutely dependent upon animal use, it surely behooves that community to address societal ethical concerns attendant upon animal use.

I have also indicated in our discussion that the concept of animal telos is presuppositional to the development of animal ethics as I have formulated it, and society seems to agree. The question then obviously arises as to whether genetic modification, which can readily be seen as effecting both major and minor changes in telos, is morally acceptable. This is certainly not the place for a full discussion of this vexatious issue. But I have argued that as long as telos changes do not negatively affect the animals' lives, i.e., that genetically modified animals are no worse off than their unmodified predecessors, and ideally are better off, there is nothing morally wrong with creating such genetic modifications. In other words, one foundational issue in genetically engineering animals is providing assurance that genetic modification occasions no harm, and ideally, as is the case in genetic modification to prevent disease or ameliorate genetic defects, the animals will be better off in virtue of such modification.

Such considerations will not be addressed until such time as the thinking of scientists is radically changed in order to assure that ethics enters into standard scientific theory and practice, and this can only be accomplished by revolutionary modification of science education.

References

1. Rollin, B. E. 1995. *The Frankenstein Syndrome: Ethical and Social Issues in the Genetic Engineering of Animals.* Cambridge Studies in Philosophy and Public Policy. Cambridge: Cambridge University Press. https://doi.org/10.1017/CBO9781139172806.
2. Animal-to-human Transplants: The Ethics of Xenoplantation. 1996. *The Nuffield Council on Bioethics.*
3. Israeli scientists 'print 3D heart using human tissue.' 2019.

4. Behnam Manesh, S., R. O. Samani, and S. B. Manesh. 2014. Ethical issues of transplanting organs from transgenic animals into human beings. *Cell Journal* 16: 353–360.
5. Xenotransplantation: The Benefits and Risks of Special Organ Transplantation. 2020. *BIO.* https://www.bio.org/articles/xenotransplantation-benefits-and-risks-special- organ-transplantation. Accessed September 2.
6. Sanders, B. 2018. Global Pig Slaughter Statistics and Charts. *Faunalytics.* October 10.
7. The Internet Classics Archive |Physics II.8,199b by Aristotle. 2020. http://classics.mit.edu/Aristotle/physics.2.ii.html. Accessed September 18.
8. Psalm 23, Old Testament. 2020. September 18.
9. Rollin, B. E. 2018. Antibiotic Use and the Demise of Husbandry. *The Journal of Ethics* 22: 45–57. https://doi.org/10.1007/s10892-017-9260-2.
10. Rollin, B. E. 2016. *A New Basis for Animal Ethics: Telos and Common Sense.* 1ed. Columbia, Missouri: University of Missouri.
11. CAST. 1981. *Scientific Aspects of the Welfare of Food Animals.* 91. Ames, IA, USA: Council for Agricultural Science and Technology (CAST).
12. Farm Animal Welfare Council. 1979. *Five Freedoms.*
13. Rollin, B.E. 1981. *Animal Rights and Human Morality.* 1st ed. Buffalo, NY, USA: Prometheus Books.
14. Bach, F. H., J. A. Fishman, N. Daniels, J. Proimos, B. Anderson, C. B. Carpenter, L. Forrow, S. C. Robson, and H. V. Fineberg. 1998. Uncertainty in Xenotransplantation: Individual Benefit versus Collective Risk. *Nature Medicine* 4: 141–144. https://doi.org/10.1038/nm0298-141.
15. Rollin, B. E. 2014. The Perfect Storm—Genetic Engineering, Science, and Ethics. *Science & Education* 23: 509–517. https://doi.org/10.1007/s11191-012-9511-3.
16. Black, K. 2004. *Scientific Illiteracy in the U.S.* Los Angeles, CA, USA: U.S. Cedars-Sinai Neurosciences Report.
17. A Look at What the Public Knows and Does Not Know about Science. 2015.
18. Somin, I. 2021. Over 80 Percent of Americans Support "Mandatory Labels on Foods Containing DNA." *Washington Post*, October 24.

19. National Center for Science Education. 2024. The latest Gallup poll on evolution | National Center for Science Education. https://ncse.ngo/latest-gallup-poll-evolution-0. Accessed May 4.
20. Gallup Inc,. 2005. Americans Weigh In on Evolution vs. Creationism in Schools. *Gallup.com*. May 24.
21. Newton, I. 1934. *Scholium to the Definitions in Philosophiae Naturalis Principia Mathematica*, ed. F. Cajori. Translated by Andrew Motte. Vol. 1. Berkeley, CA, USA: University of California Press.
22. Wittgenstein, L. 1965. *Lecture on Ethics*. Vol. 74. Philosophical Review.
23. 1989. *Michigan State News*, February 27.
24. Gaskell, G., M. W. Bauer, J. Durant, and N. C. Allum. 1999. Worlds Apart? The Reception of Genetically Modified Foods in Europe and the U.S. *Science (New York, N.Y.)* 285: 384–387. https://doi.org/10.1126/science.285.5426.384.
25. Europe ambivalent on biotechnology. 1997. *Nature* 387. Nature Publishing Group: 845–847. https://doi.org/10.1038/43051.

Chapter 9

Understanding Ethical Issues Raised by Veterinary Medical AI*

Simon Coghlan[a] **and Thomas Quinn**[b]
[a]*School of Computing and Information Systems, Centre for AI and Digital Ethics, The University of Melbourne, Melbourne, VIC, Australia*
[b]*Independent Researcher*
simon.coghlan@unimelb.edu.au, contacttomquinn@gmail.com

9.1 Introduction

Artificial intelligence (AI)—meaning digital systems that perform tasks normally requiring human intelligence [1]—may soon transform veterinary medicine [2, 3], just as it is likely to do in human medicine [4, 5]. Like human medical AI [6–8], veterinary AI raises important ethical issues. So far, these issues have not been much examined (for some work that touches on ethics and veterinary AI, see [2, 9–13]). Yet engineers who develop AI for veterinary applications will need to be aware of and even grapple with

Animals, Ethics, and Engineering: Intersections and Implications
Edited by Rosalyn W. Berne

ISBN 978-981-5129-56-4 (Hardcover), 978-1-003-65231-1 (eBook)
www.jennystanford.com

these ethical issues. In particular, these engineers must understand something about the ethical nature of professional veterinary activity, and how AI might impact it. Our aim in this chapter is to assist in furnishing such an understanding in the context of companion animal medicine.

Companion animal veterinary practice raises unique ethical issues that stem from the client–patient–practitioner relationship. (Most veterinarians work in companion animal medicine [14] and total hundreds of thousands worldwide [15]. There are tens of millions of 'pets'—including dogs, cats, and birds— in the US alone and 'pet ownership' is growing fast.) Companion animals, such as cats, rabbits, and dogs, are potentially more exposed to harm from AI than are humans because they lack the same strong social, moral and legal status. For example, the law does not effectively protect animals from wrongful injury or from clients who seek unwarranted or unjustified 'euthanasia' [16]. These conditions are relevant to the ethics of veterinary AI. At the same time, medical AI raises its own distinctive ethical issues—issues like trust, data security and algorithmic transparency—which we also discuss in the veterinary context.

AI in veterinary medicine might be used for business purposes and hospital logistics like booking appointments. Technology that affects practitioner workflow could have ethical implications, as could other AI, such as language translation apps that enable communication with linguistically diverse clients. However, AI for triage, diagnosis, prognosis, and treatment raises the most distinctive, complex, and consequential ethical questions. We concentrate on AI for such medical decision-making.

Currently, AI enjoys massive public and private investment, propelled by stories like algorithms defeating Jeopardy and Go Masters [17]. Another indication of AI's rapid ascent are recent large language models like ChatGPT and text-to-image generators that demonstrate remarkable, though sometimes strange, and biased, outputs (see Fig. 9.1). Yet most people are bewildered by the technical jargon of artificial neural networks, deep learning, computer vision, random forests and natural language processing [18]. (For explanations of AI concepts in veterinary contexts see [2, 19].) Veterinary practitioners too may not always understand, for instance, the ways in which AI learns from data and autonomously updates its algorithms to draw inferences about previously unencountered data (e.g., from patient radiographs or medical records)— and this may create uncertainty about its use in healthcare.

This issue of trust in technology is important. To some degree, medical AI remains just as much an art as a science [20], and AI developers are only now exploring how to apply modern *machine learning* (ML) methods successfully in medicine. This involves experimenting with how data are collected and pre-processed, how AI models are applied and optimized and how model performance is evaluated. Each step contains many nuances that could affect model operation in clinic settings and unintentionally harm patients and clients. While busy practitioners cannot be expected to understand all these nuances, they will increasingly need at least a basic understanding of the ethical risks and benefits of AI. This paper identifies and examines these ethical issues.

The chapter runs as follows. Section 9.2 outlines medical AI in veterinary practice. Section 9.3 introduces ethical principles of AI, human medicine and veterinary medicine. Section 9.4 identifies and examines nine ethical issues raised by veterinary AI. Section 9.5 discusses important ethical norms in veterinary medicine and AI's distinctive implications in that realm, as well as providing some practical guidance for AI's use.

9.2 AI in Veterinary Medicine

Earlier medical AI involved knowledge-based systems, such as the 1970s program MYCIN [21, 22]. These 'expert' systems involved hard-coding medical expertise from experts to generate rules and infer clinical diagnoses. However, they struggled with the inherent complexity of medical decision-making [23]. Modern ML has proved more adept. These models absorb vast data to 'learn' rules automatically in the form of mathematical functions that relate predictor variables to target variables. One very successful type of ML, *deep learning*, employs so-called 'deep neural networks' (DNNs) [24]. DNNs have layers of processing units linked together in patterns somewhat like brain neurons [1]. There can be hundreds to billions of artificial neurons in DNNs and numerous layers.

AI today often involves 'supervised' machine learning in which samples that are used to *train* models are labelled. For example, an

ML system may be trained on thousands or millions of biopsy images labelled as either cancerous or healthy tissue. Once trained, the model can be tested on new images to make predictions (e.g., about cancer) and can then undergo evaluation for diagnostic accuracy and compared with clinician performance. Ideally, the model is subjected to a clinical trial to establish efficacy and cost-effectiveness before being implemented in practice, where its effectiveness should continue to be assessed.

AI shows promise in veterinary medicine. For example, one ML algorithm for detecting canine hyperadrenocorticism had a sensitivity of 96.3% and a specificity of 97.2%, reportedly outperforming other screening methods [25]. Some models classify animal cancer, retinal atrophy, or colitis based on images [26]. Deep learning can be applied to detect fecal parasites [27] or identify canine cardiac enlargement [28]. Some models can outperform veterinary radiologists at certain tasks [29], and others predict seizures in epileptic dogs from ambulatory intracranial sensors [30]. AI might also improve veterinary surgery [31] and one day guide robotic veterinary surgeons [32, 33]. Natural language processing might usefully extract clinical information from patient records for analysis. Finally, there are direct-to-consumer AI products, such as one that predicts differential diagnoses for canine alopecia [34].

Potentially, some AI tools will be more accurate and faster than practitioners and cost-effective for clients. Perhaps, as some suggest, AI will bring "tremendous potential efficiencies and quality improvements in veterinary medicine" [2]. But it also comes with risks and ethical concerns.

9.3 Principles in AI, Medical and Veterinary Ethics

General AI ethics guidelines speak of ethical principles like transparency, accountability, data security, privacy, safety, fairness and environmental sustainability [35]. Many of these principles arise from the distinctive nature of AI and the special risks it creates. As we shall see, such AI ethics principles play a role in the ethics of

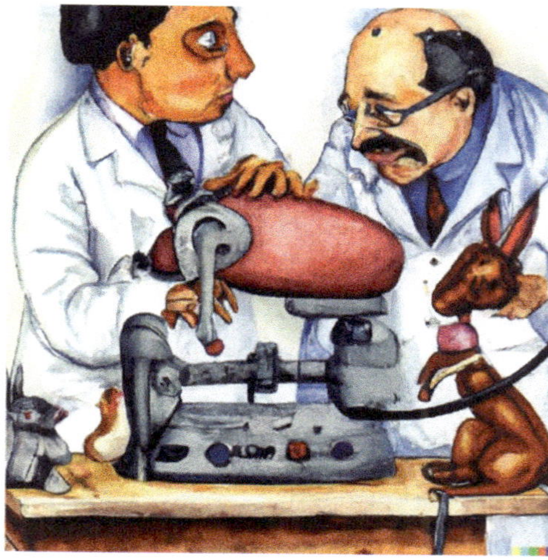

Figure 9.1 An image produced by OpenAI's DALL-E program in response to the text prompt *'a watercolor painting of veterinarians pondering the relationship between medicine, machine and animal'*. Note how the AI presents veterinarians as middle-aged white men— an example of AI bias (Image obtained via open access website https://openai.com/dall-e-2/).

veterinary AI. AI ethics also borrows from medical ethics [36] and its four widely accepted bioethical principles: nonmaleficence (do no harm), beneficence (do good), respect for autonomy (respect a person's ability to act on their own values and preferences), and justice (e.g., ensure fair distribution of medical resources) [37].

These medical ethics principles arguably apply in veterinary practice. For example, many would accept that veterinarians have responsibilities to promote patient well-being and avoid harming them and to respect the autonomy of clients. However, there are ethically relevant differences with human medicine that can affect those principles' application [38]. For example, human medical practice is mostly funded by large public or private insurance schemes, whereas veterinary medicine is mainly paid for 'out of pocket' by private individuals, who sometimes struggle to afford medical attention for their unwell animals [39] (Compared to medical practitioners, veterinary practitioners are typically less directly confronted by the principle of justice. Nonetheless, there are ethical questions about the need for public funding of veterinarian medicine and fairness for poorer clients and their animals [40]. Expensive but beneficial AI for animals could conceivably raise justice issues.) Consequently, some clients (and veterinarians) opt for cheaper and inferior diagnostics and treatment and even sometimes for 'economic euthanasia' [41].

Obviously, animal patients cannot provide autonomous consent for medical interventions (although animals might provide or withhold *assent* [42]). Hence, companion animal medicine somewhat resembles pediatric medicine (and to some degree gerontology). Medical practitioners and Boards typically endorse an *ethically patient-centered approach* [43] that prioritizes significant patient interests over the interests of other parties like parents [44]. While most parents pursue their children's best interests, pediatricians may override parental autonomy when parents refuse necessary interventions or urge harmful treatment for their children [45]. While they respect parents' interests, pediatricians see their primary duty as being to the patient.

Veterinary medicine has enjoyed comparatively less discussion—and agreement—about the right ethical principles to follow and how they should be interpreted [37, 38]. (There is also disagreement about what constitutes well-being for animals [46].) This has important implications for veterinary AI. Nonetheless, in what immediately follows, we can *generally* assume that clients and practitioners seek the best for the animals and broadly agree on what that involves. Accordingly, veterinary practitioners will broadly follow principles of *nonmaleficence* (avoid and minimize harm) and *beneficence* (do good and provide benefit) regarding patients. Furthermore, veterinarians generally *respect the autonomy* of their clients. These principles inform our identification of nine ethical issues in veterinary AI.

9.4 Ethical Issues Raised by Veterinary AI

The nine ethical issues we identify and explain below (Table 9.1) refer to situations that demand ethical judgement about AI. Such deliberation may involve moral values, principles and theories. Later, we will see how these ethical issues variously affect the three

Table 9.1 Nine ethical issues for veterinary AI with ethical issues examples

Ethical issues for veterinary AI	Examples
1. *Accuracy and reliability*	Models developed using training data from Labradors or beagles might not generalize to other dog breeds. Automated pain recognition using computer vision of animal's faces may be unreliable [47].
2. *Overdiagnosis*	Indiscriminate application of AI to blood samples for surgical candidates or fecal samples from healthy patients leading to overdiagnosis AI applied to medical developments like proteomics, genomics and metabolomics [2] could exacerbate overdiagnosis and associated unnecessary or harmful treatment, e.g., antibiotics, diet supplements
3. *Transparency*	In 'Blackbox' models, the basis of a prediction—e.g., 'patient Y has canine lymphoma with 90% probability'—is unknown or unknowable, reducing confidence in those models. Users may not know that an opaque system is partly basing its predictions on extraneous features like labels placed on X-rays rather than on relevant features like the shape, location, and density of a radiographic mass. A for-profit company might keep the workings of their medical AI secret
4. *Data security*	Hackers of an AI database could expose or sell client and patient details. Wearables attached to the outside of the animal's body could capture personal client data [48] like location and activities that may be inadvertently leaked. A domestic abuser could access information about a partner and their animal, resulting in violence against both [49].
5. *Trust and distrust*	Blackbox systems might reduce trust, although humans are also prone to over-trusting machines, which can cause harm [50]. An AI interface might 'nudge' veterinarians to overprescribe tests or drugs sold by the AI company [51].

(*Continued*)

Table 9.1 (*Continued*)

Ethical issues for veterinary AI	Examples
6. *Autonomy of clients*	Client trepidation and misunderstanding of AI may be abetted by science fiction tales, media hype and stories about AI failures and dangers, necessitating clear informed consent processes. Some clients may prefer transparent AI.
7. *Information overload*	Rapid AI in-clinic testing could reduce labor in preparing lab specimens or repeating external tests due to poor sampling. Additional job pressures may result from needing to understand, set up, maintain, and interpret unfamiliar AI systems. A temptation could exist to 'refer' cases to AI assistants, diminishing decision-making skills or encouraging less thorough histories and examinations [52].
8. *Responsibility for AI-influenced outcomes*	A practitioner may accept AI diagnoses at face value (without considering other evidence) thereby failing to take proper responsibility for decisions
9. *Environmental effects*	Training large neural networks produces significant carbon emissions. Plastics and minerals in AI devices can cause pollution [53].

parties in the central patient–client–practitioner relationship (and occasionally parties beyond it).

9.4.1 Accuracy and Reliability

Accurate and reliable AI in pathology, radiography, medicine and surgery could significantly benefit patients, including by eliminating certain human biases and misjudgments. Equally, inaccurate AI could harm patients (and clients) through misdiagnoses and poor treatment recommendations. Importantly, some AI tools may be accurate in terms of test set evaluation but unreliable in clinical practice. This may occur when the training and test sets are not representative of the intended real-world use case or contain biases. For example, an AI screening tool trained to recognize pneumonia from the audio of coughs obtained from hospitalized patients may be accurate for patients in hospital but inaccurate for outpatients [20]. And veterinary AI developed in Northern Hemisphere contexts may be less reliable in Southern Hemisphere contexts.

Even when AI is trained and evaluated on representative data and found to be accurate, this may not translate to improved clinical outcomes. Medical AI is frequently not well studied in this respect, despite the surrounding hype [54]. Although randomized clinical trials are gold-standard in evidence-based medicine, a recent systematic review found that few medical AI studies use randomization and only 9/81 nonrandomized studies were prospective [55]. Some AI is flawed by design. For example, AI purporting to diagnose emotions from photos of human faces has been criticized because expressions do not always correlate with emotional states [53]. This problem may also afflict AI for diagnosing animals' affective, pain, or welfare states [56].

9.4.2 Overdiagnosis

Overdiagnosis involves diagnosis of conditions that are harmless to the patient [57]. For example, AI might identify harmless bone defects or 'incidentalomas' [58]. Overdiagnosis is a growing but frequently overlooked concern [59] which can generate unnecessary additional testing and treatment [60]. A significant cause of overdiagnosis is large screening programs of apparently healthy individuals [61]. Veterinary AI might significantly promote overdiagnosis and on a larger scale than before, including by promoting more defensive medicine [62]. Therefore, AI-based overdiagnosis should be recognized and minimized where possible.

9.4.2.1 Transparency

Transparency broadly refers to users' knowledge of how an AI system arrived at its prediction [63]. In deep neural networks, the reasons that underlie the model's prediction can be intrinsically unknowable, due to the model's enormous complexity. Such algorithmically opaque models are dubbed 'Blackboxes'. For some AI, a trade-off may arise between model performance and intelligibility. Transparency can also be reduced by for-profit companies that conceal their AI's workings from users and competitors. Even when AI models are open source and available, busy practitioners may find it too onerous to seek and digest such information.

Some believe that Blackbox AI is not problematic if it is accurate. After all, practitioners justifiably prescribe drugs with largely unknown mechanisms. It is true that use of opaque systems can also sometimes be justified. However, algorithmic opacity can hamper the detection of inaccuracies and biases in predictions. In contrast, interpretable AI can more readily be 'caught out' making mistakes, thereby aiding quality assurance and safety. Some therefore argue that medical Blackboxes should be altogether avoided [64], or else used only when equally accurate interpretable systems are unavailable [65] or when non-transparent systems are demonstrably and significantly superior.

9.4.3 Data Security

Data used to train AI can be private, sensitive, and extensive. Data stored locally or on company servers might be leaked, sold on, or hacked. Malicious agents mounting adversarial attacks can even render AI systems unreliable [66], while anonymized health data can sometimes be matched with other data to reidentify individuals [67]. Models may suffer 'attack' and yield personally identifiable data previously used during training even after data deletion [68]. Veterinary-related data are not immune from these risks. Clients may thus have an interest in data security and in providing consent for reuse of their data, e.g., to further train AI tools.

9.4.4 Trust and Distrust

Having trustworthy technology will be important if AI is to be beneficial [69]. Unwarranted trust in AI can cause its misuse, while unwarranted distrust can cause disuse that deprives patients of benefits [70]. For example, failure to employ in-house AI that saves time on external pathology processing could cause critical time delays for sick animals. Distrust in AI by clients, perhaps exacerbated by troubling news stories or personal experiences, may even precipitate more general distrust in the veterinary profession. Distrust can rise for opaque systems [71], while excessive trust may result from 'automation bias' [72]. Conversely, humans sometimes wrongly ignore computer-based outputs, especially when outputs are obscure or prone to false alarms. AI companies may heavily promote their wares or even use medical AI to recommend their other products or tests and veterinarians invested in AI could face conflicts of interest. The veterinary profession should be aware of such commercial pressures and tactics that could influence clinical decision-making.

9.4.5 Autonomy of Clients

Respect for the autonomy of human patients standardly requires obtaining their (or their guardians') informed consent for interventions. This requires giving patients' relevant information about the nature, risks and benefits of interventions [73]. Plausibly, veterinary practitioners should similarly inform their clients of "the advantages, disadvantages and most likely outcomes for each [care] option; the possibilities of favorable and unfavorable outcomes; the likelihood that additional testing or treatment might be needed; the associated costs; and the strength of the supporting evidence" [74]. It has been argued that medical practitioners using medical AI should understand and convey its pitfalls to human patients

[75]. Respect for client autonomy at least prima facie requires that veterinarians explain to clients the broad nature, risks and benefits of chosen AI-based interventions, just as they do with other interventions. Furthermore, many clients may be ignorant, misinformed, or uncertain about AI, heightening the need for providing clear information about its pros and cons. For example, practitioners might need to explain how an AI tool can sometimes make misdiagnoses due its training data, or that it has not yet been subjected to rigorous clinical testing.

Veterinarians must normally explain to clients the general basis of their diagnoses and prognoses in ways non-medical people can understand. In Blackbox AI, however, algorithmic opacity precludes client (and practitioner) understanding of the reasons behind the machine's predictions or recommendations. That may not trouble some clients, but others may prefer transparent AI that provides such explanations [20].

9.4.6 Information Overload and Skill Erosion

Some AI might also improve life for veterinarians. Partial outsourcing of cognition to trustworthy AI 'assistants' may ease workloads [2]. Yet AI, which is a complex and ever-evolving technology, might also increase information overload for veterinarians who already endure high workplace stresses [76]. Not all technologies make our lives easier—consider the way that household appliances have not always reduced domestic labor mostly undertaken by women [77]. A recent survey found that 70% of medical practitioners believed "digital health technologies will be a challenging burden" and that they lacked "time to learn the value of the technology or foster the belief in their ability to use it... ultimately taking time away from patient care rather than improving it" [78].

Gradual erosion of medical skills through machine reliance is another theoretical possibility [79]. Some skill erosion may be overall beneficial, as when generalists refer complex patients to specialists for improved health outcomes [74], although that change has sometimes reduced accessibility to healthcare. However, over-reliance on fast and convenient intelligent decision support tools

[80] might in time weaken medical skills that veterinarians should retain.

9.4.7 Responsibility for AI-influenced outcomes

Accountability is an important idea in AI ethics because it can be unclear who is legally and ethically responsible for AI-generated harms. The difficulty of assigning or determining liability is called the 'responsibility gap' [81]. Responsible parties could include engineers, companies, practitioners, professional organizations, regulatory bodies and clinic managers and owners. Until medical AI reaches a very high degree of reliability, there is reason to say that individual practitioners must remain ethically and professionally responsible for using it. This is especially important for non-transparent AI where detection of harmful outputs can be more difficult.

9.4.8 Environmental Effects

Although the environmental effects of healthcare generally [82] and of AI specifically, are often neglected, these harms can be considerable [83]. Veterinary AI could contribute to AI's overall environmental impact [84]. While veterinarians are rightly focused on their immediate patients' well-being, there is a case for becoming more aware of veterinary medicine's increasing environmental footprint [85] and for seeking more sustainable AI tools where possible.

9.5 Veterinary AI and Ethical Responsibilities, Risks, and Guidance

9.5.1 Role and Responsibilities of Practitioners

As we have shown, AI could have both positive and negative implications for patients, clients, and practitioners. In companion animal medicine, the interests of these parties are often aligned: what benefits or harms patients often benefits or harms clients (and sometimes practitioners). Nonetheless, the interests and wishes of clients (and practitioners) and companion animal patients can sometimes conflict [86, 87]. This raises important ethical questions about veterinarians' role and responsibilities [40, 88–96] and how they relate to AI.

While many veterinarians traditionally saw their primary obligations as being to the 'owner' of the animal rather than to the patient themselves [92], this profoundly human-centered view began to shift as societal attitudes to animals evolved and the profession began to appreciate the strength of human-animal relationships [97, 98]. Nonetheless, veterinarians can still have different understandings of the strengths of their duties—differences which move to the forefront when the interests of patients and the wishes of clients or clinic managers conflict.

Most contemporary veterinarians would broadly claim to be advocates for their patients, yet 'advocate' admits of degrees. A *strong patient advocate* [99] or *ethically patient-centered* practitioner is more determined to safeguard the patient's interests and speak up on their behalf [100]. While the patient-centered practitioner will not ignore clients' perspectives and situations, such as economic insecurity [74], they will search hard for solutions that promote the patient's important interests and they may sometimes refuse to go along with harmful requests from clients. Like pediatricians [92], patient-centered veterinarians prioritize beneficence and nonmaleficence towards the patient over, say, respect for client autonomy on those key occasions of conflict. They will also seek to safeguard patient interests when they receive pressure from other parties, such as peers or clinic managers, to act counter to their patients' interests.

A veterinarian's conception of their role and responsibilities could affect their behavior toward AI. For example, some practitioners may more readily acquiesce to pressure from clients or clinic managers who are enthusiastic about AI and who urge the adoption of these tools despite the fact that those tools may lack rigorous scientific validation and/ or feature an uninterpretable and relatively risky ML model. A patient-centered practitioner would only use veterinary AI in higher stakes situations when they had grounds to believe it would be of overall benefit to the patient. Another example of how a practitioner's ethical stance could influence their use of medical AI concerns the important ethical issue of euthanasia [92]. Imagine that an AI system designed to make treatment recommendations for animals presents 'euthanasia' as an option for a patient who, despite their condition, could probably have a decent life with appropriate treatment. Although such treatment recommendations do not yet feature in AI, it is entirely conceivable that they will appear in some future veterinary AI. If that happens, it is possible that the client (and veterinarian) could be influenced by an AI recommendation for euthanasia that is not ethically justified. While *client*-centered practitioners may agree to a client's request for euthanasia based on an AI recommendation or option, an ethically *patient*-centered practitioner would strongly counsel the client to reject that aspect of the AI's recommendation. The converse situation may occur when an AI recommends onerous and futile treatment for a dying patient who would thereby be made much worse-off and so suffer what has been termed 'dysthanasia' [52, 101]. If future AI makes treatment recommendations as well as diagnoses, veterinarians will need to be aware of the potential for uncritical acceptance of such advice from machines.

9.5.2 Distinctive Risks Associated with Veterinary AI

Some risk factors are distinctive or especially salient for veterinary AI and are worth highlighting. First and perhaps most importantly, companion animals, unlike humans, are classed as legal property and enjoy relatively few social and regulatory protections [102]. Moreover, our societies remain profoundly human-centered overall, typically affording little moral consideration to animals compared to humans [103]. This pronounced ethical anthropocentrism shows itself in the fact that AI ethics has largely neglected nonhuman animals (exceptions include [104, 105])—both directly as subjects of AI itself and indirectly as subjects of the environmental impacts of AI [46]. Consequently, some AI developers and some veterinarians may devote less energy and care than they might to ensuring that AI promotes patients' interests (and may have less legal impetus to do so). Furthermore, most veterinarians work in small businesses or corporate-run hospitals; this could potentially result in pressure to increase profit and client turnover, which may overtly or subtly affect patient care [86], such as by promoting unnecessary testing and treatment.

Second, being less regulated than human medicine, veterinary medicine potentially affords more opportunities for experimenting with cutting-edge yet relatively untested treatments. Indeed, one sometimes hears the view that AI might be 'tested' on animal patients before being used on human patients. Quain et al. [52] argue that the freedom to pursue various kinds of advanced but experimental veterinary care, such as stem-cell treatment, can sometimes (though not always) pose extra risks to patients. Misguided, faulty, or insufficiently tested AI also carries risks despite being a promising cutting-edge technology. AI can be used on animal patients without the same testing and regulatory approval (e.g., by the Food and Drug Administration) that human AI requires. Additionally, veterinary medicine has fewer resources for research into medical interventions and devices [2].

Third, there is currently qualitative and quantitative data scarcity for animals compared to humans for training ML models [19]. Veterinary data records lack the requirements for consistency and standardization sometimes imposed on human medical data records [12]. These factors might make it more difficult to develop and deploy effective and reliable ML models. (Note, however, that the relatively minimal legal regulation of animal health records could sometimes *improve* data access.) Although data scarcity can be overcome through data sharing agreements, such sharing also raises risks for the privacy of medical records.

9.5.3 Ethical Guidance Concerning Veterinary AI

As we noted, the ways in which practitioners approach AI depend partly on their ethical understanding of their role and responsibilities as veterinarians (as well as on their understanding of AI and level of enthusiasm for it). Let us assume that practitioners and clinic owners and hospital managers, generally prioritize the interests of patients or act in ethically patient-centered ways. Drawing on the above analysis, we suggest the ethical principles and goals listed in Table 9.2 for governing AI use in veterinary medicine.

Table 9.2 Ethical principles/goals, examples of not meeting principles/goals, and corresponding recommendations regarding veterinary AI

Ethical principle/goal	Example of not meeting principles/goals	Recommendations
Nonmaleficence	AI produces significant false positives, causing harmful, unnecessary treatment; AI unjustifiably recommends 'euthanasia' or results in 'dysthanasia'.	Ensure AI does no harm or least harm possible and be aware of the possibility of commercial exploitation of animals and clients. Assess susceptibility to bias, data availability and quality, overfitting, confounders.
Beneficence	AI produces significant false negatives, causing missed diagnoses and untreated animals.	Be aware of dark patterns and potential for overdiagnosis. Consider harm done to animals in the development of AI tools. Establish as far as possible whether the AI tool is reliable and accurate. Determine if the AI has evidential support of clinical benefit outside labs and in relevantly similar real-life settings.
Transparency	Retrospective studies show a Blackbox DNN has made misdiagnoses across numerous hospitals.	Carefully justify use of uninterpretable or extrinsically non-transparent AI, which can hamper identification of harms to patients. Use and encourage development of transparent AI where possible.
Respect for client autonomy	Failure to explain AI results in client rejection of AI decision support tools that benefit patients.	Take client preferences seriously to avoid AI or use transparent AI, but balance this against patient well-being. Educate clients about risks and benefits of AI applications. Inform clients of financial costs of AI tools.

(Continued)

Table 9.2 (*Continued*)

Ethical principle/goal	**Example of not meeting principles/goals**	**Recommendations**
Data security and privacy	A hacker accesses a clinic's database and harasses a client or steals their ID.	Ensure client and patient data is as secure as possible from data leakage, theft.
Feasibility	Highly expensive AI substantially raises medical/surgical costs for all clients and further overburdens stressed practitioners.	Examine whether the AI is cost-effective for clients and practices. Ask if it places excessive cognitive demands on practitioners or threatens skill erosion
Accountability	A media exposé finds that veterinary bodies have not warned practitioners about poorly validated and opaque AI, causing reduced trust in the profession.	Determine who is responsible for harms and lack of benefits caused by AI tools. Ask whether there is a responsible practitioner 'in the loop' Pay heed to relevant laws and regulations about animal cruelty, professional duties, codes of conduct, consumer protection, etc.
Environmental Sustainability	Veterinarians accused of irresponsibly using emission-intensive deep learning.	Investigate magnitude of environmental harm caused by the development of AI tools prior to promoting or purchasing

Engineers working on veterinary AI will benefit from understanding these principles and goals, and the associated recommendations and examples, presented in the table.

9.6 Conclusion

Veterinary medicine is a socially valued profession that, like human medicine, is likely to be significantly affected by AI. In this chapter, we showed that veterinary AI creates risks, benefits and ethical issues that are both familiar from human medicine and unique or distinctive. Ethical responses to veterinary AI can be influenced by views about practitioner roles and responsibilities. In general, contemporary veterinarians aim to practice nonmaleficence and beneficence towards patients and to respect client autonomy. However, these principles may be interpreted differently. For example, a strongly patient-centered practitioner who prioritizes patients' vital interests may refuse to use insufficiently tested or excessively risky medical AI even when clients or clinic owners or managers improperly demand it. Equally, the patient-centered practitioner might persuade uncertain or skeptical clients that sufficiently validated and trialed AI tools can significantly benefit patients. Our discussion of the ethical nature of professional companion animal veterinary practice can inform how engineers approach AI development and its promotion. Engineers working in this area should be aware of the risks and benefits of AI for companion animals, guardians, and practitioners. In addition, they should understand the various ethical stances and professional obligations that veterinarians who use AI can have towards their patients and clients.

References

1. Norvig, P., and S. Russell. 2021. *Artificial Intelligence: A Modern Approach, Global Edition*. 4th ed. Harlow: Pearson.
2. Basran, P. S., and R. B. Appleby. 2022. The unmet potential of artificial intelligence in veterinary medicine. *American Journal of Veterinary*

Research 83. American Veterinary Medical Association: 385–392. https://doi.org/10.2460/ajvr.22.03.0038.

3. Lab, WIRED Brand. 2024. WIRED Brand Lab |Cloud to Clinic: Zoetis' vision for veterinary practices. *Wired.* https://www.wired.com/sponsored/story/cloud-to-clinic-zoetis-vision-for-veterinary-practices/. Accessed January 27.
4. Topol, E. J. 2019. High-performance medicine: the convergence of human and artificial intelligence. *Nature Medicine* 25. Nature Publishing Group: 44–56. https://doi.org/10.1038/s41591-018-0300-7.
5. Wilson, A., H, Saeed, C. Pringle, I. Eleftheriou, P. A. Bromiley, and A. Brass. 2021. Artificial intelligence projects in healthcare: 10 practical tips for success in a clinical environment. *BMJ Health & Care Informatics* 28: e100323. https://doi.org/10.1136/bmjhci-2021-100323.
6. A., Kristina, E. Peičius, and P. Astromskis. 2021. Ethical and legal challenges of informed consent applying artificial intelligence in medical diagnostic consultations. *AI & SOCIETY* 36: 509–520. https://doi.org/10.1007/s00146-020-01008-9.
7. Dalton-Brown, S. 2020. The ethics of medical AI and the physician-patient relationship. *Cambridge Quarterly of Healthcare Ethics* 29: 115–121. https://doi.org/10.1017/S0963180119000847.
8. Keskinbora, K. H. 2019. Medical ethics considerations on artificial intelligence. *Journal of Clinical Neuroscience* 64: 277–282. https://doi.org/10.1016/j.jocn.2019.03.001.
9. Ezanno, P., S. Picault, G. Beaunée, X. Bailly, F. Muñoz, R. Duboz, H. Monod, and J.-F. Guégan. 2021. Research perspectives on animal health in the era of artificial intelligence. *Veterinary Research* 52: 40. https://doi.org/10.1186/s13567-021-00902-4.
10. Steagall, P. V., H. Bustamante, C. B. Johnson, and P. V. Turner. 2021. Pain management in farm animals: focus on cattle, sheep and pigs. *Animals* 11: 1483. https://doi.org/10.3390/ani11061483.
11. Neethirajan, S. 2021. Ethics of digital animal farming. Preprints. https://doi.org/10.20944/preprints202107.0368.v1.
12. Lustgarten, J. L., A. Zehnder, W. Shipman, E. Gancher, and T. L Webb. 2020. Veterinary informatics: forging the future between veterinary medicine, human medicine, and One Health initiatives—a joint paper by the Association for Veterinary Informatics (AVI) and the CTSA One Health Alliance (COHA). *JAMIA Open* 3: 306–317. https://doi.org/10.1093/jamiaopen/ooaa005.

13. Wong, Z. S. Y., J. Zhou, and Q. Zhang. 2019. Artificial Intelligence for infectious disease Big Data Analytics. *Infection, Disease & Health* 24: 44–48. https://doi.org/10.1016/j.idh.2018.10.002.
14. U.S. Veterinarian numbers 2018 |American Veterinary Medical Association. 2022. https://www.avma.org/resources-tools/reports-statistics/market-research-statistics-us-veterinarians-2018. Accessed July 15.
15. Home. 2024. *WSAVA*. https://wsava.org/. Accessed January 27.
16. Favre, D. 2016. An International Treaty for Animal Welfare. In *Animal Law and Welfare - International Perspectives*, ed. D. Cao and S. White, 87–106. Ius Gentium: Comparative Perspectives on Law and Justice. Cham: Springer International Publishing. https://doi.org/10.1007/978-3-319-26818-7_5.
17. Mitchell, M. 2019. *Artificial Intelligence: A Guide for Thinking Humans*. 1 ed. New York: Farrar, Straus and Giroux.
18. Waljee, A. K., and P. D. R. Higgins. 2010. Machine learning in medicine: a primer for physicians. *Official Journal of the American College of Gastroenterology | ACG* 105: 1224. https://doi.org/10.1038/ajg.2010.173.
19. Appleby, R. B., and P. S. Basran. 2022. Artificial intelligence in veterinary medicine. *Journal of the American Veterinary Medical Association* 260: 819–824. https://doi.org/10.2460/javma.22.03.0093.
20. Quinn, T. P., M. Senadeera, S. Jacobs, S. Coghlan, and V. Le. 2021. Trust and medical AI: the challenges we face and the expertise needed to overcome them. *Journal of the American Medical Informatics Association: JAMIA* 28: 890–894. https://doi.org/10.1093/jamia/ocaa268.
21. Barnett, G. 1982. The computer and clinical judgement. *New England Journal of Medicine* 307: 493–494.
22. Schwartz, W. B., R. S. Patil, and P. Szolovits. 1987. Artificial intelligence in medicine where do we stand? *Jurimetrics* 27: 362–369.
23. Partridge, D. 1987. The scope and limitations of first generation expert systems. *Future Generation Computer Systems* 3: 1–10. https://doi.org/10.1016/0167-739X(87)90038-0.
24. Bengio, Y., and Y. LeCun. 2007. Scaling learning algorithms towards AI. *Large-Scale Kernel Machines* 34: 1–41.
25. Reagan, K. L., B. A. Reagan, and C. Gilor. 2020. Machine learning algorithm as a diagnostic tool for hypoadrenocorticism in dogs. *Domestic Animal Endocrinology* 72: 106396. https://doi.org/10.1016/j.domaniend.2019.106396.

26. Zuraw, A., and F. Aeffner. 2022. Whole-slide imaging, tissue image analysis, and artificial intelligence in veterinary pathology: An updated introduction and review. *Veterinary Pathology* 59: 6–25. https://doi.org/10.1177/03009858211040484.
27. Nagamori, Y., R. H. Sedlak, A. DeRosa, A. Pullins, T. Cree, M. Loenser, B. S. Larson, R. B. Smith, C. Penn, and R. Goldstein. 2021. Further evaluation and validation of the VETSCAN IMAGYST: in-clinic feline and canine fecal parasite detection system integrated with a deep learning algorithm. *Parasites & Vectors* 14: 89. https://doi.org/10.1186/s13071-021-04591-y.
28. Li, Shen, Z. W., L. C. Visser, E. R. Wisner, and H. Cheng. 2020. Pilot study: application of artificial intelligence for detecting left atrial enlargement on canine thoracic radiographs. *Veterinary Radiology & Ultrasound* 61: 611–618. https://doi.org/10.1111/vru.12901.
29. Boissady, E., A. de La Comble, X. Zhu, and A.-M. Hespel. 2020. Artificial intelligence evaluating primary thoracic lesions has an overall lower error rate compared to veterinarians or veterinarians in conjunction with the artificial intelligence. *Veterinary Radiology & Ultrasound* 61: 619–627. https://doi.org/10.1111/vru.12912.
30. Nejedly, P., V. Kremen, V. Sladky, M. Nasseri, H. Guragain, P. Klimes, J. Cimbalnik, Y. Varatharajah, B. H. Brinkmann, and G. A. Worrell. 2019. Deep-learning for seizure forecasting in canines with epilepsy. *Journal of Neural Engineering* 16: 036031. https://doi.org/10.1088/1741-2552/ab172d.
31. Souza, G. V., A. C. V. Hespanha, B. F. Paz, M. A. R. Sá, R. K. Carneiro, S. A. M. Guaita, T. V. Magalhães, B. W. Minto, and L. G. G. G. Dias. 2021. Impact of the internet on veterinary surgery. *Veterinary and Animal Science* 11: 100161. https://doi.org/10.1016/j.vas.2020.100161.
32. Esteva, A., A. Robicquet, B. Ramsundar, V. Kuleshov, M. DePristo, K. Chou, C. Cui, G. Corrado, S. Thrun, and J. Dean. 2019. A guide to deep learning in healthcare. *Nature Medicine* 25: 24–29. https://doi.org/10.1038/s41591-018-0316-z.
33. Panesar, S., Y. Cagle, D. Chander, J. Morey, J. Fernandez-Miranda, and M. Kliot. 2019. Artificial intelligence and the future of surgical robotics. *Annals of Surgery* 270: 223. https://doi.org/10.1097/SLA.0000000000003262.
34. Prevett, R. 2022. Vet AI: a pioneering platform for pets. https://foundry4.com/vet-ai-a-pioneering-platform-for-pets. *foundry4*.
35. Jobin, A., M. Ienca, and E. Vayena. 2019. The global landscape of AI ethics guidelines. *Nature Machine Intelligence* 1: 389–399. https://doi.org/10.1038/s42256-019-0088-2.

36. Mittelstadt, B. 2019. Principles alone cannot guarantee ethical AI. *Nature Machine Intelligence* 1: 501–507. https://doi.org/10.1038/s42256-019-0114-4.
37. Beauchamp, T., and J. Childress. 2001. *Principles of Biomedical Ethics.* New York: Oxford University Press.
38. Desmond, J. 2022. Medicine, value, and knowledge in the veterinary clinic: questions for and from medical anthropology and the medical humanities. *Frontiers in Veterinary Science* 9: 780482.
39. Springer, S., T. B. Lund, H. Grimm, A. T. Kristensen, S. A. Corr, and P. Sandøe. 2022. Comparing veterinarians' attitudes to and the potential influence of pet health insurance in Austria, Denmark and the UK. *Veterinary Record* 190: e1266. https://doi.org/10.1002/vetr.1266.
40. Mullan S., A. Quain, and S. Wensley. 2017. *Veterinary Ethics: Navigating Tough Cases.* 1 ed. Sheffield: 5m Publishing.
41. Boller, M., T. S. Nemanic, J. D. Anthonisz, M. Awad, J. Selinger, E. M. Boller, and M. A. Stevenson. 2020. The effect of pet insurance on presurgical euthanasia of dogs with gastric dilatation-volvulus: a novel approach to quantifying economic euthanasia in veterinary emergency medicine. *Frontiers in Veterinary Science* 7: 590615.
42. Kantin, H., and D. Wendler. 2015. Is there a role for assent or dissent in animal research? *Cambridge Quarterly of Healthcare Ethics* 24: 459–472. https://doi.org/10.1017/S0963180115000110.
43. Medical Board of Australia. Good medical practice: a code of conduct for doctors in Australia. 2024. https://www.medicalboard.gov.au/codes-guidelines-policies/code-of-conduct.aspx. Accessed January 27.
44. Fleischman, A. R. 2016. *Pediatric Ethics: Protecting the Interests of Children.* Oxford University Press. https://doi.org/10.1093/med/9780199354474.001.0001.
45. Gillam, L. 2016. The zone of parental discretion: an ethical tool for dealing with disagreement between parents and doctors about medical treatment for a child. *Clinical Ethics* 11: 1–8. https://doi.org/10.1177/1477750915622033.
46. Coghlan, S., and C. Parker. 2023. Harm to nonhuman animals from AI: a systematic account and framework. *Philosophy & Technology* 36: 25. https://doi.org/10.1007/s13347-023-00627-6.
47. Feighelstein, M., I. Shimshoni, L. R. Finka, S. P. L. Luna, D. S. Mills, and A. Zamansky. 2022. Automated recognition of pain in cats. *Scientific Reports* 12: 9575. https://doi.org/10.1038/s41598-022-13348-1.
48. van der Linden, D., A. Zamansky, I. Hadar, B. Craggs, and A. Rashid. 2019. Buddy's wearable is not your buddy: privacy implications of

pet wearables. *IEEE Security & Privacy* 17: 28–39. https://doi.org/10.1109/MSEC.2018.2888783.

49. Newberry, M. 2017. Pets in danger: exploring the link between domestic violence and animal abuse. *Aggression and Violent Behavior* 34: 273–281. https://doi.org/10.1016/j.avb.2016.11.007.
50. Kliegr, T., Š. Bahník, and J. Fürnkranz. 2021. A review of possible effects of cognitive biases on interpretation of rule-based machine learning models. *Artificial Intelligence* 295: 103458. https://doi.org/10.1016/j.artint.2021.103458.
51. Capurro, D., and E. Velloso. 2021. Dark patterns, electronic medical records, and the opioid epidemic. arXiv. https://doi.org/10.48550/arXiv.2105.08870.
52. Quain, A., M. P. Ward, and S. Mullan. 2021. Ethical challenges posed by advanced veterinary care in companion animal veterinary practice. *Animals* 11: 3010. https://doi.org/10.3390/ani11113010.
53. Crawford, K. 2021. Artificial intelligence is misreading human emotion. *The Atlantic*.
54. Kim, D. W., H. Y. Jang, K. W. Kim, Y. Shin, and S. H. Park. 2019. Design characteristics of studies reporting the performance of artificial intelligence algorithms for diagnostic analysis of medical images: results from recently published papers. *Korean Journal of Radiology* 20: 405–410. https://doi.org/10.3348/kjr.2019.0025.
55. Nagendran, M., Y. Chen, C. A. Lovejoy, A. C. Gordon, M. Komorowski, H. Harvey, E. J. Topol, J. P. A. Ioannidis, G. S. Collins, and M. Maruthappu. 2020. Artificial intelligence versus clinicians: systematic review of design, reporting standards, and claims of deep learning studies. *BMJ* 368: m689. https://doi.org/10.1136/bmj.m689.
56. Jaiswal, A., A. Krishnama Raju, and S. Deb. 2020. Facial emotion detection using deep learning. In *2020 International Conference for Emerging Technology (INCET)*, 1–5. https://doi.org/10.1109/INCET49848.2020.9154121.
57. Carter, S. M., W. Rogers, I. Heath, C. Degeling, J. Doust, and A. Barratt. 2015. The challenge of overdiagnosis begins with its definition. *BMJ* 350: h869. https://doi.org/10.1136/bmj.h869.
58. Myers, N. C. 1997. Adrenal incidentalomas: diagnostic workup of the incidentally discovered adrenal mass. *Veterinary Clinics of North America: Small Animal Practice* 27: 381–399. https://doi.org/10.1016/S0195-5616(97)50038-6.

59. McKenzie, B. A. 2016. Overdiagnosis. *Journal of the American Veterinary Medical Association* 249: 884–889. https://doi.org/10.2460/javma.249.8.884.
60. Capurro, D., S. Coghlan, and D. E. V. Pires. 2022. Preventing digital overdiagnosis. *JAMA* 327: 525–526. https://doi.org/10.1001/jama.2021.22969.
61. Woolf, S. H., and R. Harris. 2012. The harms of screening: new attention to an old concern. *JAMA* 307: 565–566. https://doi.org/10.1001/jama.2012.100.
62. Sekhar, M. S., and N. Vyas. 2013. Defensive medicine: a bane to healthcare. *Annals of Medical and Health Sciences Research* 3: 295–296. https://doi.org/10.4103/2141-9248.113688.
63. Castelvecchi, D. 2016. Can we open the black box of AI? *Nature News* 538: 20. https://doi.org/10.1038/538020a.
64. Rudin, C. 2019. Stop explaining black box machine learning models for high stakes decisions and use interpretable models instead. *Nature Machine Intelligence* 1: 206–215. https://doi.org/10.1038/s42256-019-0048-x.
65. Quinn, T. P., S. Jacobs, M. Senadeera, V. Le, and S. Coghlan. 2022. The three ghosts of medical AI: can the black-box present deliver? *Artificial Intelligence in Medicine* 124: 102158. https://doi.org/10.1016/j.artmed.2021.102158.
66. Kelly, C. J., A. Karthikesalingam, M. Suleyman, G. Corrado, and D. King. 2019. Key challenges for delivering clinical impact with artificial intelligence. *BMC Medicine* 17: 195. https://doi.org/10.1186/s12916-019-1426-2.
67. Culnane, C., B. I. P. Rubinstein, and V. Teague. 2017. Health data in an open world. arXiv. https://doi.org/10.48550/arXiv.1712.05627.
68. Carlini, N., F. Tramèr, E. Wallace, M. Jagielski, A. Herbert-Voss, K. Lee, A. Roberts, et al. 2021. Extracting training data from large language models. *Open Access to the Proceedings of the 30th USENIX Security Symposium*. 2633–2650.
69. Parasuraman, R., and V. Riley. 1997. Humans and automation: use, misuse, disuse, abuse. *Human Factors* 39: 230–253. https://doi.org/10.1518/001872097778543886.
70. Jacovi, A., A. Marasović, T. Miller, and Y. Goldberg. 2021. Formalizing trust in artificial intelligence: prerequisites, causes and goals of human trust in AI. In *Proceedings of the 2021 ACM Conference on Fairness, Accountability, and Transparency*, 624–635.

FAccT '21. New York, NY, USA: Association for Computing Machinery. https://doi.org/10.1145/3442188.3445923.

71. Ferrario, A., M. Loi, and E. Viganò. 2021. Trust does not need to be human: it is possible to trust medical AI. *Journal of Medical Ethics* 47: 437–438. https://doi.org/10.1136/medethics-2020-106922.
72. Goddard, K., A. Roudsari, and J. C. Wyatt. 2012. Automation bias: a systematic review of frequency, effect mediators, and mitigators. *Journal of the American Medical Informatics Association: JAMIA* 19: 121–127. https://doi.org/10.1136/amiajnl-2011-000089.
73. Beauchamp, T. L. 2011. Informed consent: its history, meaning, and present challenges. *Cambridge Quarterly of Healthcare Ethics* 20: 515–523. https://doi.org/10.1017/S0963180111000259.
74. Brown, C. R., L. D. Garrett, W. K. Gilles, K. E. Houlihan, E. McCobb, S. Pailler, H. Putnam, et al. 2021. Spectrum of care: more than treatment options. *Journal of the American Veterinary Medical Association* 259: 712–717. https://doi.org/10.2460/javma.259.7.712.
75. Geis, J. R., A. Brady, C. C. Wu, J. Spencer, E. Ranschaert, J. L. Jaremko, S. G. Langer, et al. 2019. Ethics of artificial intelligence in radiology: summary of the joint European and North American multisociety statement. *Insights into Imaging* 10: 101. https://doi.org/10.1186/s13244-019-0785-8.
76. Pohl, R., J. Botscharow, I. Böckelmann, and B. Thielmann. 2022. Stress and strain among veterinarians: a scoping review. *Irish Veterinary Journal* 75: 15. https://doi.org/10.1186/s13620-022-00220-x.
77. Cowan, R. S. 1983. *More Work For Mother: The Ironies Of Household Technology From The Open Hearth To The Microwave*. New York, NY: Pantheon Books.
78. Clinician of the Future: a 2022 report. 2024. *www.elsevier.com*. https://www.elsevier.com/connect/clinician-of-the-future. Accessed January 27.
79. Mittelstadt, B. D., and L. Floridi. 2016. The ethics of big data: current and foreseeable issues in biomedical contexts. *Science and Engineering Ethics* 22: 303–341. https://doi.org/10.1007/s11948-015-9652-2.
80. Kempt, H., J.-C. Heilinger, and S. K. Nagel. 2022. "I'm afraid I can't let you do that, Doctor": meaningful disagreements with AI in medical contexts. *AI & SOCIETY* 38: 1407–1414. https://doi.org/10.1007/s00146-022-01418-x.
81. Santoni de Sio, F., and G. Mecacci. 2021. Four responsibility gaps with artificial intelligence: why they matter and how to address them.

Philosophy & Technology 34: 1057–1084. https://doi.org/10.1007/s13347-021-00450-x.

82. Lenzen, M., A. Malik, M. Li, J. Fry, H. Weisz, P.-P. Pichler, L. Suveges M. Chaves, A. Capon, and D. Pencheon. 2020. The environmental footprint of health care: a global assessment. *The Lancet Planetary Health* 4: e271–e279. https://doi.org/10.1016/S2542-5196(20)30121-2.
83. Hagendorff, T. 2022. Blind spots in AI ethics. *AI and Ethics* 2: 851–867. https://doi.org/10.1007/s43681-021-00122-8.
84. Jones, R. S., and E. West. 2019. Environmental sustainability in veterinary anaesthesia. *Veterinary Anaesthesia and Analgesia* 46: 409–420. https://doi.org/10.1016/j.vaa.2018.12.008.
85. Koytcheva, M. K., L. K. Sauerwein, T. L. Webb, S. A. Baumgarn, S. A. Skeels, and C. G. Duncan. 2021. A systematic review of environmental sustainability in veterinary practice. *Topics in Companion Animal Medicine* 44: 100550. https://doi.org/10.1016/j.tcam.2021.100550.
86. Rosoff, P. M., J. Moga, B. Keene, C. Adin, C. Fogle, R. Ruderman, H. Hopkinso, and C. Weyhrauch. 2018. Resolving ethical dilemmas in a tertiary care veterinary specialty hospital: adaptation of the human clinical consultation committee model. *The American Journal of Bioethics* 18: 41–53. https://doi.org/10.1080/15265161.2017.1409824.
87. Springer, S., P. Sandøe, H. Grimm, S. A. Corr, A. T. Kristensen, and T. B. Lund. 2021. Managing conflicting ethical concerns in modern small animal practice—A comparative study of veterinarian's decision ethics in Austria, Denmark and the UK. *PLOS ONE* 16: e0253420. https://doi.org/10.1371/journal.pone.0253420.
88. Kimera, S. I., and J. E. D. Mlangwa. 2016. Veterinary Ethics. In *Encyclopedia of Global Bioethics*, ed. H. t. Have, 2937–2947. Cham: Springer International Publishing. https://doi.org/10.1007/978-3-319-09483-0_435.
89. Legood, G. 2000. *Veterinary Ethics*. Bloomsbury Publishing.
90. Magalhães-Sant'Ana, M., S. J. More, D. B. Morton, M. Osborne, and A. Hanlon. 2015. What do European veterinary codes of conduct actually say and mean? A case study approach. *Veterinary Record* 176: 654–654. https://doi.org/10.1136/vr.103005.
91. Moses, L. 2018. Another experience in resolving veterinary ethical dilemmas: observations from a veterinarian performing ethics consultation. *The American Journal of Bioethics* 18: 67–69. https://doi.org/10.1080/15265161.2017.1409825.

92. Rollin, B. E. 2006. *An Introduction to Veterinary Medical Ethics: Theory And Cases,* 2nd ed. Ames, Iowa: Wiley-Blackwell.
93. Sandøe, P., S. Corr, and C. Palmer. 2015. *Companion Animal Ethics*. John Wiley & Sons.
94. Tannenbaum, J. 1991. Ethics and animal welfare: the inextricable connection. *Journal of the American Veterinary Medical Association* 198: 1360–1376.
95. Yeates, J., and J. Savulescu. 2017. Companion animal ethics: a special area of moral theory and practice? *Ethical Theory and Moral Practice* 20: 347–359. https://doi.org/10.1007/s10677-016-9778-6.
96. Yeates, J. W., and D. C. J. Main. 2010. The ethics of influencing clients. *Journal of the American Veterinary Medical Association* 237: 263–267. https://doi.org/10.2460/javma.237.3.263.
97. Knesl, O., B. L. Hart, A. H. Fine, and L. Cooper. 2016. Opportunities for incorporating the human-animal bond in companion animal practice. *Journal of the American Veterinary Medical Association* 249: 42–44. https://doi.org/10.2460/javma.249.1.42.
98. Serpell, J. 1996. *In the Company of Animals: A Study of Human-Animal Relationships*. Cambridge; New York: Cambridge University Press.
99. Coghlan, S. 2018. Strong patient advocacy and the fundamental ethical role of veterinarians. *Journal of Agricultural and Environmental Ethics* 31: 349–367. https://doi.org/10.1007/s10806-018-9729-4.
100. Hernandez, E., A. Fawcett, E. Brouwer, J. Rau, and P. V. Turner. 2018. Speaking up: veterinary ethical responsibilities and animal welfare issues in everyday practice. *Animals: an Open Access Journal from MDPI* 8: 15. https://doi.org/10.3390/ani8010015.
101. Clark, J. D., and D. M. Dudzinski. 2013. The culture of dysthanasia: attempting CPR in terminally ill children. *Pediatrics* 131: 572–580. https://doi.org/10.1542/peds.2012-0393.
102. Sunstein, C. 2003. The rights of animals. *University of Chicago Law Review* 70: 387–401.
103. Singer, P. 1995. *Animal Liberation*. New York: Random House.
104. Owe, A., and S. D. Baum. 2021. Moral consideration of nonhumans in the ethics of artificial intelligence. *AI and Ethics* 1: 517–528. https://doi.org/10.1007/s43681-021-00065-0.
105. Singer, P., and Y. F. Tse. 2023. AI ethics: the case for including animals. *AI and Ethics* 3: 539–551. https://doi.org/10.1007/s43681-022-00187-z.

Chapter 10

The 3Rs Principles and Genetic Pain Disenhancement*

Samuel Camenzind[a] and Matthias Eggel[b]
[a]*APART-GSK Fellow (Austrian Academy of Sciences), Department of Philosophy, University of Vienna, Vienna 1090, Austria*
[b]*Philosophisches Seminar, Universität Basel, Basel 4051, Switzerland*
samuel.camenzind@univie.ac.at, matthias.eggel@unibas.ch

10.1 Introduction

The 3Rs principles—Replacement, Refinement and Reduction—were set out by the British scientists William Russell and Rex Burch, in 1959 in *The Principles of Humane Experimental Technique*. They were introduced as a measure of professional self-regulation in response to the increasing amount of animal research taking place and the invalid results that were related to inhuman treatment of animals [1; p. 3f]. Since the 1980s, the principles have become an integral part of various legal regulations. They also appear in the official statements of good practice of most European and many non-European universities and research institutions dealing with

*Reprinted from *Animal Welfare*, 31(4), 495–503, 2022, with permission from Cambridge University Press.

Animals, Ethics, and Engineering: Intersections and Implications
Edited by Rosalyn W. Berne

ISBN 978-981-5129-56-4 (Hardcover), 978-1-003-65231-1 (eBook)
www.jennystanford.com

animal experimentation [2], as well as playing a crucial role in the ethical evaluation and authorization of animal experiments in several countries.

The fact that the 3Rs have been successfully implemented across a range of legislative domains based on different ethical theories (themselves based on different moral traditions and schools) indicates their compatibility with a wide range of ethical stances.

However, what is the original meaning, or content, of the 3Rs? Furthermore, what is their normative groundwork? These questions are important for two reasons. First, the 3Rs were developed prior to the emergence of animal ethics as a scientific discipline in the 1970s [3] and before many ground-breaking developments in modern biotechnology, such as somatic cell nuclear transfer cloning and genetic engineering. Second, notwithstanding more than fifty years of developments in the content and methodology of animal ethics, we claim that ethical evaluation of the presuppositions informing the 3Rs principles and the changing ethical context continues to lag behind scientific and practical efforts to implement the 3Rs [4].

We therefore aim to analyze the normative foundations of the 3Rs principles in contemporary terms and concepts, and to explain their relevance for contemporary animal research. After establishing the normative groundwork of the 3Rs, we will look at the practical implications of this groundwork. To this end, we shall investigate, from the normative perspective of the original 3Rs, whether the genetic disenhancement of research animals to limit their ability to feel pain and suffer (hereafter "GPD" for genetic pain disenhancement) would comply with the 3Rs principles. We use GPD as a practical example, since it is being discussed today as a promising way of solving one of the key moral issues raised by animal research, i.e., animal pain and suffering. By discussing GPD within the context of the 3Rs we aim not only to gain an insight into whether GPD is compatible with one or more of the 3Rs, but to also develop a better understanding of the specific normative foundation of the 3Rs principles and the conceptual limitations and practical implications of that foundation.

Lastly, from a contemporary standpoint, we reflect critically on the normative position of the 3Rs principles and their underlying concept of harm. We argue that reducing moral concerns about animal research to those that are intelligible within a sentientist framework (e.g., harm and suffering), as the 3Rs do, represents an oversimplification of the moral issues involved. We suggest that interference with abilities, instrumentalization, flourishing, and death are all important aspects of animal ethics requiring consideration—and all the more so in light of the possibilities offered by modern genetic biotechnologies, including GPD.

10.2 Overview of The 3Rs Principles

The UK's National Centre for the Replacement, Refinement and Reduction of Animals in Research (NC3R) distinguishes two forms of *Replacement*. "Full Replacement" ("absolute Replacement" by Russell and Burch) aims at avoiding "the use of any research animals. It includes the use of human volunteers, tissues and cells, mathematical and computer models, and established cell lines" (NC3R 2021). "Partial Replacement" includes "the use of some animals that, based on current scientific thinking, are not considered capable of experiencing suffering", e.g., *Drosophila*, nematode worms and embryonic and fetal forms of vertebrates. Also included here are the use of primary cells (and tissues) taken from animals killed solely for this purpose. Others understand the use of animals that are alleged to be less sentient as "relative Replacement" [5].

Reduction aims to reduce the number of animals used for a specific experiment to the minimum required to guarantee sufficient statistical power. It also includes methods to maximize knowledge gain per animal used (e.g., longitudinal imaging). The rationale for this principle is that using either too many or too few animals in a research project is morally problematic. Using too many animals is wrong because no additional knowledge is gained while using too few is problematic because it may jeopardize the validity of a study.

Refinement aims to minimize the amount of harm and suffering inflicted on research animals to the scientifically necessary minimum. This not only entails minimization of negative impacts such as pain and suffering but also includes promotion of the positive well-being of animals (e.g., through environmental enrichment).

10.3 The Normative Groundwork of the 3Rs Principles

Unfortunately, Russell and Burch do not consider the groundwork of the moral theory their principles are based on in detail, nor do they attempt to justify these principles. For various reasons, this is unsurprising. For one thing, Russell and Burch were not ethicists. They had a background in zoology, psychology and microbiology. They developed a normative framework to assess animal research from within the field of animal research [4, 6]. This internal approach may have helped the 3Rs to become a success. The 3Rs principles were not experienced as an external interference restricting scientific freedoms. Instead, they appeared to be a measure of professional self-regulation and a necessary means to progress and develop scientific fields involving laboratory animal use.

From a historical perspective, the lack of moral theory is also not very surprising, given that animal ethics as an academic discipline evolved ten years after publication of *The Principles of Humane Experimental Technique*.

Nevertheless, an outline and justification of the moral framework underpinning the 3Rs would have been valuable. First, as we will try to show, the 3Rs principles represent a unique moral position in comparison with the other ethical approaches being explored in the middle of the twentieth century. Second, such an outline and justification might have precluded different interpretations of the 3Rs and their long-term purpose. For example, it would arguably have been clearer whether the 3Rs are a method to minimize the distress of animals within an anthropocentric scientific framework or instead an abolitionist tool to refrain from animal research entirely (as a final goal). Third, a moral groundwork would have provided normative guidelines allowing us to assess the compatibility of particular practices (such as GPD) with the 3Rs.

Although Russell and Burch do not themselves examine the groundwork of the moral theory their principles are based on, the moral theory they implicitly relied upon can be deduced to a degree. This theory can be described as one that is hierarchical and sentientist. It involves deontological principles.

The first two chapters of *The Principles of Humane Experimental Technique* explain that the 3Rs are based on the ethical principle of humanity. This principle can be articulated in the form of an imperative: "Treat animals as humanely as possible." "Humanely" here is not to be understood in the sense of social co-operation, but rather in the sense of humanness. The notion of humane treatment is explained with reference to subjectively negative experienced mental states such as pain and fear, and distress more generally [1; pp. 14, 15]. Noticeably, and in contrast with later discussions [7], the main focus in the 1959 book was not on animal welfare as a phenomenon including both positive and negative mental states. It was on the negative welfare states of pain, fear and distress [8; p. 267]. (This will become relevant when we turn to GPD specifically.)

The second chapter of *The Principles of Humane Experimental Technique* also presents a discussion of the nature of the pain that animals are able to experience and asks how distress can be measured. The overall aim of the authors is to examine the "concept of humanity (or inhumanity) as an objective assessment of the effects of any procedure on the animal subject" [1; p. 14]. The concept of inhumanity is further distinguished into two forms: *direct* and *contingent* inhumanity. The former concerns "the infliction of distress as an unavoidable consequence of the procedure employed," the latter includes "the infliction of distress as an incidental and inadvertent by-product of the use of the procedure, which is not necessary for its success" [1; p. 54]. Imperfect husbandry conditions and harsh handling would count as indirect in this sense, as would unanticipated post-operative complications and sub-optimal hygiene conditions.

Although Russell and Burch mention death, and the killing and euthanasia of animals in several places, the topic is not discussed extensively from a moral perspective. What is also missing in the first two chapters, and indeed throughout the rest of the book, is a comprehensive consideration and justification of the grounds on which we should treat animals as humanely as possible. Why should we do this? Should we treat animals humanely for their own sake or for other reasons, such as economic or scientific reasons? Or should we show a "friendly and constructive attitude to the lower animals" because cruelty to animals tends to make the perpetrator callous [1; p. 14], a view that is associated with Kant, whose work anticipates the 3Rs principles [9; p. 62f]. Both interpretations are possible: Russell and Burch open their survey with a confident assertion that society owes "many if not most of the benefits of modern medicine and countless advances in fundamental scientific knowledge" to animal experimentation, and with the further statement that good animal welfare promotes successful science [1; p. 3f].

While the first, historical comment on the contribution of animal experimentation to medical progress has been challenged by medical historians [10], the second claim, that "good quality science depends upon good animal welfare," is still widely accepted today [11, chapter 7.8]. In effect, the second claim represents the major theme of Russell and Burch's book. From this claim, we might be tempted to infer that animal welfare has only an instrumental value for qualitatively better, more efficient science. Such an inference would signal a moral anthropocentrism in which animals lack moral status and in which animal welfare is considered, at most, indirectly.

However, it seems unlikely that this was Russell and Burch's intended meaning. This is indicated by the prominent role given in the book to Charles Darwin. Each chapter is introduced with a quote from Darwin, and in the final paragraph of the foreword Darwin is presented as a scientist who was very concerned about the welfare of experimental animals and more active in "furthering the progress of humane experimental technique" than most other scientists [1; p. xiv]. In addition, the claims the authors make about relative and partial Replacement (see below), and their focus on sentient animals—all vertebrates [1; p. 6]—are indicators that Russell and Burch assume the moral status of all sentient animals. So, treating animals in the most humane way, besides being good scientific practice, instrumentally speaking, may also be of direct, moral value. This is a good reason for thinking that Russell and Burch were not working in an anthropocentric paradigm: they were promoting a sentient position that grants all sentient animals moral status [4, 8].

Even if this interpretation is correct, there are a number of reasons to think that the moral significance attributed to animals by Russell and Burch was not very high. First of all, it is obvious that Russell and Burch accept the lexical priority of human interests over animal interests [12; p. 6]. The infliction of harm such as pain, or suffering, or stress on animals is generally permitted for scientific purposes as long as it is done in a "humane" manner. The authors do not appear, then, to see animal research as a severe moral problem, or a moral dilemma. Secondly, no harm-benefit-analysis is recommended by Russell and Burch. This means the animal's interest in avoiding suffering (harm scale) cannot outweigh the scientific benefits (benefit scale). In other words, using the 3Rs principles alone results in the position that any scientific goal—even trivial ones—trumps severe harms inflicted on animals.

To conclude, the 3Rs approach essentially aims to promote animal welfare while at the same time striving for scientific validity. Importantly, however, when these two goals conflict with each other, or at least appear to do so, priority is given to scientific validity. We refer to this position as "hierarchical sentientism" or "human priority position" [13; p. 16].

Despite the marginal consideration they give to the animal's own interests, one has to acknowledge the strict bindingness of the 3Rs principles. The 3Rs are categorical duties—they are neither supererogatory ideals nor *prima facie* duties that can be balanced against other moral requirements. Certainly, Replacement and Reduction seem to be perfect (negative) duties leaving little if any room for discussion of how they are to be complied with, although Refinement leaves more room here, allowing us to ask how the distress of animals is best avoided or minimized, and how the positive welfare of animals is to be promoted.

Given the theories, or moral attitudes, that were common currency when Russell and Burch were writing, we can see that their ethical position, in setting out the 3Rs approach, was somewhat new and unique, and far from self-explanatory. Firstly, they departed from the historically dominant anthropocentric paradigm, which excludes animals from the moral community completely. Giving animals moral status, however, they followed the paths of neither Leonard Nelson's [14] interest-based ethics, which considered animal and human interests to be equal, nor Henry Salt's [15; p. 1] animal rights account, which claimed, famously, that animals have rights if humans do. Moreover, as mentioned above, against the dominant utilitarian tradition in Great Britain [16], the 3Rs principles do not allow us to conduct a harm-benefit analysis.

Finally, in contrast with Albert Schweitzer [17], who expanded the moral community in his ethics of "reverence for life" to all living beings, including non-vertebrates and plants, Russell and Burch focus only on sentient animals. We can summarize this by saying that the 3Rs represent a hierarchical, sentientist position associated with deontological, categorical, absolute principles to promote the humane treatment of animals.

Having provided some insight into the normative foundation of the 3Rs principles, we will now examine a biotechnological innovation that is currently being discussed as a promising 3Rs method. This is genetic engineering to reduce animals' ability to feel pain and suffer. We wish to discuss animal disenhancement of this kind for two reasons. On one hand, we want to assess whether, and if so how, it complies with the principles of Replacement, Reduction and Refinement. On the other, we believe the assessment will throw light on the ethical foundations of the 3Rs principles, their conceptual focus, and also their limitations, as well as their practical implications for modern biotechnology and animal research.

10.4 Animal Disenhancement

Animal research causes pain and suffering to millions of sentient research animals every year [18]. There is a wide consensus among the general public, politicians, scientists and ethical theorists that this is morally problematic.

The consensus on the moral relevance of animal pain and suffering, and on the desirability of avoiding unnecessary pain, is evident in ongoing efforts to improve implementation of the 3Rs in practice—e.g., by reducing the suffering of laboratory animals by providing better housing conditions and better pain treatment (e.g., analgesia and anesthesia), and through developments in surgical technique, science education and training, and so on. Notwithstanding these efforts to minimize pain and stress, scientific research continues to inflict significant suffering on laboratory animals. This is partly because the opportunities to reduce animal suffering are limited for economic and practical reasons. For example, methods designed to better satisfy the 3Rs will only be implemented if they are affordable, do not impair scientific validity and can feasibly be applied to a large number of individual animals. It is also because pain, suffering, distress, and anxiety, in different grades of severity, are integral accompaniments to animal research procedures. As a result of these inherent practical and economic limitations, recent discussion of how we can improve animal welfare further, in practice, has turned to genetic options, including genetic disenhancement.

The term "animal disenhancement" was first coined by Paul Thompson to refer to the alteration of animals to better suit their environment, either by breeding or via "biotechnological reduction or elimination of capabilities in non-human animals to mitigate animal welfare problems" [19]. The idea of altering research and farm animals for specific human use emerged in the course of the 1980s and 1990s [20–26]. The basic idea is to adapt the phenotype of an animal so that it is better suited to life in the (otherwise harmful) conditions involved in a specific use, such as animal experimentation or livestock farming, through breeding or genetic engineering.

To explain the rationale behind animal disenhancement we need to refer to three historical developments. First, consider the historico-cultural use of animals for human purposes. The intensive use of animals in agriculture and scientific research involves harms to the animals some of which are intrinsic to the procedures and housing systems while others are only contingent side-effects. As a strongly institutionalized practice, animal research that involves stress for the animals is highly likely to continue in the future. This fact becomes morally relevant in view of the second historical development.

Second, the emergence of animal ethics and the rising moral status of animals in academia and society over the last fifty years have significantly altered our understanding of the human-animal relationship. For example, the (juridical) rights of animals and our moral duties towards them have changed. With this development, procedures involving animal harm have been considered increasingly morally problematic—more and more, they have come to be regarded as requiring justification. The ongoing intensive use of animals creates a tension between the rising moral standing of animals and consequent desire to protect them, on one side, and the benefits (e.g., dairy products, research-based knowledge) acquired through our use of animals, on the other.

The third historical development revolves around the proposal that the tension described above can be eased using our recently acquired biotechnological ability to manipulate the genome of animals (e.g., DNA-microinjection, embryonic stem cell mediated gene-transfer, SCNT-cloning and genetic engineering with nuclease-systems such as CRISPR/Cas) to create animals that are "better suited" for human purposes. Hence, scientific progress, and the advent of modern biotechnology, has not just given us new opportunities for animal use. It has also introduced an opportunity to reduce animal suffering in various areas of human—in other words, our—animal use. For example, it has been suggested that we should use congenitally blind chickens in order to reduce injurious pecking in the flock [27]. Hornless cows have been engineered to render the painful cutting off their horns via de-horning obsolete [28]. These measures are designed to reduce farm animals' infliction of pain on each other. By contrast, genetic animal disenhancement aims, not to modify external traits and the behavioral tendencies and capabilities of animals, but rather to reduce or eliminate the animals' ability to experience pain and suffering itself: it reduces the animals' experience of negative emotional states by limiting their *capacity* to have those states [29–32].

10.5 Paradigm Shift with the Latest Technological Developments: From Thought Experiment to Realistic Possibility?

According to some authors, GPD to reduce animal suffering seems feasible following recent technological developments in genetic engineering such as clustered regularly interspaced short palindromic repeats-CRISPR associated proteins (CRISPR-Cas), zinc finger nucleases (ZFN) and transcription activator-like effector nucleases (TALENs) [33].

These breakthroughs in gene editing technology have added significantly to the options available where genetic alteration of animal genotypes and phenotypes is concerned. They function like "molecular scissors" (i.e., custom-made designer nucleases) capable of targeting a specific DNA sequence and inducing a precise DNA break. Compared with previous biotechnological methods, the new techniques are claimed to be easier to use, cheaper and very precise (although this is also a feature of older methods). Perhaps most importantly, unlike previous biotechnological methods, and their limited application to different species, the new generation nuclease systems have been applied in several species and have so far proven adaptable [33, 34]. Thus, they appear to offer unprecedented possibilities.

Importantly, according to an account of recent advances in genetic engineering given by Shriver, particular enzymes and peptides have been identified that relate directly to what is considered the "affective dimension of pain"—in other words, to "caring about" the painful sensation. For example, studies in mice have shown a link between the affective dimension of pain and the AC1 and AC8 enzymes, the peptide P311 [33]. Interfering with these pathways could potentially reduce the affective dimension of pain and/or chronic pain symptoms while leaving the acute pain response intact [33]. It must be noted that despite the euphoric announcements, as of the year 2025, no ideal animal model has been bred in which the experience of pain could be eliminated. The following considerations, therefore, still have the status of a thought experiment.

10.6 GPD in the Light of the 3Rs Principles

Would GPD comply with the principles of Replacement, Refinement and Reduction? The implications of GPD for animal research have previously been considered [30], and in the course of this discussion some attention was given to the 3Rs principles [29, 31, 32]. However, we propose to systematically analyze the relationship between GPD and each principle separately. We shall use the ethical position of Russell and Burch as a normative standard of evaluation. Approaching the issues in this way, we aim to gain an insight into whether GPD is compatible with (one, two, or all three of) the 3Rs principles. We will also seek to obtain a better understanding of the normative underpinning of the 3Rs principles, and the conceptual and practical limitations that follow from this in the context of contemporary biotechnological developments.

10.6.1 GPD and Replacement

According to Russell and Burch [1; p. 70.] sentient animals can be replaced in two different ways. That is, they distinguish between *absolute* and *relative* Replacement. Absolute Replacement is described by Russell and Burch as the "absolute ideal" [1]. Today, it includes the substitution of animal models with non-animal methods such as organs-on-a-chip, high-throughput systems (HTC), mathematical models and computational modelling, systematic literature reviews or the use of human volunteers (among many other methods). Partial and relative Replacement (and also comparative substitution) involve animal-derived material such as organoids, cells or fluids and "animals that are lower on the phylogenetic scale" [35]. The latter may be invertebrates (e.g., *Drosophila melanogaster* or the amoeba [*Dictyostelium discoideum*]) or lower vertebrates. Additionally, relative Replacement involves "non-recovery experiments on living and intact but completely anesthetized animals" and the use of painlessly killed animals [1].

The rationale for using lower vertebrates is that these animals are believed to be "less sentient," thereby possessing a lower capacity to suffer compared to the higher vertebrates (the same distinction could be made among higher and lower vertebrates). Here, we should note that Russell and Burch did not explicitly discuss exactly what "less sentient" means, or the evidence on which ascriptions of it are based [36; p. 656]. From a sentientist perspective it is reasonable to substitute animals which display clear evidence of sentience with animals whose possession of sentience is less well evidenced. However, Russell and Burch have cautioned against the use of the evolutionary *scala naturae* as a norm from which guiding moral principles of animal research can be derived [35, 37; p. 28]. We also know now, from the case of cephalopods and crustaceans, that the emergence of sentience did not follow a single evolutionary lineage [38], so the idea that determinations of sentience can be based on a *scala naturae* is highly questionable.

More importantly, relative Replacement cannot be accurately said to solve the moral problem of animal distress. Rather, it shifts it to the replacement animals. Although they are (allegedly) less sentient than the animals they replace, they remain sentient. Thus, they are capable of experiencing positive and negative mental states. The degree to which GPD affects other negative mental states, such as anxiety, fear and boredom, is crucial in any serious assessment of GPD as a relative Replacement method. Since it still involves the use of animals, GPD will be compatible only with relative or partial Replacement, not absolute Replacement, however among the relative Replacement methods GPD is unique, because it assumes that genetically disenhanced animals are unable to experience pain but capable, nonetheless, of experiencing positive mental states.

10.6.2 GPD and Reduction

As originally conceived by Russell and Burch, Reduction is not concerned with reducing the total number of animals in animal research *per se*, but only tries to minimize the number of animals for a specific experiment with a specific goal. Regarding this, a sufficient number of animals is the number that allows us, with the required statistical power, to answer a specific study question. We can call this "relative Reduction." An alternate interpretation would be to say that Reduction aims to reduce the total number of animals in animal research *per se* [29]. We call this "absolute Reduction."

Where relative Reduction is concerned, the generation of GPD animals neither promotes nor obstructs Reduction, since relative Reduction is only concerned with the correct number of animals to reach a specific study aim.

The claim that GPD research (and GMO research generally) will lead to greater numbers of animals being used in research, and the inference that it therefore introduces a conflict between the goals of Reduction and Refinement, rests on a particular understanding of Reduction as "absolute Reduction." Only from an absolute Reduction perspective (one that goes beyond the original 3Rs framework) can it be questioned whether an effort to minimize harm and suffering (i.e., Refinement) which at the same time could, theoretically, lead to increased animal use is morally acceptable.

10.6.3 GPD and Refinement

Refinement is defined as "any decrease in the incidence or severity of inhumane procedures." Russell and Burch add that "its object is simply to reduce to an absolute minimum the amount of distress imposed on those animals that are still used" [1; pp. 64, 134]. It is the *ultima ratio* method within the 3Rs principles and is applied after the other two Rs have been applied.

Refinement is usually taken to refer to animal housing and husbandry conditions, experimental procedures and the handling of the animals, experimental design, conduct and reporting, and also animal models [39]. *Prima facie*, with its potential to significantly reduce animal suffering, GPD is morally desirable from a Refinement point of view. However, this depends on the assumption that GPD animals are not more likely to sustain increased injuries, such as bruising, which may adversely affect the animal's well-being, than animals without GPD. Here, the situation is not entirely straightforward. Studies have shown, on the one hand, that eliminating the affective pain dimension (i.e., the negative subjective experience of pain) does not impair acute responses to noxious stimulation [40]. Thus, GPD animals are likely to continue to exhibit normal guarding behavior (i.e., behavior involving the avoidance of activities normally causing pain), making an increase in self-injuries unlikely. Moreover, even if bruises should result from such guarding behavior, the bruises themselves would not cause any substantial suffering. From a purely sentientist point of view (where only negative subjective experiences are considered morally relevant) bruising will not be morally problematic anyway, even if the animals injure themselves more often, and even if the injuries have negative consequences, such as broken bones, limping and so on, as long as these consequences do not cause pain or suffering [31].

Offsetting these considerations is the fact that humans with pain insensitivity disorder are significantly more vulnerable to physical injuries—they have a higher morbidity and mortality rate than the average for the population as a whole [41]. Where GPD animals (e.g., mice or rodents) are social animals, the disenhancement may also have an impact on the social behavior of the animals [32]. How GPD affects the organism as a whole and the social behavior of animal groups is yet to be properly investigated. What is clear is that in order to comply with Russell and Burch's sentientism, GPD must not have any negative (subjective) effects on the animals. Thus, the technology used to create GPD animals needs to be able to effectively manipulate the pain perception of animals without causing negative subjective side-effects.

Taking all of the above considerations together, we can see that within the science-centered, sentientist normative framework of the 3Rs principles, the generation of GPD animals is acceptable: in principle, the use of such animals is compatible, at least, with relative Replacement and Refinement, while being value-neutral as regards Reduction. This conclusion is sound provided that GPD actually works and so long as the GPD animals do not suffer more than their unaltered kin and are still able to have positive emotions.

Of course, this raises the question as to whether the sentientist understanding of harm is an oversimplification, or questionable reduction of the moral issues at stake. Therefore, we shall now reflect on the ethical foundations of the 3Rs principles and elaborate what we consider to be the shortcomings of the 3R principles as a contemporary framework for morally acceptable conduct in animal research.

10.7 The 3Rs under Ethical Scrutiny

10.7.1 Limited Moral Significance of Animals

One criticism concerns the moral significance ascribed to animals within the hierarchical sentientism of Russell and Burch. Notwithstanding our argument—and *pace* Vorstenbosch—animals do have a moral standing of their own within the 3Rs. However, we share Vorstenbosch's assessment, that the decisions about animal use in science are primarily a matter of science and scientists [4; p. 343]. This means that even trivial scientific interests could justify severe stress for a large number of animals which will eventually be killed. This criticism is not new and has led to the implementation of a harm-benefit analysis (HBA) in the approval process for animal research in many countries around the world. While, according to the 3Rs principles, all experiments that are scientifically sound and necessary (i.e., instrumentally essential) are morally permissible, the implementation of HBA represents a paradigm change.

Here, experiments that are scientifically sound and considered necessary can still be morally prohibited if the harm they inflict is disproportionate to the expected knowledge gain (goal-related essentiality). The extent to which an HBA, over and above mere compliance with the 3Rs, improves ethical evaluation critically depends, of course, on the underlying concept of harm used in the HBA, and on the moral significance given to animals. This means that for an HBA to actually represent a paradigm change in practice, not just theory, significantly more moral weight has to be given to animals than they were accorded within the 3Rs.

10.7.2 The Limited Harm Concept

In the ongoing discussion of GPD, a central issue concerns the concept of "harm" and the question whether GPD animals are "worse off" than normal species members [19, 29, 33, 42, 43]. This issue arises because there is a fundamental ethical distinction between sentientist and non-sentientist notions of harm. The question of morally relevant goods is also relevant to the moral assessment of the 3R principles.

While GPD will potentially reduce negative subjective experience in animals, it also raises questions about objective forms of harm and the possibility of harmless wrongs. Non-sentientists claim that the subjectivist harm concept employed in the 3Rs principles is too narrow, and that other morally relevant properties exist going beyond merely subjective harms. For example, altering an animal's *telos* [23, 24] or violating its integrity [44] is a morally significant act. Unlike sentientist harms, non-sentientist harms include harms that do not cause subjective experience in the affected individual [45]. An example of non-sentientist harm and harmless wrongdoing can also be found in the concept of dignity that is used in the Swiss Animal Welfare Act [46]. In this act, negative impacts on animals are defined to include not only pain, suffering and anxiety, but also major interference with the animal's appearance or abilities (non-sentientist harms) or excessive instrumentalization and humiliation (harmless wrongdoing).

The word "instrumentalization" here refers to ways of using animals that fail to reflect the fact that the animals have value in themselves and are more than mere means to human ends. GPD could potentially be an example of impermissible instrumentalization because it changes the abilities of animals in ways that do not take the animal's inherent value sufficiently into account. It is important to stress at this point that we are not claiming that the Swiss dignity concept is flawless or superior to the 3Rs in general. Our goal is merely to point out that, with recent developments in bioethics, the 3Rs have been rightly criticized as being too narrow in their conception of harm and assumptions about what is morally relevant.

The problem of instrumentalization arises especially clearly in connection with so-called "waste," or "surplus," animals. It is to this matter that we now turn.

10.7.3 Death and Flourishing

Russell and Burch mention the death, killing and euthanasia of animals several times in their book. However, the topic is not discussed from a moral perspective and hence the death of research animals does not seem to represent a moral problem according to the authors. Humane (i.e., pain-free) killing is compatible with their sentientist approach. As long as the animals are killed with the least harm necessary there is no moral problem. However, this position has been objected to by critics who claim that death as such, and premature death especially, frustrate flourishing and hence represent a harm by deprivation. The moral importance of killing animals, and of animal death more generally, is clear, because in animal research the number of animals that are killed without ever having been used in an experiment (e.g., waste animals, breeding animals and surplus animals) is significant. Hence, we believe that the 3Rs, in Russell and Burch's formulation of them, neglect an important aspect of animal research.

We can distinguish two different types of surplus animal: those used exclusively for breeding (this includes animals used in the generation of GPD animals) and those which, after being bred, are not used in experiments.

Although this problem was recognized a long time ago in the debate about animal disenhancement [20], it is neglected in the current discussion of GPD and animal research in general. Hence, the actual debate over GPD underestimates the complexity of ethical issues raised by animal research and modern biotechnology. It is difficult to estimate the numbers of animals required for research into GPD and the eventual generation of GPD animals, but we think it is safe to say that this number would be considerable.

To neglect the deaths of these animals is problematic, because within the sentientist approach of the 3Rs the pain and suffering of every animal should count. Furthermore, flourishing is an important element in modern interpretations of the Refinement principle. Whereas Russell and Burch were only concerned with the avoidance of negative subjective experiences, modern interpretations of Refinement include the promotion of positive experiences for research animals (e.g., play made possible by environmental enrichment).

10.7.4 Refinement for Its Own Sake?

Within Russell and Burch's hierarchical sentientism, human interests have lexical priority over animal interests. The instrumentalization of animals for the benefit of humans is morally permitted, but whether we are permitted to use an animal for another animal's sake is not clear at all. From Russell and Burch's perspective nothing can be concluded about the relations between animals themselves. Given this, we might ask: Is it morally permissible to cause pain and suffering to animals during the development of GPD to prevent future pain and suffering in GPD animals?

There seems to be no clear-cut answer to this question, even outside the framework of the 3Rs. Interestingly, however, in Switzerland a project proposing to investigate the unknown harms of a planned procedure to improve future study designs was refused a license with the explanation that "studies solely to improve the design of future experiments were unwarranted" [47]. Here, the officials decided, in effect, that Refinement for Refinement's sake is unjustified.

10.8 Conclusion

GPD, and modern biotechnology more generally, interferes with animals' lives and abilities in ways going well beyond what was possible in 1959 when the 3Rs principles were developed by Russell and Burch. Applications that could not have been foreseen then are now being developed, and ethical theories have been developed to respond to these developments. Our aim has been to investigate whether the 3Rs are able to deal with issues raised by modern animal research. To this end, we have used GPD as paradigmatic proxy for modern biotechnology—an approach that seemed especially suitable given that GPD is believed to be a promising method to further improve scientists' ability to satisfy the 3Rs. We have also analyzed the normative groundwork of the 3Rs principles and applied it to the question of whether GPD is morally permissible.

We have shown that GPD could potentially be regarded as a relative Replacement and Refinement method, while being value-neutral as regards relative Reduction. Consequently, we conclude that the 3Rs principles favor GPD, and that only absolute Reduction provides a potential basis, within the 3Rs framework, for moral misgivings about GPD.

Importantly, we have argued that the 3Rs principles, with their sentientist concept of harm, oversimplify the complexity of the ethical issues raised by animal research and modern biotechnology. This is highlighted in the discussion on GPD, where there is a tendency to reduce the ethical issue to minimizing pain and suffering, and thus to oversimplify the moral concerns there are about animal research.

In moral philosophy, when scrutinizing a problem, it is often impossible to take every aspect into consideration. One has to define the most important issues and ignore (at least, for the time being) others. However, it is important that the moral aspects one focuses on still adequately represent the initial problem. That is, formulated solutions need to do more than fit the simplified problem. They should be equally valid when applied to the initial problem. Where this is not the case, and the formulated solution only fits the simplified model, a moral-philosophical artifact is produced.

In other words, the 3Rs principle, if applied to GPD alone (as a proxy for modern biotechnology), runs the risk of generating moral artifacts. Here, the problem with the 3Rs principles lies with their excessively narrow, sentientist concept of harm. This concept of harm limits moral concerns to those about the negative subjective experiences of animals. It ignores the fact that animals can also be wronged by excessive instrumentalization, and interference with their abilities, and also the fact that animals' deaths as a harm of deprivation must also be taken into account. Finally, the 3Rs principles fail to take into consideration the numbers of animals (and their deaths) required to generate GPD animals.

Inadequacies in the 3Rs framework have been recognized previously. These have been sought to be remedied (at least, partially) by complementing the 3Rs with HBA. However, while HBA introduces some improvements—e.g., in bringing about paradigmatic change in not only the evaluation of scientific necessity, but also the proportionality of inflicted harms and expected knowledge gains—it often continues to suffer from many of the same problems as the 3Rs framework. According to the critique we mentioned above, HBA only represents an improvement to the extent that it takes non-sentient harms and harmless wrongdoing more fully into account than the 3Rs principles do—and in practice it often fails to do this to an appreciable extent.

In our interhuman relations, the avoidance of pain and suffering plays a crucial role, but our moral obligations are not simply a matter of not (subjectively) harming our fellow humans. Similarly, we might have a moral responsibility to also consider other aspects in our relations with research animals.

References

1. Russell, W. M. S., and R Burch. 1959. *The Principles of Humane Experimental Technique*. London, UK: Metheun.
2. Bayne, K., G. S. Ramachandra, E. A. Rivera, and J. Wang. 2015. The Evolution of Animal Welfare and the 3Rs in Brazil, China, and India. *Journal of the American Association for Laboratory Animal Science* 54: 181–191.

3. Grimm, H., S. Camenzind, and A. Aigner. 2016. Tierethik, 78–96.
4. Vorstenbosch, J. M. G. 2005. The Ethics of the Three Rs principle: A reconsideration. *Animal Welfare* 14: 339–345. https://doi.org/10.1017/S0962728600029675.
5. Olsson, I., N. Franco, D. Weary, and P. Sandøe. 2012. The 3 Rs principle—mind the ethical gap!
6. Grimm, H. 2013. Ethik im Kontext des Tierversuchs. 21–54. Baden-Baden, Germany: Nomos. https://doi.org/10.5771/9783845250328-21.
7. Russell, W. M. S. 2005. The Three Rs: Past, Present and Future. *Animal Welfare* 14: 279–286. https://doi.org/10.1017/S0962728600029596.
8. Ferrari, A. 2008. Genmaus & Co. Gentechnisch veränderte Tiere in der Biomedizin.
9. Camenzind, S. 2020. Instrumentalisierung: Zu einer Grundkategorie der Ethik der Mensch-Tier-Beziehung. In Instrumentalisierung. Brill mentis.
10. Lafollette, H., and N. Shanks. 1996. *Brute Science. Dilemmas of Animal Experimentation*. London, UK: Routledge.
11. Terrestrial Code Online Access. WOAH—World Organisation for Animal Health.
12. Rippe, K. P. 2009. Güterabwägungen im Tierversuchsbereich: Anmerkungen zu einem ethischen Paradigmenwechsel. ALTEX: 3–10.
13. Brody, B. A. 1998. *The Ethics of Biomedical Research: An International Perspective*. Oxford University Press. https://doi.org/10.1093/oso/9780195090079.001.0001.
14. Leonard N. 1923. *System of Ethics*. Yale University Press.
15. Salt, H. 1980. *Animals' Rights. Considered in Relation to Social Progress*. Clarks Summit: PA, USA: Society for Animal Rights.
16. Bentham, J. 1996. *An Introduction to the Principles of Morals and Legislation*. Oxford, UK: Claredon Press.
17. Schweitzer, A. 1923. *The Philosophy of Civilization*. NY, USA: Prometheus.
18. Knight, A. 2019. Critically Evaluating Animal Research. In *Animal Experimentation: Working Towards a Paradigm Change*, 321–340. Brill. https://doi.org/10.1163/9789004391192_015.
19. Thompson, P. B. 2008. The Opposite of Human Enhancement: Nanotechnology and the Blind Chicken Problem. *NanoEthics* 2: 305–316. https://doi.org/10.1007/s11569-008-0052-9.
20. Macer, D. 1989. Uncertainties About 'Painless' Animals. *Bioethics* 3: 226–235. https://doi.org/10.1111/j.1467-8519.1989.tb00342.x.

21. Comstock, G. 1992. What Obligations Have Scientists to Transgenic Animals? College Station, TX, USA: Center for Biotechnology, Policy and Ethics.
22. Comstock, G. L. 2000. On the Ethical Case Against Agricultural Biotechnology. In *Vexing Nature? On the Ethical Case Against Agricultural Biotechnology*, ed. G. L. Comstock, 139–173. Boston, MA: Springer US. https://doi.org/10.1007/978-1-4615-1397-1_5.
23. Rollin, B. 1986. On Telos and Genetic Manipulation. *Between the Species* 2. https://doi.org/10.15368/bts.1986v2n2.9.
24. Rollin, B. E. 1996. Bad Ethics, Good Ethics and the Genetic Engineering of Animals in Agriculture. *Journal of Animal Science* 74: 535–41. https://doi.org/10.2527/1996.743535x.
25. Ryder, R. 1995. Animal Genetic Engineering and Human Progress. In *Animal Genetic Engineering: of Pigs*, Oncomice and Men, P. Wheale and R. McNally, 1–16. London, UK: London: Pluto Press.
26. Sandøe, P., B. Forsman, and A. K. Hansen. 1996. Transgenic Animals: The Need for Ethical Dialogue. In *Welfare Aspects of Transgenic Animals*, ed. L. F. M. Van Zutphen and M. Van Der Meer, 90–101. Berlin, Heidelberg: Springer. https://doi.org/10.1007/978-3-642-60589-5_8.
27. Sandøe, P., P. M. Hocking, B. Förkman, K. Haldane, H. H. Kristensen, and C. Palmer. 2014. The Blind Hens' Challenge: Does it Undermine the View that Only Welfare Matters in Our Dealings with Animals? *Environmental Values* 23: 727–742. https://doi.org/10.3197/096327114X13947900181950.
28. Carlson, D. F., C. A. Lancto, B. Zang, E.-S. Kim, M. Walton, D. Oldeschulte, C. Seabury, T. S. Sonstegard, and S. C. Fahrenkrug. 2016. Production of Hornless Dairy Cattle from Genome-Edited Cell Lines. *Nature Biotechnology* 34: 479–481. https://doi.org/10.1038/nbt.3560.
29. Ferrari, A. 2006. Genetically Modified Laboratory Animals in the Name of the 3Rs? ALTEX 23: 294–307.
30. Shriver, A. 2015. Would the Elimination of the Capacity to Suffer Solve Ethical Dilemmas in Experimental Animal Research? In *Ethical Issues in Behavioral Neuroscience*, ed. G. Lee, J. Illes, and F. Ohl, 117–132. Current Topics in Behavioral Neurosciences. Berlin, Heidelberg: Springer. https://doi.org/10.1007/7854_2014_318.
31. Devolder, K., and M. Eggel. 2019. No Pain, No Gain? In Defence of Genetically Disenhancing (Most) Research Animals. *Animals* 9: 154. https://doi.org/10.3390/ani9040154.

32. Eggel, M., and S. Camenzind. 2020. Authorization of Animal Research Proposals—a Comparison of Harm Concepts in Different European Regulations. Berliner und Münchener Tierärztliche Wochenschrift.
33. Shriver, A. 2009. Knocking Out Pain in Livestock: Can Technology Succeed Where Morality has Stalled? *Neuroethics* 2: 115–124. https://doi.org/10.1007/s12152-009-9048-6.
34. Shriver, A., and E. McConnachie. 2018. Genetically Modifying Livestock for Improved Welfare: A Path Forward. *Journal of Agricultural and Environmental Ethics* 31: 161–180. https://doi.org/10.1007/s10806-018-9719-6.
35. National Research Council (US) Committee for the Update of the Guide for the Care and Use of Laboratory Animals. 2011. *Guide for the Care and Use of Laboratory Animals.* 8th ed. The National Academies Collection: Reports Funded by National Institutes of Health. Washington (DC): National Academies Press (US).
36. Redmond, C. 2019. When Is an Alternative Not an Alternative? Supporting Progress for Absolute Replacement of Animals in Science. In *Animal Experimentation: Working Towards a Paradigm Change*, 654–672. Brill. https://doi.org/10.1163/9789004391192_028.
37. Blumer, V. K. 2004. Ethische Aspekte zu Tierversuchen und das Solidaritätsprinzip, 27–29. Tierversuche in der Forschung. Bonn, Germany: Lemmens Verlag- und Mediengesellschaft.
38. Godfrey-Smith, P. 2016. *Other Minds: The Octopus, the Sea, and the Deep Origins of Consciousness.* First Edition. Farrar, Straus and Giroux.
39. Herrmann, K. 2019. Refinement on the Way Towards Replacement: Are We Doing What We Can? In *Animal Experimentation: Working Towards a Paradigm Change*, 3–64. Brill. https://doi.org/10.1163/9789004391192_002.
40. Ploner, M., H.-J. Freund, and A. Schnitzler. 1999. Pain Affect without Pain Sensation in a Patient with a Postcentral Lesion. *PAIN* 81: 211. https://doi.org/10.1016/S0304-3959(99)00012-3.
41. Majeed, M. H., M. Ubaidulhaq, A. Rugnath, and I. Eriator. 2018. Extreme Ends of Pain Sensitivity in SCN9A Mutation Variants: Case Report and Literature Review. *Innovations in Clinical Neuroscience* 15: 33–35.
42. Palmer, C. 2011. Animal Disenhancement and the Non-Identity Problem: A Response to Thompson. *NanoEthics* 5: 43–48. https://doi.org/10.1007/s11569-011-0115-1.

43. Fischer, B. 2020. In Defense of Neural Disenhancement to Promote Animal Welfare. In *Neuroethics and Nonhuman Animals*, ed. L. S. M. Johnson, A. Fenton, and A. Shriver, 135–150. Advances in Neuroethics. Cham: Springer International Publishing. https://doi.org/10.1007/978-3-030-31011-0_8.
44. Rutgers, B., and R. Heeger. 1999. Inherent Worth and Respect for Animal Integrity, 41–51. Van Gorcum: Assen, Germany: Recognising the Intrinsic Value of Animals.
45. Rippe, K. P. 2008. Ethik im Ausserhumanen Berich. In *Ethik im außerhumanen Bereich*. Paderborn, Germany: Brill mentis.
46. Animal Welfare Act (SR 455). 2017.
47. Swiss Researchers Struggle to Get Animal Experiments Approved. 2019. The Scientist Magazine®.

Chapter 11

Animals and AI: The Role of Animals in AI Research and Application—An Overview and Ethical Evaluation*

Leonie Bossert[a] and Thilo Hagendorff[b]
[a]*University of Vienna,*
Universitätsstraße 7, 1010 Vienna, Austria
[b]*University of Stuttgart*
Universitätsstraße 32, 70569 Stuttgart, Germany
leonie.bossert@univie.ac.at, thilo.hagendorff@iris.uni-stuttgart.de

11.1 Introduction

Artificial intelligence (AI) technologies and their fields of application are one of the most discussed developments of recent times. They are said to have a high potential and are associated with the desire to use these technologies for improvements in social systems like medicine, mobility, or the sciences. At the same time, they are met with criticism and fear of an alleged "loss of

*Reprinted from *Technology in Society*, 67, 101678, 2021, with permission from Elsevier.

Animals, Ethics, and Engineering: Intersections and Implications
Edited by Rosalyn W. Berne

ISBN 978-981-5129-56-4 (Hardcover), 978-1-003-65231-1 (eBook)
www.jennystanford.com

sovereignty" of humanity. In this context, a demarcation between humans and machines is intensively discussed, whereby animals are often used for comparison with humans or machines alike [1]. This phenomenon is not new to the history of philosophy. Within the academic debate around AI, however, this is almost the only context in which animals appear. In our opinion, this is a problematic shortcoming of the sciences and humanities, since animals—or animal brains, to be more specific—are not just used as a model for AI; they are also massively affected by various AI applications. Within AI research, development, and application, the role of animals is either not debated at all or it is discussed in the same manner as data usage, thereby implying that animals are nothing more than living resources. This is problematic from an ethical point of view since ethical particularities regarding living beings are ignored. We argue that it is a mistake to neglect certain ethical challenges that exist related to a "new" technology, especially a technology becoming increasingly important for most societies in the world. Therefore, we want to provide an overview of how animals are affected within AI research as well as how animals are affected by applying AI technologies on them. This chapter, however, is neither a systematic literature review nor an exhaustive in-depth analysis of the field. Due to the fact that—at least to the best of our knowledge—this chapter breaks new ground, the considerations we are presenting are still in the early stages of maturation. This chapter is the first to give a comprehensive overview and to normatively explore the interconnection of animals and AI research as well as its application. The chapter aims to provide comments and tentative explorations of a new field, sometimes speculative first steps towards possible blind spots, which AI research as well as animal ethics communities might have overlooked. We want to encourage the communities to further investigate the field more in-depth in the future.

In this overview, we highlight the profound role of animals in AI research (Section 11.2), breaking it down to basic AI research (Section 11.2.1) and AI model testing (Section 11.2.2). We discuss the ethical problems arising from this (Section 11.2.3), followed by proving that animals are not only important for researching and developing AI, but that AI is also important for surveilling animals (Section 11.3). The final section (Section 11.4) pinpoints our conclusion from this for further AI research and development. The term AI represents an umbrella term for various types of machine learning methods, especially artificial neural nets, in all chapters. Here, the term is not supposed to include symbolic AI, whole-brain emulation, neuroenhancements, human–machine interfaces and the like.

11.2 Animals in AI Research

11.2.1 Neurobiology and AI Research

AI research differentiates between symbolic and subsymbolic AI. While symbolic AI aims at representing information through symbols, deductive conclusions from existing knowledge and manipulating symbols in a way intelligent machine behavior occurs, subsymbolic AI is dedicated to inductive conclusions based on implicit rules and patterns. Symbolic AI is inspired by mathematical logic, subsymbolic AI by neuroscience. Nowadays, subsymbolic AI research and the related methods, especially deep neural networks [2], are way more prevalent than symbolic AI. Out of the two competing paradigms—"making a mind" (symbolic AI) versus "modelling a brain" (subsymbolic AI) [3]– the latter approach seems to succeed. The path that led to the success of brain-inspired, deep neural networks traces back to Frank Rosenblatt's idea of the "perceptron" [4]. Here, insights into the functionality of neurons, which were heavily dependent on animal experiments, were used to develop new techniques for information processing, where neurons are simulated by computer programs. Perceptrons receive, similar to biological neurons, particular inputs, where the sum of the inputs determines whether the perceptron meets a threshold value and gives an output in the form of 1 or 0. This principle is the very fundamentum for today's multi-layered neural nets. Especially the development of convolutional neural networks was originally inspired by single-cell recordings of mammalian visual cortex [5]. Moreover, current research in "AI attention" is inspired by the primate visual system [6]. AI attention is a methodology where artificial neural networks do not process entire images but focus on certain areas of an image instead. Not least, reinforcement learning, another method where intelligent agents act in a way that maximizes certain rewards, was partly inspired by studies in animal psychology [7]. Moreover, current research aims at mapping the layout of neurons of a cubic millimeter of rat brain, using cutting-edge brain imaging tools. Said tools are for instance infrared lasers scanning the brain of fixed living animals and microscopes to scrutinize slices of rat and mice brain, unravelling information about neural circuitries in order to be able to build more brainlike artificial neural nets [8].

All in all, research on AI and animal experiments in neuroscience have an intertwined history. Leading scientists in the field argue that a "better understanding [of] biological brains could play a vital role in building intelligent machines" and that "ideas from neuroscience will become increasingly indispensable" [9]. On the one hand, such arguments indirectly promote further animal experiments as a way of inspiring and validating new as well as existing AI algorithms and architectures, neglecting approaches that can usher more capable AI systems that do not resemble structures of brains. On the other hand, current AI research acknowledges the numerous differences between artificial and biological neural nets. Although the former was inspired by the latter, the composition of artificial neural nets embraces a number of mathematical techniques that do not possess any analogies to biological systems. First and foremost, one can refer to missing biological correlates to backpropagation in this context [10]. AI researchers recognize the stark differences between machine and human or animal learning, but at the same time they demand that machine learning shall be aligned with the latter [11]. Up to now, human or animal thinking is very different from intelligent machines [12], upholding historical claims about what computers cannot do [13]. Even sophisticated neural networks with billions of interconnections represent only a tiny cubic millimeter of brain tissue. Moreover, artificial neural networks need a lot of labeled training examples. Animals or humans, on the contrary, are capable of "few-shot learning", needing only some examples of a given type of object to be able to recognize further examples. This also works in machines [14], but only to a limited extent. Hence, scaling up current deep learning techniques will not suffice to reach or surpass cognitive and sensorimotor capabilities even of invertebrate animals like insects.

According to some AI researchers, self-learning systems should not just structurally mimic biological brains, but also encompass "innate mechanisms" or "behavior programs" of living beings which are encoded in the genome and are already present at birth [15] in order to bring machine learning closer to human or animal learning. That means AI research is again supposed to draw upon animals or animal experiments, respectively, only that this time the experiments do not focus on dissecting the brain, but the genome. Either way, progress in AI research and AI capabilities depends on the study of and experimenting on living animals—that is the assumption at least. Advanced neuroimaging or single-cell recording techniques are used in in vivo experiments, visualizing processes in the brain on different levels of magnification, ranging down to single synapses [16]. In vivo methods often mean that brains of living animals are penetrated, which comprises stereotaxis surgery, a "survivable procedure" (Carter and Shieh 2010, p. 74) where animals are held in place via clamps, after which a craniotomy (a surgical removal of a part of the cranial bone to expose the brain) is performed with a drill to access the brain. Despite brain imaging or recording techniques, further "tools" from genetic bioengineering granted more and more insight into the inner "functionality" of mammalian brains. To increase the statistical power of experiments, preferably large sample sizes, meaning a great number of animals such as rodents, are used. Despite animals having to be anesthetized via gas or pharmacological agents, it must be assumed that the animals subjected to neurobiological experiments suffer severe and prolonged post-surgery pain as well as distress from captivity and various other sources [17].

This raises significant ethical questions (cf. 2.3). These questions can be detached from methodological ones. In this vein, animal experiments can be (ethically) opposed as well as (methodologically) embraced at the same time. However, with regard to the latter, one can draw analogies to discussions on the methodical reliability of animal experiments in other research fields like clinical research [18, 19]. Nevertheless, and purely viewed from a methodological stance, a lot of progress has been made in the field of AI by drawing on ideas from neurobiology, which obtained most of its insights from experiments on animals. At the same time, however, it is an entirely speculative question whether future AI research can benefit from or be accelerated by focusing on animal experiments and whether coming AI applications resemble or differ from biological brain structures, much like airplanes resemble or differ from birds, while both are able to fly. Notwithstanding the above, when assessing the nexus of AI and animal testing, one should not withhold that AI-based systems for safety inspections of medicinal substances do in many cases yield more reliable results than animal experiments [20]. In the long run, this can lead to a drastic reduction in animal experiments due to advances in AI development and use. The same could hold true in case biologically accurate models of neurons can someday be employed to replace laboratory animals in brain research at all.

11.2.2 Animals as Benchmark for AI

Apart from dissecting animal brains in a way that helps to improve artificial neural nets, animals are used as a benchmark for AI. Is current AI "smarter" than a worm, a fly, a mouse, a monkey? By comparing animal and machine behavior, it is assumed that animals can be reduced to particular performance indicators that can also be used to measure AI capabilities. This approach may lead to a dismissal of the idea that, on the one hand, artificial intelligent software is only a mimicry of intelligence, and that, on the other hand, the salient attribute of animals is their sentience (cf. 2.3), which machines lack. Notwithstanding that, some articles state that currently no AI application is "smarter" than any animal [21]. While such a broad and abstract claim may be true, it is unquestionably possible to single out specific cognitive tasks and compare animals and artificial intelligent computer software with each other. This is done as part of the "Animal-AI-Olympics", a competition where AI researchers try to determine who is closest to achieving artificial general intelligence [22]. While renowned AI tools like Pluribus, AlphaGo, AlphaZero or Agent57 achieve superhuman levels in playing particular games [23–26], these systems possess very narrow capabilities that only succeed in regard to particular tasks

[27]. This is where animals come into play. By comparing AI systems with animals' cognitive capabilities, with their behavioral flexibility in various domains, ability for transfer learning, planning and problem-solving skills, ability to synthesize new information or with their general intelligence [28, 29], respectively, animals are serving as a suitable benchmark to measure whether AI technologies can handle complex tasks. Human intelligence, depicted as a contrast to animal intelligence, is described as a long-term goal that, according to many machine learning scientists, is currently out of reach [30]. Animal intelligence in turn is supposed to be replicable by machines. In this discourse, at least two problems emerge.

Firstly, intelligence is commonly seen as an isolated abstract problem-solving capability that an individual agent or a sophisticated machine can possess. The mechanisms of this capability have to be unraveled to be able to reproduce them using technology. However, the animal (human) mind cannot be seen as some kind of processor that "digests" data coming from the senses. Such isolated perspective reproduces problematic assumptions about (animal as well as human) intelligence, reducing it to ration instead of valuing aspects like, e.g., social intelligence. Moreover, the mind cannot be separated from its body. Intelligence cannot be reproduced by disembodied computers. This is why more and more AI researchers argue in favor of the embodiment-hypothesis, stressing that machine intelligence needs some kind of body that interacts with the world and makes experiences [31]. This approach is more open towards the technological representation of social cognition, affective responses and engagement in a social environment. Here, intelligence does not just equal problem-solving skills, where a particular dimension of intelligence is stipulated as its ontological foundation. In fact, intelligence is a rich concept, comprising the ascription of meaning to experiences, the actions by which an agent behaves in a social environment as well as concepts of self-identity [32]. In animal benchmarks for AI, on the other hand, behavioral studies, where animals' problem-solving abilities are tested by giving them a reward in the form of food, are used to measure and quantify intelligence, discounting its many dimensions beyond specialized, goal-directed, mostly vision-based cognition tasks. This involves problematic assumptions about animals and consequentially also about human intelligence—a challenge already addressed within AI ethics in regard to its impact on the *conditio humana* [33].

The second problem arising in the discourse on animals as benchmarks for AI is a rather apparent one. Animals are not depicted as a categorically different entity than machines, but as being machines themselves. This becomes evident when for example looking at the logo of the "Animal-AI-Olympics". Animals like an octopus, a fox, a pigeon or a mouse are interwoven with cogs, levers, pumps, tubes, etc. These animals are natural and artificial beings at the same time (for a detailed discussion of this aspect, cf. [34]), thus closing the gap between fully artificial intelligent machines. In this context, an entirely obsolete philosophy and perspective on animals that reaches back to Descartes' mechanistic view on animals is reawakened [35]. Needless to say, with current, ample biological and behavioral knowledge on animals at hand [36–39], it cannot be stated that they are solely Cartesian machines or non-sentient automata. But the discourse on AI is retrieving the idea that animal behavior follows mechanistic principles. In this vein, animals serve as models for a gradual approach to achieve general intelligence. However, not animals, but artificial animal models called "animats" are used to simulate animal behavior and to study animal intelligence on a "primitive level" [40, 41]. These "animats" are digital animals who are supposed to learn motion and decision-making for the purpose of satisfying their "physiological" needs in a virtual environment such as the Minecraft universe. However, "animat" intelligence, even when pretending to use animal intelligence as a template for machine intelligence, does by far not manage to grasp natural intelligence in all its complexity and dimensions.

11.2.3 Ethical Implications

As we have shown in Section 11.2.1, animal experiments can play a crucial role in the development of AI models. Obviously, many ethical issues concerning animal experiments and their normative implications exist. However, the many researchers in the AI community that directly or indirectly make recourses to animal experiments do not point at these ethical issues or discuss them critically. The difficulty herein lies in the fact that numerous animals are bred, subjected to painful procedures and killed. These procedures are connected to ethical problems, one of them being the suppression of animals' interests (cf. below). AI basic research does not possess a normative dimension, so the notion goes. Here, research on animals is purely propelled by viewing animals as carriers and suppliers of data. Animals are not seen as awe-inspiring living beings, which would still be a rather (presumably "biophile," thus nature-loving) anthropocentric perspective when it comes to direct moral value, meaning that only human beings are treated as directly morally considerable beings. Going beyond an anthropocentric perspective, many animal ethicists argue that one has to think about animals as individuals with an interest in (at least) not being harmed and hence as individuals who are directly morally considerable [42–45].

The latter perspective is "common sense" within animal ethics. Regardless of whether one argues that animals should simply not be harmed [46] or whether one grants them rights [43, 44, 47], completely ignoring their interests is considered ethically wrong. The argument revolves around the assumption that sentience is the prerequisite for having interests and that those who have interests have at least an interest in freedom from pain and feeling positive emotions. Interests thereby are seen as strong interests (related to one's own well-being), meaning that something is not only *in the* interest of someone or something (so-called weak interests), but that someone *has* this interest themself. Building animal-ethical arguments on sentience is controversially discussed since it is yet unclear which animals can be considered as sentient and which cannot. With regard to AI research, however, there is no need to go into this controversy in more detail, as the animals used within the experiments mentioned above are almost without any exception sentient vertebrates. Experiments on flies, ants or other invertebrate organisms must be discussed separately.

Sentient animals, as numerous studies have shown, also have a number of other abilities that have long been ascribed to humans only. They show interests in very particular food or toys, prefer some things or some accompaniment over others, use tools, follow (mostly short-term) plans, have different traditions among the same species and communicate [29, 48–50]. Many scientists [50, 51] and philosophers [42] take this as evidence that animals do not only possess interests in not being harmed, but are also interested in following their plans, being able to choose between various objects, live in their particular social group, etc.

Linking the described ethical assumptions to AI research, one can see that on the one hand, the access to animal brains made artificial neural nets possible, but on the other hand, the "animal in the machine" is neither perceived nor acknowledged within AI research. Animals are seen as batches of data. While especially mice and rats, who are widely used in brain research, are typically carriers of various (partly problematic) meanings—they are, for instance, heavily associated with filth and disease, in AI research, they are nothing more than parts of larger machines that produce data, cleansed of all "filth" [52]. Obviously, with such a perspective on animals, one does not consider their interests or their individuality. Animal testing has a long tradition of ethical considerations with regard to laboratory animal welfare, though [53–55]. Animal welfare is often significantly connected to the quality of the derived research data. The literature on animal welfare discusses mostly the implementation of the Three Rs (reduce, replace, refine), but also other issues like training of veterinarians, the harmonization of efforts in animal care, possibilities of environmental enrichment for laboratory animals, etc. Obviously, these discussions have ramifications for animal testing in the AI field. Thus, we will not claim that animal-based research in the AI context is devoid of all ethical considerations. However, research publications in the AI field typically do not trace back the origin of the animal data they are using—which would comprise to mention stereotaxis surgery and other measures to access brains of living animals. Hence, it holds true that animals are part of AI research, but just in the form of data which is stripped off of ethical implications surrounding animal interests.

From a non-anthropocentric perspective, which sees animals as directly morally considerable individuals, not considering their interests must be challenged and rethought. On the one hand, animal experiments are assessed critically within animal ethics, since in most cases they are associated with pain, distress and captivity. On the other hand, it is ethically problematic to ignore the fact that the animals used for AI research as laboratory animals are sentient beings, not non- sentient resources. We think that when taking ethics seriously, (at least) the latter aspect must be critically evaluated from an anthropocentric perspective as well. A technical, ethical evaluation which does not take particular ethical challenges under consideration is falling short of being complete. We argue that ethical aspects regarding animals could also be taken into account when discussing "AI ethics" for demonstrating *all* consequences appearing within this field of technology. This applies to both, the scientific as well as the public debate about AI.

11.3 Applying AI on Animals

We have already taken a look at the crucial role animals or, to be more precise, their brains, play in the development of more advanced machine learning architectures. Moreover, animals or their cognitive capabilities, respectively, are used to test the performances of AI applications. But animals play a further role in the AI world as they are also controlled by AI. AI builds the foundation of modern surveillance technologies, where sensors produce too much data for human observers to sift through. These surveillance technologies are not solely directed towards humans [56], but also and maybe even more excessively towards animals, especially farmed animals. This (the so-called "animal-industrial complex" [57]) is quite a big business. One has to bear in mind that the biomass of all existing terrestrial vertebrates is distributed among 3% wild animals, 32% percent humans and 65% farmed animals [58]. Further, 60% of all mammals are farmed animals, 36% are human beings and only 4% are wild animals [59]. Similarly, only 30% of all birds live in the wild. The remaining 70% are kept as farmed animals [59]. The confinement of billions of farmed animals requires technology that can be employed to monitor, restrict and suppress the animal's agency (for details on animal agency cf. [60]). Farmed animals, sentient vertebrates like their "wild ancestors," possess the above-mentioned capabilities which lead animal ethicists to demand taking them morally directly into account. The "animal-industrial complex" of metal cages, breeding technologies, hormones, fences and surveillance systems testifies that animals are not able to perform these capabilities within the complex humans forced them into (despite one being able to observe forms of resistance of animals to the industrial practices [61, 62]).

This has recently been reinforced by AI technologies [63–68]. Animals are analyzed via camera vision-based systems for herd "management" purposes like tracking, fattening control, or facial recognition [69]. Advanced closed-circuit television (CCTV) systems are used to automatically detect and track pigs, allowing the automatic detection of potential health problems without human observation. Robots drive through poultry houses and collect "layers", which are floor eggs not placed in the designated nest boxes. Acoustic data analytic tools listen to "poultry" birdcalls and process the audio files so that they can be used as early warning systems to detect diseases. Echo-sounder technologies at aquacultures can detect whether caged fish are swimming near the water surface and use this as a signal for feeding times. Feeding sensors in combination with predictive analytics are used for "performance" prediction. Support Vector Machine classifiers in combination with sensor devices are used to detect early symptoms of lameness in cows. On cockroach farms, AI systems analyze sensor data for features like the insects' temperature, food intake and humidity in order to ensure an optimal growth environment. Robot vehicles driving around in factory farms for "broilers" are supposed to move the animals around to improve the feed conversion ratio. In "processing plants", intelligent robot arms equipped with cameras are used for "poultry deboning", cutting animal bodies apart with greater efficiency than human butchers. Besides AI applications that directly aim at animal control, various machine learning methods are used to optimize fodder compositions, design vaccines and to analyze animals' genetics.

All mentioned AI-based animal surveillance technologies have the purpose of increasing efficiency, paring down personnel and increasing farmers' profit margins in a conflicting relationship between care and control. In some cases, for instance in early detection of lameness in cows [67], measures to avoid herdsmen's financial losses can go hand in hand with actual animal welfare. However, the envisaged steps to increase animal welfare uncritically operate in the wider framework of factory farming, which is not questioned as such. Notwithstanding that, apart from surveillance measures directed at people, animal surveillance in general is—with very few exceptions [70–72]—neither a public nor academic topic. The same holds true for AI-based methods for animal control in factory farms, respectively. Since farmed animals are not seen as subjects but as material resources that must effectively be bred, fattened and processed, they can also not be seen as subjects of or subjected to surveillance. Animals—in contrast to humans subjected to surveillance measures—do not change their behavior due to surveillance, though. They lack awareness of being surveilled and are since not succumbed to chilling effects [73]. Nevertheless, surveillance and AI tools pave the way for not just semi-automated, but fully automated factory farms. Automation then leads to even greater emotional distances between the human perception of animal suffering. Thus, the application of AI on animals is carried out for the benefits of humans, not the former.

Many ethical problems are associated with these applications. First of all, an evaluation of an increased automation is ethically important when dealing with sentient animals. We assume two entangled problems to be the greatest ones from an animal ethics point of view. The first problem is creating an even greater distance between humans and the agricultural animal industry through almost full automation. This makes the violent treatment of animals even further invisible to even more people, perpetuating it rather than modifying it towards a more respectful treatment of animals [74]. Here, one could object that in an automated factory farm individual cases of misconduct of animal abuse cannot be brushed aside by blaming and simply replacing minimum wage workers, at least not when the cases are clearly documented by (AI-based) surveillance techniques. Hence, the company responsible for the affected facilities would have to take responsibility. However, in practice, many cases of intra-industry documentations of severe animal welfare and abuse issues exist [75], but these do very rarely have real consequences in favor of the animals [76]. Similar outcomes would have to be expected in case AI-based systems reveal that animal welfare metrics are not met. We assume that it is questionable whether such incidents reach the public at all, and if so, whether the public's reaction to that would cause perceptible changes for the affected animals.

The second problem is related to the first one, being the commodification of animals per se and the fact that here too their interests are not only lowered compared to human interests, but simply ignored. Moreover, one could ask the highly philosophical question if farmed animals like pigs and cows should have a right to privacy [77], despite not being aware of the surveillance. One could also go into detail about ethical claims against further increasing the efficiency logic in agriculture, because among others it leads to t only monetary output being of importance when dealing with other living beings, viewing them as resources only. All of these issues are highly relevant. Within this paper, however, we do not want to discuss the very concrete ethical challenges associated with the application of AI technologies on animals. Instead, we want to recall our observation that within current debates on AI (ethics), the role of animals is being ignored. This also holds true for practices in which AI is applied on animals as we described them. We argue that AI practitioners as well as public media and policy debates should include these aspects when discussing the role of AI and the consequences of AI for societies in order to be complete in their assessments.

This also holds true for debates about nature conservation. AI surveillance technologies for animals are not only used in agricultural systems, but also for nature conservation, to be more precise in species conservation. Examples are projects like "BirdVision", "Wildbook" or the "Elephant Listening Project". "BirdVision" aims at constructing windmills equipped with AI, which recognizes nearby flying birds and thus regulates the rotor blades speed, lowering the speed or stopping the rotation of blades when a bird is coming too close [78]. The AI used in the "Wildbook" project is able to identify individual wild animals of particular species, e.g., giraffes, sperm whales or humpback whales [79]. Thereby, the technology helps with setting up wildlife inventory and with collecting data of animal migration, supporting researchers and nature conservationists in order to improve their knowledge about the life of these animals. Moreover, the "Elephant Listening Project" shall protect elephants from poaching [79]. Rangers are informed about the elephant's location or promptly informed about alarming noises such as shots, so that poachers can be tracked down at the scene.

Beyond that, AI technologies are also developed for helping stray and shelter animals [80]. Profiling shelter animals should lead to faster and happier adoptions, since AI is assumed to help find the individual animal suiting human preferences best and thus prevents abandonment of the animal. Two other promises regarding shelter animals are the earlier detection of malnutrition through a "malnutrition screening algorithm" as well as AI technologies predicting disease breakout in environments where many animals are living in confined space with an increased likelihood of disease outbreaks. Moreover, researchers hope that AI will make it possible to decode animal language in a way that linguistic communication with animals becomes possible via AI. This is currently being researched, but whether this "Dr. Dolittle" dream will come true remains uncertain [81].

This type of AI technology use is undoubtedly beneficial to animal individuals and contributes to their protection and well-being. However, it should not be overlooked that in regard to the conservation practice, in which the above-described use of these technologies is embedded, a practice is addressed, which focuses on species protection instead of protection of individuals. From an animal ethics perspective, it would be desirable to add a dimension to nature conservation practice in which the interests of affected wild animals are deemed relevant as well [82]. This would at least provide the option of using AI technologies for the benefit of specific animal individuals—as it is the case with helping individual animals within shelters—in contrast to their current use in the animal industry.

11.4 Conclusion

We have shown that animals, respectively animal brains, play a crucial role in the development of AI technologies. This development is partly rooted in research results gained in animal experiments (cf. Section 11.2.1) and tests its technologies by comparing them to animal capabilities (cf. Section 11.2.2). Within this research, animals are seen as mere resources and providers of data only. Animals' interests are ignored, which we pointed out as an ethical issue since animals in this context are sentient vertebrates who are able to feel pain and pleasure. They also possess many capabilities similar to human ones (cf. Section 11.2.3). But animals are not only used for (the development of) AI technologies, AI technologies are also used for (surveilling) animals (cf. Section 11.3). This happens in the animal-industrial complex, where surveillance technologies are used for perfecting the outcome of animal industrial farming while reducing the amount of work done by humans, striving for an automated animal industry and perpetuating a violent treatment of animals, which we also pointed out to be ethically problematic. In addition, AI surveillance technologies are used for species conservation when tracking or identifying individual wild animals, when constructing, e.g., windmills equipped with AI to protect birds, for helping stray or shelter animals with a faster adoption by human caretakers, etc.

As the use of AI in these domains illustrates, the technology can be used in ways that can benefit individual animals, protect them and educate people about their way of life (for example, by studying the previously unknown migration routes of whales). Increasing this practice would be desirable from an animal ethics perspective—but also from a biophile anthropocentric view since wild animal and species protection is a highly important endeavor for many humans. On the other hand, changing human-animal-relationships in regard to farmed animals can also benefit humans in manifold ways (e.g., through building more sustainable futures [83]). AI may help with achieving that when it is used alternatively to currently existing and predominant ways. For this to happen, however, it is necessary to take animal interests seriously at the research and developmental stage of AI technologies. This is a long way to go and AI researchers and developers need to be trained not only in regard to ethical human issues, but also regarding other living beings. We believe that this is a path worth taking and that it will ultimately benefit animals *and* people.

References

1. Suddendorf, T. 2013. *THE GAP — The Science of What Separates Us From Other Animals*. New York.
2. LeCun, Y., Y. Bengio, and G. Hinton. 2015. Deep learning. *Nature* 521. Nature Publishing Group: 436–444. https://doi.org/10.1038/nature14539.
3. Dreyfus, H. L., and S. E. Dreyfus. 1991. Making a mind versus modelling the brain: artificial intelligence back at the branchpoint. In *Understanding the Artificial: On the Future Shape of Artificial Intelligence*, ed. M. Negrotti, 33–54. Artificial Intelligence and Society. London: Springer. https://doi.org/10.1007/978-1-4471-1776-6_3.
4. Rosenblatt, F. 1958. The perceptron: a probabilistic model for information storage and organization in the brain. *Psychological Review* 65: 386–408. https://doi.org/10.1037/h0042519.
5. Hubel, D. H., and T. N. Wiesel. 1959. Receptive fields of single neurones in the cat's striate cortex. *The Journal of Physiology* 148: 574–591. https://doi.org/10.1113/jphysiol.1959.sp006308.
6. Moore, T., and M. Zirnsak. 2017. Neural mechanisms of selective visual attention. *Annual Review of Psychology* 68: 47–72. https://doi.org/10.1146/annurev-psych-122414-033400.
7. Sutton, R. S., and A. G. Barto. 2018. *Reinforcement Learning: An Introduction*. Cambridge, MA, USA: A Bradford Book.
8. Strickland, E. 2020. AI Designers Find Inspiration in Rat Brains - IEEE Spectrum. https://spectrum.ieee.org/ai-designers-find-inspiration-in-rat-brains. Accessed July 30.
9. Hassabis, D., D. Kumaran, C. Summerfield, and M. Botvinick. 2017. Neuroscience-Inspired Artificial Intelligence. *Neuron* 95: 245–258. https://doi.org/10.1016/j.neuron.2017.06.011.
10. Bengio, Y., D.-H. Lee, J. Bornschein, T. Mesnard, and Z. Lin. 2016. Towards biologically plausible deep learning. arXiv. https://doi.org/10.48550/arXiv.1502.04156.
11. Lake, B. M., T. D. Ullman, J. B. Tenenbaum, and S. J. Gershman. 2017. Building machines that learn and think like people. *Behavioral and Brain Sciences* 40: e253. https://doi.org/10.1017/S0140525X16001837.
12. Hagendorff, T., and K. Wezel. 2020. 15 challenges for AI: or what AI (currently) can't do. *AI & SOCIETY* 35: 355–365. https://doi.org/10.1007/s00146-019-00886-y.

13. Dreyfus, H. L. 2014. What Computers Can't Do: A Critique of Artificial Reason. In *20. What Computers Can't Do: A Critique of Artificial Reason*, 90–100. Princeton University Press. https://doi.org/10.1515/9781400848393-021.
14. Fabi, S., S. Otte, J. G. Wiese, and M. V. Butz. 2020. investigating efficient learning and compositionality in generative LSTM networks. arXiv. https://doi.org/10.48550/arXiv.2004.07754.
15. Zador, A. M. 2019. A critique of pure learning and what artificial neural networks can learn from animal brains. *Nature Communications* 10: 3770. https://doi.org/10.1038/s41467-019-11786-6.
16. Nishiyama, Jun, and Ryohei Yasuda. 2015. Biochemical computation for spine structural plasticity. *Neuron* 87: 63–75. https://doi.org/10.1016/j.neuron.2015.05.043.
17. Morgan, K. N., and C. T. Tromborg. 2007. Sources of stress in captivity. *Applied Animal Behaviour Science* 102: 262–302. https://doi.org/10.1016/j.applanim.2006.05.032.
18. van der Worp, H. B., D. W. Howells, E. S. Sena, M. J. Porritt, S. Rewell, V. O'Collins, and M. R. Macleod. 2010. Can animal models of disease reliably inform human studies? *PLoS Medicine* 7: e1000245. https://doi.org/10.1371/journal.pmed.1000245.
19. Greek, R., and A. Menache. 2013. Systematic reviews of animal models: methodology versus epistemology. *International Journal of Medical Sciences* 10: 206–221. https://doi.org/10.7150/ijms.5529.
20. Luechtefeld, T., D. Marsh, C. Rowlands, and T. Hartung. 2018. Machine learning of toxicological big data enables read-across structure activity relationships (RASAR) outperforming animal test reproducibility. *Toxicological Sciences: An Official Journal of the Society of Toxicology* 165: 198–212. https://doi.org/10.1093/toxsci/kfy152.
21. Colagrossi, M. 2018. What animals is A.I. currently smarter than? https://bigthink.com/technology-innovation/what-animals-is-ai-currently-smarter-than/
22. Crosby, M., B. Beyret, and M. Halina. 2019. The Animal-AI Olympics. *Nature Machine Intelligence* 1: 257–257. https://doi.org/10.1038/s42256-019-0050-3.
23. Badia, A. P., B. Piot, S. Kapturowski, P. Sprechmann, A. Vitvitskyi, D. Guo, and C. Blundell. 2020. Agent57: Outperforming the Atari Human Benchmark. arXiv. https://doi.org/10.48550/arXiv.2003.13350.
24. Silver, D., J. Schrittwieser, K. Simonyan, I. Antonoglou, A. Huang, A. Guez, T. Hubert, et al. 2017. Mastering the game of go without human knowledge. *Nature* 550: 354–359. https://doi.org/10.1038/nature24270.

25. Vinyals, O., I. Babuschkin, W. M. Czarnecki, M. Mathieu, A. Dudzik, J. Chung, D. H. Choi, et al. 2019. Grandmaster level in StarCraft II using multi-agent reinforcement learning. *Nature* 575: 350–354. https://doi.org/10.1038/s41586-019-1724-z.

26. Brown, N., and T. Sandholm. 2019. Superhuman AI for multiplayer poker. *Science* 365: 885–890. https://doi.org/10.1126/science.aay2400.

27. Marcus, G. 2018. Innateness, AlphaZero, and Artificial Intelligence. arXiv.

28. Hauser, M. D. 2000. *Wild Minds: What Animals Really Think.* New York: Henry Holt.

29. Bekoff, M., and J. Pierce. 2010. *Wild Justice: The Moral Lives of Animals.* Illustrated ed. Chicago: University of Chicago Press.

30. Brockman, J., ed. 2019. *Possible Minds: Twenty-Five Ways of Looking at AI.* 1 ed. New York: Penguin Press.

31. Mitchell, M. 2019. *Artificial Intelligence: A Guide for Thinking Humans.* 1 ed. New York: Farrar, Straus and Giroux.

32. Clocksin, W. F. 2003. Artificial intelligence and the future. *Philosophical Transactions. Series A, Mathematical, Physical, and Engineering Sciences* 361: 1721–1748. https://doi.org/10.1098/rsta.2003.1232.

33. Nida-Rümelin, J., and N. Weidenfeld. 2018. *Digitaler Humanismus: Eine Ethik für das Zeitalter der Künstlichen Intelligenz.* 4th ed. München: Piper.

34. Karafyllis, N. C. 2015. Growth of biofacts: the real thing or metaphor? In *Growth of Biofacts: The Real Thing or Metaphor?*, 141–152. transcript Verlag. https://doi.org/10.1515/9783839405185-011.

35. Descartes, R. 2003. *Treatise of Man.* Prometheus Books.

36. Bateson, P. 1991. Assessment of pain in animals. *Animal Behaviour* 42: 827–839. https://doi.org/10.1016/S0003-3472(05)80127-7.

37. Gentle, M. J. 1992. Pain in birds. *Animal Welfare* 1: 235–247. https://doi.org/10.1017/S0962728600015189.

38. Stevens, C. W. 2011. Analgesia in amphibians: preclinical studies and clinical applications. *The Veterinary Clinics of North America. Exotic Animal Practice* 14: 33–44. https://doi.org/10.1016/j.cvex.2010.09.007.

39. Webster, J. 2006. Animal sentience and animal welfare: what is it to them and what is it to us? *Applied Animal Behaviour Science* 100: 1–3. https://doi.org/10.1016/j.applanim.2006.05.012.

40. Strannegård, C., N. Svangård, and B. R. Steunebrink. 2017. The Animat Path to Artificial General Intelligence. In *Proceedings of IJCAI-17 Workshop on Architectures for Generality & Autonomy.*

41. Wilson, S. W. 1986. Knowledge growth in an artificial animal. In *Adaptive and Learning Systems*, ed. K. S. Narendra, 255–264. Boston, MA: Springer US. https://doi.org/10.1007/978-1-4757-1895-9_18.
42. Nussbaum, M. C. 2006. *Frontiers of Justice: Disability, Nationality, Species Membership*. Harvard University Press. https://doi.org/10.2307/j.ctv1c7zftw.
43. Palmer, C. 2010. *Animal Ethics in Context*. New York: Columbia University Press.
44. Regan, T. 2004. *The Case for Animal Rights*. 1 ed., Updated with a New Preface. Berkeley: University of California Press.
45. Singer, P. 2011. *Practical Ethics*. 3rd ed. New York: Cambridge University Press.
46. Thompson, P. B. 2007. Ethics on the frontiers of livestock science. *Redesigning Animal Agriculture: the Challenge of the 21st Century*. CABI Books: 30–45. https://doi.org/10.1079/9781845932237.0030.
47. Donaldson, S., and W. Kymlicka. 2013. *Zoopolis: Eine Politische Theorie der Tierrechte*. Translated by J. Schulte. 1st ed. Berlin: Suhrkamp Verlag.
48. Bradshaw, G. A. 2010. *Elephants on the Edge: What Animals Teach Us about Humanity*. New Haven, Conn.: Yale University Press.
49. Meijer, E. 2019. *Animal Languages: Revealing the Secret Conversations of the Living World*. London: John Murray.
50. Sommer, V. 2014. Zoologie: Von "Mensch und Tier" zu "Menschen und andere Tiere." In, 359–386. https://doi.org/10.14361/transcript.9783839425183.359.
51. Waal, F. B. M. de. 1997. *Good Natured: The Origins of Right and Wrong in Humans and Other Animals*. Revised ed. Cambridge, Mass.: Harvard University Press.
52. Birke, L. 2003. Who–or what–are the rats (and mice) in the laboratory. *Society & Animals: Social Scientific Studies of the Human Experience of Other Animals* 11: 207–224. https://doi.org/10.1163/156853003322773023.
53. Prescott, M. J., and K. Lidster. 2017. Improving quality of science through better animal welfare: the NC3Rs strategy. *Lab Animal* 46: 152–156. https://doi.org/10.1038/laban.1217.
54. Bayne, K., and P. V. Turner, ed. 2013. *Laboratory Animal Welfare*. 1st ed. Amsterdam; Boston: Academic Press.
55. The principles of humane experimental technique. 1960. *Medical Journal of Australia* 1: 500–500. https://doi.org/10.5694/j.1326-5377.1960.tb73127.x.

56. Andersen, R. 2020. The panopticon is already here. *The Atlantic*, July 29.
57. Noske, B. 1989. *Humans and Other Animals: Beyond the Boundaries of Anthropology*. Pluto Press.
58. Pachauri, R. 2008. Eating our future: the environmental impact of industrial animal agriculture. In.
59. Bar-On, Y. M., R. Phillips, and R. Milo. 2018. The biomass distribution on Earth. *Proceedings of the National Academy of Sciences* 115: 6506–6511. https://doi.org/10.1073/pnas.1711842115.
60. McFarland, S., and R. Hediger. 2009. Animals and agency: an interdisciplinary exploration. In *Animals and Agency*. Brill.
61. Amir, F. 2018. *Schwein und Zeit. Tiere, Politik, Revolte*. 1., Originalveröffentlichung edition. Hamburg: Edition Nautilus GmbH.
62. Hribal, J., and J. St Clair. 2011. *Fear of the Animal Planet: The Hidden History of Animal Resistance*. Edinburgh: AK Press.
63. Connolly, A. 2019. Is artificial intelligence right for poultry production?, WATTAgNet. *WATTPoultry.com*. August 22.
64. Stine, L. 2019. French poultry tech startup Tibot Technologies raises €3m seed round for health-boosting robot. *AgFunderNews*. October 16.
65. Zhang, L., H. Gray, X. Ye, L. Collins, and N. Allinson. 2019. Automatic individual pig detection and tracking in pig farms. *Sensors* 19: 1188. https://doi.org/10.3390/s19051188.
66. Chong, Z. 2020. AI helps grow 6 billion roaches at China's largest breeding site. *CNET*. https://www.cnet.com/science/ai-helps-grow-6b-roaches-at-chinas-largest-breeding-facility/. Accessed August 10.
67. Haladjian, J., Z. Hodaie, S. Nüske, and B. Brügge. 2017. Gait anomaly detection in dairy cattle. In *Proceedings of the Fourth International Conference on Animal-Computer Interaction*, 1–8. ACI '17. New York, NY, USA: Association for Computing Machinery. https://doi.org/10.1145/3152130.3152135.
68. Carpio, F., A. Jukan, A. I. M. Sanchez, N. Amla, and N. Kemper. 2017. Beyond production indicators: a novel smart farming application and system for animal welfare. In *Proceedings of the Fourth International Conference on Animal-Computer Interaction*, 1–11. ACI '17. New York, NY, USA: Association for Computing Machinery. https://doi.org/10.1145/3152130.3152140.
69. Hansen, M. F., M. L. Smith, L. N. Smith, M. G. Salter, E. M. Baxter, M. Farish, and B. Grieve. 2018. Towards on-farm pig face recognition using convolutional neural networks. *Computers in Industry* 98: 145–152. https://doi.org/10.1016/j.compind.2018.02.016.

70. Braverman, I. 2012. Zooveillance: Foucault goes to the zoo. SSRN scholarly paper. Rochester, NY.

71. Donaldson, A.. 2012. Surveillance and non-humans. In. Routledge. https://doi.org/10.4324/9780203814949.ch3_1_d.

72. Bolinski, I. 2015. Animal tagging: Zur Ubiquität der smar ten Tiere. In *Animal Tagging: Zur Ubiquität der smar ten Tiere*, 205–222. transcript Verlag. https://doi.org/10.1515/9783839430460-009.

73. Staben, J. 2016. *Der Abschreckungseffekt auf die Grundrechtsausübung*. 1st ed. Vol. 6. Internet Und Gesellschaft. Tübingen: Mohr Siebeck.

74. Was sich am Fleisch entscheidet: Über die politische Bedeutung von Tieren: Hagendorff, Thilo: Amazon.de: Books. 2024. https://www.amazon.de/Was-sich-Fleisch-entscheidet-politische/dp/3963172371. Accessed February 9.

75. Große, E. 2017. Untersuchungen an verendeten/getöteten Schweinen in Verarbeitungsbetrieben für tierische Nebenprodukte—Deutsche Digitale Bibliothek.

76. Bülte, J. 2018. Zur faktischen Straflosigkeit institutionalisierter Agrarkriminalität 2018: 36–56.

77. Pepper, A. 2020. Glass panels and peepholes: nonhuman animals and the right to privacy. *Pacific Philosophical Quarterly* 101: 628–650. https://doi.org/10.1111/papq.12329.

78. Jungblut S.-I. 2020. BirdVision: Intelligenter Vogelschutz an Windrädern. *Digital for Good |RESET.ORG.*

79. Jungblut S.-I. 2019. Künstliche Intelligenz—Können wir mit Rechenleistung unseren Planeten retten? *Digital for Good | RESET.ORG.*

80. What can artificial intelligence do for stray animals | the vet futurist. 2024. https://vetfuturist.com/what-can-artificial-intelligence-do-stray-animals. Accessed February 9.

81. #author.fullName}. 2021. Dr Dolittle machines: how AI is helping us talk to the animals. *New Scientist*. https://www.newscientist.com/article/mg24833133-500-dr-dolittle-machines-how-ai-is-helping-us-talk-to-the-animals/. Accessed June 8.

82. Aitken, G. 2018. *A New Approach to Conservation*. Loundon: Routledge.

83. Poore, J., and T. Nemecek. 2018. Reducing food's environmental impacts through producers and consumers. *Science* 360: 987–992. https://doi.org/10.1126/science.aaq0216.

Chapter 12

Touchscreen Uses for Animals: Teaching Discrimination Tasks

Jennifer Taylor-O'Connor,[a,b] **Corinne C. Renguette,**[c] **and Sierra Ouimette**[d]

[a] *Department of Veterinary Medicine and Biomedical Sciences, Texas A&M University, College Station, Texas, USA*
[b] *Parrot Kindergarten, Florida, USA*
[c] *Department of Technology Leadership and Communication, Indiana University Purdue University Indianapolis, Indianapolis, Indiana, USA*
[d] *Pet Kindergarten, Texas A&M University, College Station, Texas, USA*
jen@parrotkindergarten.com

12.1 Background

The fields of engineering and technology have influenced research with non-human animals (hereafter, "animals") in a variety of ways including designing technology and systems to monitor animal welfare [1], studying animal locomotion across species to inform biomechanical design [2], improving farm and zoo animal enrichment and welfare [3, 4], and enhancing enrichment and welfare for animals in research labs [5] and private homes [6], to name only a few examples. Whether to advance current technologies

Animals, Ethics, and Engineering: Intersections and Implications
Edited by Rosalyn W. Berne

ISBN 978-981-5129-56-4 (Hardcover), 978-1-003-65231-1 (eBook)
www.jennystanford.com

for humans or to design new technologies for both human and non-human animals, engineers and scientists have the potential to contribute to the welfare and lives of animals and the responsibility to stay informed about protocols for ethical use of technology in research with and for animals.

One technological option for conducting research with and for animals is to use touchscreen devices. Touchscreen devices such as tablets and computer screens are easily accessible and offer flexibility with use, making them ideal for research data collection as well as for mental stimulation and enrichment for animals in the care of humans [7, 8].

Discrimination tasks are frequently used in science and for enrichment, as animals can interact with multiple targets on the screen (e.g., match-to-sample, choose between visual stimuli). In this chapter, we review research conducted using touchscreen devices in animal interactions to begin framing best practices for training animals on discrimination tasks. In line with the ethical practices of the field of Animal-Computer Interaction (ACI), we will focus most especially on animal-centered methods that allow animals the most agency and the least aversive consequences in the learning process. Finally, we will provide our recommendations for best practices across species.

This work has the potential to continue to inform the engineers and scientists who design and develop technology that can be used with animals. Advances in this area can help researchers inform their own design practice and provide quality and ethical enrichment while also collecting usable data.

12.1.1 Uses for Touchscreen Devices in Animal Research

Touchscreen devices may improve animal welfare and aid our understanding of animal behavior. Captive animals often suffer from stereotypies due to lack of mental enrichment. To this end, touchscreen devices can be used to improve animal psychological wellbeing with cognitive enrichment interactions such as use with primates [8–10], dogs and wolves [11, 12], cats [13], macaques [7], pigs [14], parrots [6], and whales [15]. Touchscreen devices can enable animals to signal requests with picture symbols to enhance choice and autonomy in captive environments by pressing a symbol to ask for specific foods or activities [6, 16–18]. Nevertheless, touchscreen devices are most widely used in behavior and cognition research in labs, sanctuaries, and natural environments [19–23].

12.1.2 Discrimination Tasks and Touchscreen Devices

For decades, researchers have advanced our knowledge of animal cognition and behavior with technology-based testing. A leading use of touchscreen testing with animals is discrimination learning, or teaching animals to distinguish similarities and differences between multiple stimuli [21]. Moreover, discrimination tests can form the basis of both research data collection and mental enrichment for animals in captivity.

Researchers frequently use discrimination tasks to test animal behavior and cognition [24]. Discrimination behavior is a natural behavior in animals, especially as it relates to food [24] and to evaluate danger [24]. Discrimination tasks on touchscreen devices have also been used by comparative psychology researchers to explore animals' cognitive abilities, such as inferential reasoning by exclusion in birds such as pigeons [25], cockatoos [26] and African greys [27], as well as canines [25], and a variety of primates including marmosets [28] among others [29].

Results from discrimination tasks are generally also compared against human performance on similar tasks [25]. Researchers have tested dogs' ability to discriminate emotional expressions on human faces [30] and pigeons on whether they can recognize familiar human faces [25, 30]. Discrimination tasks include investigation of animals' abilities to understand abstract concepts, often through matching-to-sample tests [31] and quantity discrimination skills [24, 32]. More recently, researchers have used discrimination tasks on touchscreen devices to explore animal preferences in order to improve welfare for captive animals [16–18].

12.1.3 Ethics in Animal-Computer Interactions

While research generally benefits humans and science, it is not always beneficial for the animals being tested, and indeed, can be very harmful for animals [33, 34]. The field of animal-computer interaction is leading improvements in the ethical treatment of captive animals, especially with regard to their participation in research. Mancini's manifesto [35] outlines an ethical framework for the use of technology in animal research. Under this framework, animal participants are viewed as stakeholder-users in every aspect of the interaction, and the research and use of technology should benefit the animal-user. This framework is opening new pathways for much research in agency-oriented, animal-centered technology design and interactions. Innovations have emerged including tangible and physical, haptic and wearable, olfactory, tracking technology, and screen technology [36]. A wide field of animal-centered technology now exists across species, including technology for farm animals such as chickens [37] and cows [38]; for domestic animals such as dogs [39] and cats [13, 40]; and for undomesticated animals such as elephants [41] and deer [42]. These studies are designed around the animal as a stakeholder and are in alignment with these ethical principles.

Within Mancini's Manifesto framework [35], touchscreen research has also evolved from forced task engagement with punishment and coercion (in rats, for example, [43]) to more voluntary, animal-centered cognition and behavior studies [44, 45]. As ethical considerations require specific benefits to individual user-animals, research tasks have also often been turned into touchscreen-based "games" to collect data from animals in a manner that utilizes consent and autonomy while providing enrichment. Such research has been conducted using dogs and wolves [11], cockatoos [16], red-footed tortoises [46], ring-tailed lemurs [47], bears [18, 24], great apes [24, 45], macaques [7], and whales [15], among others.

Control is a basic need for both humans and animals [45]. Furthermore, research demonstrates that enhancing the degree of control an animal has over their environment can lead to beneficial effects on their behavior [48–50]. Mancini suggests considering a more animal-centered approach [33–35] that includes animals' *mediated and contingent* consent to technology interactions so they have choice and control in the research process. *Mediated* consent is when someone determines for the animal whether or not they give consent. *Contingent* consent includes an animal's voluntary interaction with technology [39] without coercive food restrictions, as for example, birds flying into the lab for tests as demonstrated by cockatoos [26] or pigeons [44]; or primates interacting with computers within their enclosures as demonstrated by baboons [10] and white-face sakis [51, 52]. *Actual* consent comprises an animal's use of symbols to indicate their choice to participate in research [16, 53].

12.1.4 Review of the Literature

We surveyed the literature for examples of ethical touchscreen discrimination training in a variety of species; most particularly, we selected those that offered the greatest amount of agency to the animal participants. Here we will provide an overview of literature on touchscreen training used in discrimination tasks for animals. Next, we will recommend best practices based both on the literature and on our experiences, and we will present an expanded framework for training animals on discrimination tasks, as this forms the basis for much research and is also foundational for touchscreen-based enrichment and symbol communication training.

A difficulty in reviewing the literature in this area is that training methods that include a degree of punishment and coercion, contrary to ethical principles, are pervasive in animal research, and most particularly with rodents, pigeons, and some primates. Positive punishment is the application of aversive stimuli to decrease a behavior, for instance, turning on a bright light when the animal chooses the incorrect answer. Likewise, coercion may be defined as compelling animals to engage in a behavior to meet a basic need such as food or water, after being deprived of such. Conspecific housing and socialization opportunities are also crucial for animal welfare [44, 54]. Although there is no unanimous agreement defining agency and consent in animals [33, 34, 55], and as choice and control are highly favored by animals [45], our review comprises research that most closely conformed to the ethical principles of agency and voluntary participation [33] with low or absent applications of punishment, weight control, and coercion, and studies that included access to conspecific housing and socialization.

In some species, there were no ideal studies. Studies on rats and pigeons, for example, almost always use weight control [43, 44], so we selected what we felt was the best among competing options [56]. This review also included research on discrimination task training methods and outcomes in animals. Recent publications reflect the increasing ethical standards in research related to animal-computer interaction. For example, ethical touchscreen dog training methodologies have been created and published [57, 58]. There have also been many improvements in socializing animals during testing. For example, one study included group-housed pigeons that were able to access cognition task tests from within an aviary where they could come and go [44]; another study included enclosures that allowed animals to voluntarily enter adjoining chambers for research tasks [59]. Although we want to acknowledge that these types of studies are valuable to the advancement of the field, they were not included here, as we selected only research that included discrimination training and testing results.

Ultimately, we assigned ratings to papers concerning the particular species based on (1) minimal (or absent) weight control, (2) minimal (or absent) punitive measures, (3) voluntary (to the extent possible) involvement by the animal, and (4) chances for social interaction or housing with conspecifics.

As depicted in Table 12.1, numerous animals have undergone training to engage with touchscreen devices for discrimination tasks. In this section, we will discuss touchscreen training methods and protocols, device selection, shaping protocols, and outcomes for discrimination tasks. Because many of the studies do not fully align with our articulated application of acceptable ethical principles, we provide a framework with guidance and best practices for future work that includes agency and voluntary touchscreen interaction in research across species.

12.2 Touchscreen Protocols in Research

As touchscreen computers become more prevalent and interest in animal-computer interaction grows, researchers, facility personnel and human caregivers of captive animals have begun developing

Table 12.1 Selected discrimination training studies and their alignment with ethical principles

Species	Study	Discrimination task	Weight control	Incorrect response	Voluntary initiation	Socialization
Baboons	[10]	Two alternative forced choice and match-to sample	N/A	3 sec pause, green screen	Yes	Yes
Bears	[18, 24]	• Food preferences • Quantity • Enrichment preference	N/A N/A N/A	• No beep, dark screen 0.5 sec • Buzz tone and 750 ms pause • No tone, no food reward	No Yes Yes	No Yes Yes
Bonobos	[60]	Pictures for emotional Stroop task	N/A	3500 ms pause	Yes	Yes
Cockatoos	[16]	Pexigrams for preference corroboration	N/A	None	Yes	Yes
Dogs/Wolves	[11]	Images	N/A	3 sec pause, red screen, buzz sound	No	Yes
Gorillas	[24]	Quantity	N/A	Buzz sound and 750 ms pause	Yes	Yes
Horses	[61]	Shapes with curvatures; vertical and horizontal lines	N/A	None	Yes	Yes

(*Continued*)

Table 12.1 (*Continued*)

Species	Study	Discrimination task	Weight control	Incorrect response	Voluntary initiation	Sociali-zation
Macaques	[7]	Three-choice enrichment preference	N/A	Different tone, no reward	Yes	Yes
Marmosets	[28]	Images	Food provided after tests	3 sec pause and red screen	Yes	Yes
Rats/mice	[56]	Multiple match-to-sample	85% free-feeding	5 sec pause and extinguished house light	No	Yes
Red footed tortoises	[46]	Spatial tasks	N/A	3 sec pause	No	Yes
Whales	[15]	Match-to-sample	N/A	No whistle, no reward	Partial (invitation with 95% engagement participation rate)	No

methods to train animals to interact with touchscreen devices, e.g., [11, 16]. However, they often rely on trial-and-error methods [62], persisting with the same training even after experiencing undesired outcomes. The trial-and-error method involves waiting for a correct result to reinforce the desired behavior; however, these methods can lead to inefficiencies and frustration for both the humans and the animals [62]. Nevertheless, insights from literature focusing on touchscreen device usage can be valuable for exploring common themes across species in discrimination task training on touchscreen devices.

In this section, we outline some touchscreen and discrimination task training protocols found in the literature shown in Table 12.1. This research covers aspects from environmental setup and device specifications to habituation and training procedures. This information spans multiple varied animal species. If we were unable to find information for an animal, the table is marked N/A. This is meant to be a sampling (rather than a comprehensive list) to help create a generalizable training framework for discrimination tasks on touchscreen devices.

12.2.1 Automated Training vs. Social Learning

Researchers employ three main operant learning methods to train animals to use technology: automated training systems (AUTs), social learning methods, and a hybrid of both.

AUTs [63] do not include human presence during learning or testing (except as needed to transfer the animals to the testing area). AUTs entail animals interacting with technology through a sequential series of repetitive motor tasks (usually paired with food reinforcers for correct responses) progressing in difficulty based on certain pre-set benchmarks [10]. Advantages include a reduced human workload, location in the animals' habitats, the potential for environmental enrichment, and a standardized learning experience for more reliable comparative behavior and cognition data [10].

Social learning methods (SLMs) include human training and positive interaction during any portion of the training process. Human interaction can involve encouragement and pointing [26], Model-Rival interactions [64], and positive-reinforcement using a bridge and food rewards [11, 16]. Although social learning methods require a greater human workload [10], the benefits may produce faster results [65], better meet ethical standards for animal-computer interactions, and may be better suited to ensure each animal participant receives training at a pace that matches its individual needs. Additionally, technology stations can be placed with automatic food dispensers in the animals' home environment after training is complete for continued data collection and environmental enrichment [8, 66].

Animals trained through SLMs require fewer repetitions to learn [65], while animals trained through AUTs can require thousands of trials to achieve similar results [10]. Given that novelty, choice, and control are crucial for animal psychological well-being [8, 9], the extensive repetitions needed for animals to grasp even basic computer-interaction training tasks, especially in cases like traditional rodent and pigeon studies where animals are hungry or thirsty [10], seemingly violate the ethical standards related to animals' voluntary consent.

When the goal of the animal-computer interaction is psychological wellbeing and cognitive challenge, social learning provides the benefit of a tailored learning experience to each animal as an individual, utilizing training techniques that best meet the needs for the specific animal-user. Many non-human primates are initially trained by humans to use computer technology [8]. After the basic skills have been taught, the animals have been observed playing "computer games" in their home environments, adapting to challenges and novelty, and experiencing cognitive enrichment [8].

A *hybrid* method includes the presence of a human who is not involved in the training or testing except to offer food reinforcers. Table 12.2 presents a list of methods used to train specific species from the literature in Table 12.1.

12.2.2 Environmental Setup

Before using touchscreen devices for enrichment games, communication, or research, protocols around screen interactions should be considered. These interactions exhibit similar setup and training for environmental factors such as time of day, lighting conditions,

Table 12.2 Methods for discrimination task training in different species

Species	Training method
Baboons [10]	Automated learning
Bears	
• [18]	• Hybrid
• [24]	• Hybrid
• [67]	• Automated learning
Bonobos [60]	Automated learning
Cockatoo [16]	Social
Dogs/wolves [11]	Social
Gorillas [24]	Hybrid
Horses [61]	Social
Macaques [7]	Automated
Marmosets [28]	Automated learning
Orangutans [45]	Hybrid
Pigeons [21]	Automated learning
Rat/mice [56]	Automated learning
Red footed tortoise [46]	Social
Whales [15]	Social

auditory stimuli, and a selection of conditioned reinforcers. What follows are some examples of touchscreen environmental practices found in the literature.

12.2.3 Time of Day

Time of day is an important consideration in discrimination task training. Fatigue can impact an animal's engagement in the training session. For instance, rats have a reverse-light cycle, so nocturnal training may yield better results than training during the daytime., Additionally, aligning training sessions to complement the animal's feeding schedule and peak activity cycles may yield positive results. Table 12.3 outlines some examples of the times of day the animals in the selected literature review were trained and tested.

12.2.4 Lighting

Lighting conditions can impact the activity level of the animal, and can potentially serve as aversive stimuli [68, 69]. Species have

Table 12.3 Time of day examples for touchscreen discrimination training

Animal	Time of day
Baboons [10]	24 h availability
Bears	
• [18]	• 1:00 p.m.
• [24]	• N/A
• [67]	• Sampling from 10 a.m.–12 p.m. and 2–4 p.m.
Bonobos [60]	12p.m.–3p.m.
Cockatoos [16]	8–11a.m. or 4–7p.m.
Dogs/wolves [11]	N/A
Gorillas [24]	7a.m.–10a.m.
Horses [61]	N/A
Macaques [7]	8a.m.–6p.m. (lights off from 7p.m.–7a.m.)
Marmosets [28]	11a.m.–3p.m.
Orangutans [45]	N/A
Pigeons [21]	N/A
Rats/mice [56]	9a.m.–5p.m.; reverse day/night (lights are off from 7a.m.–7p.m.)
Red footed tortoises [46]	9a.m.–5p.m.
Whales [15]	N/A

different reactions and preferences around lighting conditions, for example, rats and mice are active in the dark and sleep and hide in the light, while most parrots go to sleep in the dark. Table 12.4 presents a survey of the lighting conditions in the literature.

12.2.5 Sounds as Conditioned Reinforcers

Sudden noises or undesirable sounds may startle animals, creating an aversive association with the apparatus and decreasing training accuracy. Some sounds, such as a human's voice offering encouragement when they are present during training or quiet soothing music, may become a positive reinforcer (e.g., [11]).

Auditory tones or beeps may be intentionally conditioned as reinforcers. Conditioning a sound begins by introducing it at a low-level and pairing the sound with a food reinforcer to create a conditioned positive reinforcer. Here we survey the sounds as

Table 12.4 Lighting condition examples for touchscreen discrimination training

Animal	Brightness
Baboons [60]	N/A
Bears [18, 24]	N/A
Bonobos [10]	N/A
Cockatoos [16]	Standard indoor house lighting
Dogs/Wolves [11]	N/A
Gorillas [24]	N/A
Horses [61]	Outdoor
Macaques [7]	Natural and artificial, outdoor and indoor
Marmosets [28]	Between 40–30 LUX in cage
Orangutans [45]	N/A
Pigeons [21]	2-watt house light in chamber
Rats/mice [56]	3-watt house light in chamber
Red footed tortoises [46]	25-watt fluorescent tube lights
Whales [15]	N/A

conditioned reinforcers in the selected touchscreen discrimination training literature.

12.2.6 Devices, Hardware, Software, and Feeding Mechanisms

Perhaps the single most important consideration influencing an animal's performance in discrimination tasks is the compatibility of the device to the animal's biology and physiology. While the devices used for each species can vary widely, common considerations include the relative size of the device to the size of the animal, a safe device enclosure or case, a responsive screen, software adapted for the animal's biology, and a reward system tailored to the specific animal's preferences. This might also include a feeding mechanism that responds to successful performance on tasks.

Special attention should be placed on choosing the screen size of the device (and its enclosure) in relation to the animal's body (e.g., [12, 70]), the safety provided by the enclosure to prevent harm to both the animal and the device, and the size of the pixel stimuli on

Table 12.5 Sounds as conditioned reinforcers for touchscreen discrimination training

Animal	Shaping	Correct answer	Incorrect answer
Baboons [10]	None	None	None
Bears			
• [18]	• Pleasant auditory beep	• Pleasant auditory beep	• None
• [24]	• Melodic tone	• Melodic tone	• Buzzer
• [67]	• Tone	• Tone	• Different Tone
Bonobos [60]	N/A	Tone	None
Cockatoos [16]	"Good" (verbal)	"Good" (verbal)	None
Dogs/wolves [11]	Clicker for shaping; beep for food delivery	Tone	Buzzer
Gorillas [24]	Melodic tone	Melodic tone	Buzzer
Horses [61]	Beep and chime	Chime	Buzzer
Macaques [7]	Acoustic Tone ('ding')	Acoustic tone ('ding')	Different tone ('error')
Marmosets [28]	Acoustic tone	Melodic tone	None
Orangutans [45]	Auditory bridge	Auditory bridge	None
Pigeons [21]	Acoustic signal	Acoustic signal	None
Rats/mice [56]	Tone	Tone	None
Red footed tortoises [46]	None	None	None
Whales [15]	Whistle	Whistle	None

the screen. Hardware and software may be developed internally or available open source/commercially so it can be adapted as needed. The most important consideration is adapting the setup to the species' physiology and biology [12, 70, 71], as well as Umwelt—the sensory world as it is experienced by an animal [72].

12.2.7 Screen Size

Screen size can impact training success. A screen that is too small demands advanced motor skills as the target stimuli size may be too small, and a screen that is too large demands more body movement to navigate between discrimination stimuli. For example, research has studied touchscreen interaction development and target stimuli size and spacing for dogs [12] and rodents [43]. Ideally, an animal should be able to center in front of the screen and reach multiple options from a stationary position. Table 12.6 presents the available pixel and screen sizes used for specific animals during discrimination training.

12.2.8 Hardware and Screen Sensitivity

In addition to the issue of the optimal screen size, screens must have an appropriate level of sensitivity so they will react to the animal contact. Often, a more sensitive screen will require less motor-skill training from the animal to learn to engage in the discrimination tasks. Screens may utilize infrared panels or capacitive touch or may need modification for a given species. For example, some researchers made significant modifications by creating a simulated touchscreen using a computer and a monitor behind glass since whales are very large and remain underwater during training [15]. This required the trainers to be able to see where the whale was touching, but opened up many new underwater opportunities. This review includes a list of the hardware mentioned in the selected literature that was either acquired through commercial laboratory suppliers, built internally, or commercially available.

Just as hardware needs to be tailored to specific species, software also can be adapted for animals [33, 39] (Hirskyj-Douglas and Read, 2018), (Mancini, 2017). Important considerations in software

Table 12.6 Screen sizes and pixel sizes for touchscreen discrimination training

Animal	Symbol pixel size	Screen size
Baboons [10]	300 × 300	29 cm × 29 cm
Bonobos [60]	N/A	56 cm
Bears • [18] • [24] • [67]	 • N/A • 2268 × 2419 • N/A	 • 48 cm • 48 cm • N/A
Dogs/Wolves [11]	200–300	16 cm × 27 cm
Cockatoos [16]	144 × 144	26 cm
Gorillas [24]	2268 × 2419	48 cm
Horses [61]	300 × 300 initially, then reduced	107 cm
Macaques [7]	N/A	38 cm
Marmosets [28]	200 × 200	27 cm × 15 cm
Orangutans [45]	N/A	53 cm
Pigeons [21]	72 × 72	38 cm
Rats [56]	208 × 208	38 cm
Red footed tortoises [46]	94 × 94	38 cm
Whales [15]	N/A	94 cm × 53 cm

include sensitivity to input (touch) as well as spacing and size of stimuli. Like hardware, the software below includes internally developed, open-source, and commercially available apps.

12.2.9 Food Reward Device

In automated learning systems [10, 21, 56], feeding mechanisms affix to the device itself and in hybrid and social learning systems [16, 24, 45] the food reward is delivered by the trainer. The location can assist with increasing animals' task success. For mice and rats, the feeder device is often located in the back of the chamber so they must run to the back [56]. They then have space to observe the new stimuli on the screen as they approach. Researchers with dogs relocated the device from the front to the side as the dogs were too distracted by the initial location and did not pay attention to the screen [11]. In hybrid and social learning systems [24, 60] the food is systematically hand-delivered, generally over the touchscreen and

Table 12.7 Hardware and software samples for touchscreen discrimination training

Animal	Apparatus	Hardware/software name
Baboons [10]	19-in. LCD touch monitor	Developed using eprime language
Bears • [18] • [24] • [67]	• Panasonic CF-19 Toughbook or Asus Aspire One Laptop projected wo a 19" VarTech Armorall capacitive touchscreen monitor affixed to a rolling LCD cart • Panasonic Toughbook laptop computer and a 19-inch, Vartech Armorall Capacitive, touch-screen monitor welded to the front of a rolling computer cart • Infrared screen (Touhmonitor, CarrollTouch touch technology, dual Serial/USB touch interface	• Real practice or inquisit 3.0 (millisecond.com) • Visual basic for windows • N/A
Bonobos [60]	22" ViewSonic TD2220 touch-sensitive monitor (1920 × 1080 resolution) connected to the researcher's (DWL) computer on an adjustable cart	OpenSesame
Cockatoos [16]	Samsung Galaxy Tab A, children's plastic case	CommBoard by Shmootz, Ltd.
Dogs/wolves [11]	15" laptop, TFT computer monitor mounted behind infrared touchframe	Vienna comparative cognition technology
Gorillas [24]	Panasonic Toughbook Laptop CF19 computer and 1900 VarTech Armorall capacitive touch-screen monitor welded inside a rolling LCD panel cart encased with top and sides	Visual basic for windows
Horses [61]	42-inch LCD touchscreen monitor controlled by the Surface Acoustic Wave system on a portable stand	N/A

(Continued)

Table 12.7 (*Continued*)

Animal	Apparatus	Hardware/software name
Macaques [7]	eXperimental Behavioral Instrument, MacBook Air (MacOS Catalina 10.15), microcontroller (Teensy 3.5), 15" touchscreen, 2 peristaltic pumps, custom-made mouthpiece 24 cm from the touchscreen	Multiple-choice interface programmed by MWorks (http://mworks-project.org) v 0.10, 1 July 2023
Marmosets [28]	Mini laptop PC with touch-sensitive screen (Model SC, Japan), acrylic case	Microsoft direct X
Orangutans [45]	HP Desktop 260-A129 PC and a 21" color PC computer monitor with a Keytec Magic Touch touchscreen	Custom touchscreen-delivered program written in Java
Pigeons [21]	38 cm thin film transistor display 5 cm behind the pecking key (key with a switch that registers a single peck)	N/A
Rats [56]	Flat screen monitor, Craft Data Ltd.	ELO Touchsystems, displaze
Red footed tortoises [46]	15-inch IR "CarrollTouch" touchframe (Model D87587-001, 15 in., without frame)	Vienna Comparative Cognition Technology Cognition Lab 1.9
Whales [15]	43" Monitor (I-O DATA, EX-LD4K432DB) 40 cm behind an underwater acrylic window—monitor controlled by a laptop	Custom program written in Microsoft Visual Studio 2019 (Microsoft Corporation, Redmond, WA, USA)

Table 12.8 Feeding mechanisms or reinforcer devices for touchscreen discrimination training

Animal	Food reward device
Baboons [60]	Homemade food dispenser
Bears	
• [18]	• PVC chute
• [24]	• Trainer/human
• [67]	• Pellet dispenser
Bonobos [60]	Manual via PVC tube
Cockatoos [16]	Trainer/human
Dogs/wolves [11]	30-treat capacity food dispenser, Messerli Research Treat and Train from PetSafe Pet Tutor
Gorillas [24]	Trainer/human
Horses [61]	Tray below screen
Macaques [7]	Mouthpiece and pump for liquid delivery
Marmosets [28]	Unavailable
Orangutans [45]	Trainer/human
Pigeons [30]	Grain feeder
Rats [56]	Pellet receptacle attached to a 45 mg pellet dispenser
Red footed tortoises [46]	Polyoxymethylene plate (diameter 47 cm) with 16 small, indented placeholders
Whales [15]	Trainer/human

centered to avoid side bias development. Reinforcers should also be adapted to the devices. For instance, [11] noted that soft dog food got stuck in the feeder device.

12.3 Introducing Animals to Touchscreen Learning Systems

The literature suggests that introducing animals to touchscreen learning often begins with identifying animal preferences to find the best food reinforcers, delivering them during optimal times when the animal is most motivated to learn, and optimizing the environment for successful learning. Here we will explore the food reinforcers used in the reviewed literature, the kinds of weight

Table 12.9 Food reinforcer practices for touchscreen discrimination training

Animal	Reinforcer
Baboons [10]	Grains of dry wheat
Bears • [18] • [24] • [67]	• Almonds, biscuits, raisins, grapes • Fruits, vegetables, honey roasted peanuts, banana pellets, dried banana chips • yogurt-covered raisins and wafer cookies • Flavored pellets
Bonobos [10]	DK zoological trainings biscuit
Cockatoos [16]	Sunflower seeds, almonds, pine nuts
Dogs/wolves [11]	Liver sausage in a tube, cream cheese, peanut butter, normal dry or semi-dry food, cubed hard cheese, hot dogs
Gorillas [24]	Regular zoo diet of fruits, vegetables, and Mazuri Primate Leafeater Chow
Horses [61]	Carrots
Macaques [7]	Diluted juice
Marmosets [28]	Gum powder, soybean powder, marshmallow, sponge cake, egg cookie, nuts, sweet potato, cheese
Orangutans [45]	Preferred food reward (i.e., blueberry)
Pigeons [21]	0.2g grain
Red footed tortoise [46]	Mushrooms, strawberries, and sweet corn among other preferred fruits and vegetables
Rats/mice [56]	Cheerios, yogurt, meat baby food, Formula P, research diets
Whales [15]	Fish

control often used in studies (1), animal touchscreen contact style, session frequency, and methods to habituate animals to the screen.

12.3.1 Weight Control and Reinforcers

Many (but not all) automated learning systems have included some form of weight control [21, 56], and both automated and social learning systems surveyed here use food reinforcers when the animal answers correctly. Here we will review the kinds of reinforcers and weight control (1) for each species surveyed [8].

Reinforcers can vary from animal to animal, from day to day, and from minute to minute and can be chosen by the animal [60]. Some systems use multiple reinforcers [16, 24]. (See Step 2 in Section 12.5, "Recommended Best Practices" for more information on reinforcers).

12.3.2 Contact Style

Contact style refers to the way an animal is best able to 'touch' or make contact with the screen and is often determined by the animal's body type and movement style. For example, birds, mice, rats, and dogs tend to use a nose-poke or beak/tongue-touch on the screen, and marmosets are taught to touch with their fingers. Whales, on the other hand, use their rostrum (similar to a beak, it is the crest on their upper jaw). Table 12.10 lists contact style for animals from the selected articles.

Table 12.10 Touchscreen contact styles for discrimination training

Animal	Contact style
Baboons [10]	Hand
Bears	
• [18]	• Nose, rarely claw
• [24]	• Claw and nose
• [67]	• Tongue, claw (rarely)
Bonobos [60]	Hand
Cockatoos [16]	Tongue/beak touch
Dogs/wolves [11]	Nose, paw
Gorillas [24]	Hand
Horses [61]	Nose
Macaques [7]	Unspecified
Marmosets [28]	Hand
Orangutans [45]	Wooden dowel
Pigeons [59]	Peck at pecking key (key with a switch that registers a single peck)
Rats/mice [59]	Nose poke, paw
Red footed tortoises [46]	Touch with nose
Whales [15]	Touch with rostrum (crest on upper jaw)

Table 12.11 Session duration and frequency samples for touchscreen discrimination training

Animal	Session duration/frequency
Baboons [10]	24 h availability
Bears • [18] • [24] • [67]	• 3 × /week, 10–15 min per day for 4–5 sessions per day • 3 × /week, 8–10 sessions/day • 1 × /week, February through October, varied by voluntary engagement
Bonobos [60]	4–5 × /week/varied by voluntary engagement
Cockatoos [16]	5 × /week, 12 min/session
Dogs/wolves [11]	1 × /week, 30 min/session
Gorillas [24]	3 × /week, 4–5 sessions/day
Horses [61]	4–7 sessions/day
Macaques [7]	Available 8–10 h per day on 6 consecutive Saturdays
Marmosets [28]	5 × /week, 60–90 min/day
Orangutans [45]	3–4 × /week, 1–2 sessions/day
Pigeons [21]	5 × /week, 40 trials
Rats/mice [56]	30 min/session
Red footed tortoises [46]	5 × /week, up to 30 min/session
Whales [15]	~5 sec between trials, 12 trials=1 session, 1–4 sessions per day, 5 min between sessions

12.3.3 Session Duration and Frequency

The frequency and length of training sessions will vary widely depending on the individual animals and their preferences should be considered. Ideally, voluntary engagement by the animal may also impact the duration and frequency.

Table 12.11 lists some of the training schedules in the selected articles, although they may not reflect animals' duration capacity or preference as training frequency may reflect trainers' and experimenters' availability.

12.3.4 Habituation to the Device

To avoid eliciting a fear response, repeating the exposure, or habituating the animal to the device so it becomes a regular part of

Table 12.12 Samples of habituation methods for touchscreen discrimination training

Animal	Habituation method
Baboons [10]	None needed; in housing area
Bears [18, 24, 67]	None needed; in housing area
Bonobos [60]	None needed; in housing area
Cockatoos [16]	1 session of desensitization training and introduction of touchscreen using food reinforcers
Dogs/wolves [11]	Luring first to touchscreen then to treat dispenser; adapting to feeder device sound
Gorillas [24]	None needed; in housing area
Horses [61]	None needed; in housing area
Marmosets [28]	None needed; in housing area
Orangutans [45]	None needed; in housing area
Pigeons [21]	Grain dispenser beneath tablet illuminated for 3 sec when grain is delivered, several sessions in testing enclosure
Rats/mice [56]	15 min habituation in testing enclosure with pellet dispenser
Red footed tortoises [46]	30 min habituation in testing enclosure with food in dispenser until all food eaten for 3 sessions
Whales [15]	Progressively increased complexity of target behavior beginning with real objects and then adding the screen

their routine is important so the setting remains highly enriching. Devices that are contained in the animal's housing areas do not typically need separate habituation procedures and, therefore, did not include details regarding habituation except to mention that the devices were placed in the housing area, which indicates that they were already part of the animal's regular routine. More about habituation is included in Section 12.5, "Recommended Best Practices."

12.4 Discrimination Training

Discrimination training can be broken into multiple phases and is a somewhat recursive, iterative process, not necessarily a linear process where testing will confirm success or prompt additional preparation. Before discrimination training, shaping breaks the more complex tasks into smaller tasks and helps condition them to prepare the animal for the training. Shaping can then be followed by more complex discrimination tasks and testing, additional shaping, more discrimination tasks, and re-testing. Table 12.13 lists examples of the initiation and steps of the shaping portion of the training found in the selected research.

12.4.1 Targeting and Stimuli Discrimination Training

Targeting (or target training) is the practice of training an animal to touch or indicate by looking at a desired target, such as a toy, a dowel, or a dot on a screen, for example. Targeting is a precursor to discrimination training [73]. Table 12.14 lists some of the examples of discrimination training from the literature including how the test was initiated, highlights from the process, what the animal experienced with correct vs. incorrect responses, and what performance success looked like in that particular study.

12.4.2 Stop Criteria

It is crucial to establish criteria to determine the point at which the session or experiment will end. Ideally, this would be driven at least partially by the animal, although motivation can wane over time and may be reinvigorated by high-value reinforcers, e.g., [15]). Some studies require a given number of trials or sessions. Table 12.15 displays examples of the criteria for stopping a given portion of the training or study. In some instances (marked with an asterisk) the engagement was voluntary or some of the subjects did not advance.

12.5 Recommended Best Practices

Frameworks for touchscreen interactions have been developed with some species utilizing positive reinforcement and force-free training. For instance, [70] drew from Don Norman's model of "How Humans Do Things" in his work with canines. The seven-stage model

Table 12.13 Shaping for touchscreen discrimination training

Animal	**Shaping training**
Baboons [10]	Initiation: Presenting hand through arm-port resulted in showing texture on screen and food reward Step 1: Presenting hand with sequentially 1 sec, 15 sec and 40 sec delay or touching screen resulted in food reward Step 2: Presenting hand displayed 300 × 300 stimulus for 15 sec; touching resulted in food reward
Bears • [18] • [24] • [67]	• Initiation: Stationed in front of touchscreen, rewarded for touching blank screen. Step 1: two-alternative forced choice task to reinforce blue circle (not yellow) to represent "like". • Unpublished • Initiation: Approaching and interacting in any manner. Step 1: Dots of honey placed on screen to encourage interaction. A touch anywhere on screen (red) resulted in food reward. Step 2: Smaller red space to target for food reward. Continued to shape to specific stimuli.
Bonobos [60]	Initiation: N/A Step 1: Shaped to touch small targets on screen Step 2: N/A
Cockatoos [16]	Initiation: Accept an offered treat Step 1: Touch anywhere on the screen for a low-value reinforcer Step 2: Touch a target dot on the screen for a high-value reinforcer
Dogs/wolves [11]	Initiation: Lured to position in front of touchscreen with food reward Step 1: Rewarded for touching screen Step 2: Rewarded for touching circle on screen
Gorillas [24]	Unpublished
Horses [61]	Initiation: Nose-touch with successive approximation to physical circle Step 1: Nose-touch with successive approximation to touchscreen circle with physical angle bar guides Step 2: Nose-touch and pixel size adjustment based on horse behavior

(Continued)

Table 12.13 *(Continued)*

Animal	Shaping training
Macaques [7]	Initiation: Voluntarily touched the task bar at bottom of screen Steps not present—previous experience using the same device for other tasks
Marmosets [28]	Initiation: N/A Step 1: Touched a mashed banana on the screen atop a colorful square stimulus with visual and auditory reinforcers Step 2: Received a reward for touching the square without the mashed banana
Orangutans [45]	N/A
Pigeons [21]	Initiation: Voluntarily entered test area Step 1: Trained to touch a disk image centered on the screen for a reward Step 2: Trained to touch a disk image in different locations on the screen for a reward
Rats/mice [56]	Initiation: Nose touch to initiate contact before each trial in pre-training Step 1: Trained to collect treats in a 30 sec interval schedule Step 2: Required to respond to a single stimulus on the screen to receive food reward
Red footed tortoises [46]	Initiation: Presentation of strawberry on screen combined with manually shaping approximations toward screen Step 1: Manual training to touch stimulus on screen Step 2: Manual training to multiple touches of different stimuli on screen
Whales [15]	Initiation: Previously trained hand signal invited whale to come to window, showed sample, and waited for whale to approach window before beginning. Step 1: Target trained to touch the window with rostrum for a whistle (used in prior training) and food reward. Then using various match-to-sample objects Step 2: Physical objects first, then photos, screens, geometric shapes, mirrored shapes

Table 12.14 General discrimination training examples

Animal	Test initiation	Stimuli process	Correct response	Incorrect response	Performance success
Baboons [10]	Hand presentation	Presentation of S+ and S- images on touch-screen with random left-right variations	Food reward	3 sec pause	80% accuracy in 200 consecutive trials
Bonobos [60]	N/A	Selecting the box outline colored red (S+) vs. blue (S-)	S+ resulted in a food reward; additional reward every 5th correct response and additional rewards for completing trials	S- resulted in 3500 ms pause	80% accuracy was achieved within 1–2 24-trial sessions
Bears • [18] • [24] • [67]	Touching screen with nose for food reward N/A Approach apparatus and activate screen	Stimuli were photos of preferred and disliked foods associated with colored circles Two arrays of differing dot quantities (S+ and S-), larger quantity assigned S+ Presentation of S+ and S- images on touch-screen with random left-right variations	Beep and food reward Food reward Food reward and tone	No beep, no food reward, 500 ms pause Buzz sound and 750 ms pause Brief timeout and different tone	80% accuracy was achieved for preferred food selection, varied for more complex tasks 5 out of 6 consecutive trials in a session of 20; 80% accuracy within 4 consecutive 20-trial sessions Selected touchscreen over traditional enrichment

(*Continued*)

Table 12.14 (*Continued*)

Animal	Test initiation	Stimuli process	Correct response	Incorrect response	Performance success
Cockatoos [16]	Accepts a food reward	Two stimuli conditioned one at a time with 4–7 repetitions followed by discrimination tasks	Audible bridge; food reward provided	2 sec pause; trial repeated	70% accuracy or better
Dogs/wolves [11]	Licks paste off screen; feeding device offers reward	Presentation of S+ and S- images on touchscreen with random left-right variations	Tone bridge; food reward	Buzz sound; 3 sec pause	66% accuracy in 30 trials
Gorillas [24]	N/A	Two arrays of differing dot quantities (S+ and S-), larger quantity assigned S+	Food reward	Buzz sound and 750 ms pause	80% accuracy within 30 sessions of 80-trials
Horses [61]	N/A	Two black circles presented on the screen with S+ and S- size-dependent	Food reward	None	Maintaining accuracy rate above 70% across 12 sessions of 12 trials
Macaques [7]	Animal initiates by touching taskbar at bottom of screen	Multiple choice with three stimuli: two motor tasks and one picture-viewing task	Acoustic feedback and fluid (diluted juice) reward for motor tasks, no reward for picture task	Different acoustic feedback and no fluid reward	Selections analyzed to determine preference

Marmosets [28]	Red square presented at the center of the screen until touched	S+ and S- pictures randomly distributed on screen in a two-choice force task	Food reward, 3 sec interval	5 sec interval, repeat and recue	90% in 100 trials.
Orangutans [45]	Approaching testing apparatus	Preference for free or forced choice analyzed via S+ (forced choice) and S- (free choice)	S+ resulted in an immediate food reward	S- resulted in several additional free-choice selection keys and then same food reward	Selections analyzed to determine preference for forced or free choice
Pigeons [21]	Pecking at screen for a food reward	Multiple matching: 10 on screen, 5+, 5- peck to select stimulus image Go/no go: one stimulus behind pecking key (key with a switch that registers a single peck)	Stimulus disappears when pecked; reinforcer given after all 5 are gone; food reward provided	Nothing	Completing 50 40-trial training sessions
Rats/mice [56]	Nose poke to screen for reward	Two stimuli presented an equal # of times per session, alternate sides, a given stimulus does not appear on the same side more than 3 × consecutively.	Tone, light, pellet	5 sec pause + repeat until correct	percent of correct choices per session of 100 trials

(*Continued*)

Table 12.14 (*Continued*)

Animal	Test initiation	Stimuli process	Correct response	Incorrect response	Performance success
Red footed tortoises [46]	Touches triangle to begin test	Two blue circles presented on the screen with S+ and S- location-dependent and counter-balanced	Stimuli disappear and reward presented	3 sec pause with a blank screen; trial repeated	10 completed 20-block trials above chance
Whales [15]	Trainer hand signal invites whale to testing area. Whale touches simulated screen with tip of rostrum to initiate test	Match-to-sample stimuli in underwater window.	Whistle and food reward	No whistle, no reward	80% accuracy, for real objects, photographs, and monitor presentations, 88% for normal geometric patterns, 75% for mirror images of geometric patterns

Table 12.15 Examples of criteria for stopping the training sessions

Animal	Discrimination test stop criteria
Baboons [10]	N/A
Bonobos [60]	24 trials or animal disengagement*
Bears	
• [18]	• Upon consuming "an appropriate number of rewards as determined by the care staff" (p. 5)
• [24]	• 80 trials or animal disengagement*
• [67]	• Autonomous—available for at-will engagement once per week
Cockatoos [16]	N/A
Dogs/wolves [11]	30 trials
Gorillas [24]	80 trials or animal disengagement*
Horses [61]	12 or 24 trials per session or animal disengagement*
Macaques [7]	Autonomous ~8 h sessions available for at-will engagement (6 Saturdays)
Marmosets [28]	60 min; failure to meet next level criteria within 5000 trials
Orangutans [45]	64 trials or animal disengagement for 10 min
Pigeons [21]	40 trials
Rats/mice [56]	100 trials or 60 min
Red Footed Tortoises [46]	20 trials or animal disengagement*
Whales [15]	1–4 sessions of 12 trials per session per day

of action includes two phases: an execution phase and an evaluation phase. Within those phases are the following steps:

- Execution Phase: (1) Identify a goal; (2) formulate a plan of action; (3) specify each stage in sequence; and (4) perform the action sequence.
- Evaluation Phase: (5) Perceive the consequential state of the world; (6) interpret that perception; and (7) compare that interpretation to the desired goal.

This two-phase framework can be used to evaluate how well a system works and where it may fall short in a cycle [70]. This is illustrated in his work on touchscreen interfaces [12, 70], in which training with back-chaining (training the last step of a complex task first) on a canine touchscreen task resulted in dogs performing the sequence without distraction, while luring them (guiding with treats) through the steps caused the dogs to show distraction between the steps and without fluid execution of the task (2). Using this framework to assess the complex task, it seems that between steps three and four, the animal seemed confused about the expected action at that particular stage of the sequence and the point at which the sequence of actions was supposed to end, resulting in an execution error.

Freil et al. (2017) [70] distinguish between two types of execution errors: slips and mistakes. These categories are particularly relevant in scenarios where animals interact with screens, necessitating both motor skills and cognitive planning and execution abilities. Slips are essentially motor-control or mechanical errors, such as when a parrot accidentally hits the wrong stimulus because they are positioned too closely, or when a dog fails to pull a rope with sufficient force to activate an electronic device. Mistakes, on the other hand, arise from a mismatch between the plan and the desired outcome, like failing a discrimination task because of pursuing an incorrect goal or formulating a flawed plan. Understanding these subcategories in animal-computer interaction is crucial; slips indicate the need for either an environmental adjustment or further motor control training, whereas mistakes suggest a potential misunderstanding or cognitive limitation of the animal in completing the task.

Engineers, computer scientists, and researchers may be able to use this framework to work toward designing discrimination task systems that can reduce accidental touches to nearby stimuli and to measure points of breakdown in animal error rates, especially at stages 4 and 5. Scaffolded training in which the animal advances only after demonstrating the ability to successfully complete the task reduces errors at the planning levels, stages 1 and 2. A signal such as a tone or a light that is classically conditioned with food reinforcers to indicate accurate completion of a task, or a pause and corrective trial for an error, informs the animal in the evaluation phase, leading to a new plan in task execution.

Cunha, the primary author, is an expert in parrot research and training with over a decade of expertise in symbolic representation and concept training for parrots. She is co-founder of an online educational platform for parrot caregivers and has made significant scientific contributions through her research on parrot communication using tablets, enrichment games, and analog symbols. She is an international speaker, and her methods focus on choice and consent, which have been translated and shared around the world. Renguette, the second author, is both a university researcher and an associate professor in technical communication with experience in assessment of innovative educational technologies, STEM education, and applied linguistics. Ouimette, another co-author, co-founded and works as a trainer in Pet Kindergarten. Drawing from both our personal experiences and from the research literature, we know that the individual preferences of each animal shape their approach to learning discrimination tasks. Nevertheless, we have compiled a set of recommended steps to guide the implementation of training:

- **Step 1:** Set up the environment for choice and species-specific ideal learning conditions
- **Step 2:** Determine reinforcers and habituate the animal to the training area and apparatus
- **Step 3:** Shape behavior using positive reinforcement to initiate contact with the touchscreen device
- **Step 4:** Teach stimuli discrimination tasks with positive reinforcement

Step 1: Set Up the Environment for Choice and Species-Specific Ideal Learning Conditions

The external environment plays a large role in determining an animal's success in learning tasks. Most animals will choose to engage in training sessions voluntarily, without the need for weight control or force, if conditions have been thoughtfully created to be sensitive to species-specific circadian rhythms and dietary preferences. To promote choice, select a neutral location where the animal can easily choose to engage or disengage.

To best identify individual animal preferences and optimal times, observe the animal throughout the day and identify the most behaviorally active times (e.g., mornings and evenings for birds, lights-off periods for rodents). Schedule training sessions during those periods to ensure the animal is alert. Likewise, engage in training with reinforcers between mealtimes to allow for free access to a base diet while choosing to train for 'bonus' food rewards.

Here are some examples of certain time preferences for some animals:

- Parrots are most active in the morning and late afternoon times, when in the wild they would be foraging for food and interacting with other parrots. Training is ideally during these windows of activity and in areas with minimal distractions.
- Dogs and cats exhibit periods of activity interspersed with rest, during which they are visibly alert and engaged. Optimal learning environments for them include low-distraction settings, such as small quiet rooms.
- Rodents and small exotic animals are most active during dawn and dusk, preferring subdued lighting conditions. They thrive in environments with minimal noise and movement.
- Reptiles require semi-dark lighting and a heated space maintained at an ideal temperature. They benefit from tranquil surroundings with minimal disturbances.

For equines, learning sessions are best conducted in the morning and late afternoon to avoid lethargy caused by excessive heat in warm climates. These sessions should take place in environments with low distractions, allowing for focused, primarily individual learning experiences. Select apparatus based on the animal's ability to touch stimuli from a stationary position centered in front of the screen and with stimulus pixel sizes and distribution that have been ideally tested for optimal sizing for the species (see Table 12.6). Use larger icons and symbols for bigger animals due to their greater error margin. Smaller animals can use smaller icons and screens. Oversized icons and devices may prevent the animal from reaching all the icons or symbols without moving, which is not preferable. For training stations, we recommend:

- Birds: Utilize a centrally positioned perch parallel to the screen at eye level.

- Dogs and Cats: Offer a raised flat platform at a height allowing for safe independent access.
- Rodents and Small Exotics: Provide a colored mat, ensuring it is low enough for them to independently jump on and off securely, with optional elevation if needed.
- Reptiles: Offer a colored mat without elevation ideally.
- Equine: Supply a colored mat with minimal elevation, if any, recommended for use.

Touchscreens should be very sensitive and responsive to animal touches to reduce frustration and shorten motor skill training periods.

Lighting can also be adjusted to the species' natural preferences, as non-preferred lighting can be an unintentional aversive stimulus.

When setting up equipment in environments where animals are socially housed, care should be taken to provide multiple training stations to reduce the chances of aggression and guarding behaviors over resources. Additionally, for animals housed either individually or in groups, a method that encourages voluntary participation in research involves allowing them to enter and exit testing areas at twill. Trials should be offered for as long as the animal is present, and data collection methods should ensure that data isn't compromised even if an animal decides to leave mid-trial (e.g., [45]). In such setups, non-human primates have been observed voluntarily engaging with "video games" for more than an hour daily [8], and parrots have been willing to train for up to 40 min daily without requiring weight management and with the freedom to exit training sessions whenever they choose [16].

In an environment that is sensitive to species-specific needs and preferences, touchscreen training and discrimination tasks may provide excellent enrichment, in addition to providing research data.

Step 2: Determine Reinforcers and Habituate the Animal to the Training Area and Apparatus

To create a low-frustration, high-enrichment environment for a force-free, positive learning experience, selecting optimal food reinforcers and an optimal schedule of reinforcement will help the animal habituate to the environment and apparatus.

Ideally, the reinforcer is tailored to the animal's preference. This may be done by offering a selection of foods and observing the order in which the animal eats them. A one or two-value food reinforcement system may be employed, depending on the learning system. During two-value reinforcement systems, a high-value and a low-value food reinforcer are used. When the animal engages in the target behavior, a *bridge*—generally a sound—alerts them that they have engaged in the target behavior. Immediately after the bridge, a high-value reinforcer is provided. When the animal attempts the target behavior without success the stimuli is withdrawn for a 3 sec pause and they are offered a lower-value reinforcer. In this way, the animal has a high rate of reinforcement and motivation and, in time, will often voluntarily train for long durations. We add a note here that some high-value rewards may be too high and cause over-arousal in some animals.

In single-value reinforcement systems, the animal is rewarded for target behavior with a tone followed by a food reinforcer. After two unsuccessful attempts, some animals may benefit from being offered a simpler behavior (such as touching a large dot on the screen) to reduce frustration and maintain motivation. Upon successful completion of the simpler target behavior, the animal is rewarded with a tone and a food reinforcer and the target task resumes. This ensures a high rate of reinforcement with opportunities for success after unsuccessful attempts at a target behavior.

Higher value food reinforcers placed near the device (or manually delivered) may help with voluntary engagement during habituation until the animal displays non-vigilant comfort behavior in the presence of the device. When the animal is comfortable with the device, behavior shaping for touchscreen training may begin. At the beginning of each of the subsequent training sessions – shaping and discrimination tasks– the animal is given a "free treat" to build behavioral momentum and encourage their participation at the beginning of the session.

Step 3: Shape Behavior Using Positive Reinforcement to Initiate Capacitive Contact with the Touchscreen Device

Using positive reinforcement and behavior shaping, animals can be trained to interact with a touchscreen device by following a sequence of steps. Initially, in both social and automated learning setups, the animal receives a food reward for making any contact with the touchscreen. Applying soft foods like nut butter or cheese (depending on the animal's dietary needs) directly to the screen can entice the animal to interact with the touchscreen. Once the animal purposefully touches the screen, a quiet tone or other non-aversive signal is emitted, and they are rewarded with more reinforcers. This method encourages the animal to slowly progress towards engaging with the touchscreen.

Next, to guide the animal towards the desired behavior of interacting with smaller stimuli on the screen—which is generally part of discrimination task trials—a large circle is displayed on the screen. If the animal touches this circle, it is reinforced with both a tone and a food treat. After the animal consistently achieves the target behavior with a level of accuracy considered appropriate for its species across multiple sessions, the size of the circle can be incrementally reduced to match that of the stimuli used in discrimination tasks. This circle may also be moved to different locations on the screen. Once the animal consistently and accurately engages with this target stimulus, it is appropriate to begin discrimination training.

Step 4: Teach Stimuli Discrimination Tasks with Positive Reinforcement

Prior to using the touchscreen for discrimination tasks, several "pre-steps" must be completed including analyzing the location, selecting a successful easy backup target behavior option, and ensuring options for the animal-participant to choose to discontinue the training.

The first step of analyzing the location of the task is important because many animals have behavioral or visual preferences, they could have a dominant side because of brain lateralization (similar to human left- or right-handedness), or they can be easily influenced by cues or repeated reinforcements on one side. Therefore, the target stimuli should be positioned in a location that will minimize side biases in successive presentations. Although random presentations can be generated automatically in some applications, when a stimulus is shown in the same location for three or more successions, even randomly, a bias may be associated with that location. This bias can be minimized by dividing the touch screen into quadrants or intentionally designing the application to allow for a minimum number of times per side consecutively. In this way, random conditions would limit the possibility of the target stimulus being presented in the same location consecutively.

The second step for single-value reinforcement systems is selecting an easy fallback successful target behavior to reduce frustration during discrimination task training (as mentioned in the habituation section). Animals may get frustrated and lose motivation to continue if they select incorrectly multiple times. For example, if an animal selects incorrectly twice in a row, the large, original target circle from the shaping procedure may appear in the center of the screen. When the animal touches the circle, they receive a food reinforcer. This continued positive reinforcement aids in motivation and reducing frustration.

Finally, to promote consent, choice, control, and autonomy, the animal should be able to discontinue the activity at will. We recommend a design that includes an 'exit' button on the testing page that will trigger an auto-save so the animal may exit the test without discarding the data. For animals that have access to touchscreen-based enrichment games, they may be given high-value reinforcers for discrimination task sessions and lower-value reinforcers for other tablet-based activities.

Training on discrimination tasks starts by displaying two different stimuli on the touchscreen. Under the least-reinforcing scenario method [74], a correct selection is rewarded with a habituated sound and a food reinforcer, whereas an incorrect choice leads to a 3 sec screen-timeout and delay before the next attempt. This can be followed by the presentation of either a new task or repeat trials for correction. Errorless learning progressively introduces the two stimuli, ensuring minimal opportunities for mistakes.

In both social and automated learning frameworks, an initial phase involves making the correct choice unmistakably clear, often through guiding the selection or highlighting the correct stimulus while pairing that correct response with a sound and a food reward. Incorrect responses are gently corrected by either temporarily halting the interaction with a brief pause or by not providing the usual rewards, rather than through negative consequences.

As the training progresses, the distinction between the stimuli becomes gradually more nuanced, yet is designed in such a way that the likelihood of error remains low. This might involve adjusting the contrast or size of the stimuli to help the animal distinguish between them more easily. The touchscreen software can aid this process by introducing a short pause with a blank screen or introducing a slightly altered task that again encourages the correct response without penalty.

In social learning sessions, trainers may initially use physical cues or position the touchscreen in a manner to ensure the animal targets the correct stimulus. This fosters a direct and clear association from the outset. This guidance is slowly phased out, increasing the animal's reliance on its own judgment and understanding of the task and fostering independence in a supportive, error-minimized environment.

For automated learning systems, the approach similarly involves providing immediate rewards for correct responses to reinforce the desired behavior, gradually increasing task complexity in such a way that the animal can successfully make the correct associations with reduced assistance. Throughout this process, the focus is on creating a learning environment where errors are less likely to occur, supporting a smooth progression towards independent successful task completion.

12.6 Future Directions

This chapter presents a review of studies of touchscreen use for discrimination task training by 14 species that met our criteria for animal-centered ethics. Future work could include a more robust discussion of the technological design elements around touchscreen devices for optimizing discrimination tasks including considerations for object shape, color, depth perception, and the animals' specific visual abilities, Fitts' testing for each species, animal size-to-screen size ratios, and whether capacitive of infrared screens are more effective for each species' touchscreen interactions. Future work could also include other types of touchscreen tasks and games and more detailed insights and nuances into species-specific shaping and training. In addition, it would be valuable to have a more comprehensive discussion of ethical frameworks and how they could be relevant to animal use of touchscreen devices while also contributing to animal wellbeing.

12.7 Conclusion

Advances in technology and design of touchscreen devices have contributed to captive animal welfare in many ways, including providing novel opportunities for enrichment and ethical data collection for research endeavors related to animal-computer interactions. Emerging trends position animals as crucial stakeholders in the research that impacts them. This comprehensive review of the literature around training animals to use touchscreen devices for discrimination tasks can inform researchers, engineers, and computer scientists about methods and approaches for using touchscreens for social interactions, enhanced enrichment, and research purposes within the framework of animal-centered ethics. This work spans 14 species of animals and includes aspects such as environmental setup, touchscreen device hardware and software, habituation and training, as well as discrimination task outcomes.

Moreover, we present an expanded framework of recommendations and guidelines for best practices for implementing discrimination training across diverse species encompassing (1) environmental setup conducive to species-specific learning, (2) animal-specific reinforcer selection and habituation, (3) behavior shaping with positive reinforcement, and (4) teaching stimuli discrimination tasks with positive reinforcement. It is our hope that this will help other researchers better understand methods for training animals to use touchscreens and that through this work, engineers and computer scientists are able to stay informed about interdisciplinary advances in the area of animal-computer interaction so they are better able to contribute to designing technology that improves the wellbeing of captive animals.

References

1. Jukan, A., X. Masip-Bruin, and N. Amla. 2017. Smart computing and sensing technologies for animal welfare: a systematic review. *ACM Computing Surveys* 50: 10:1–10:27. https://doi.org/10.1145/3041960.
2. Ramdya, P., and A. J. Ijspeert. 2023. The neuromechanics of animal locomotion: from biology to robotics and back. *Science Robotics* 8: eadg0279. https://doi.org/10.1126/scirobotics.adg0279.
3. Brando, S., A. Vitale, and M. Bacon. 2023. Promoting good nonhuman primate welfare outside regular working hours. *Animals: An Open Access Journal from MDPI* 13: 1423. https://doi.org/10.3390/ani13081423.
4. van Weeghel, H. J. E., A. P. Bos, M. H. Jansen, W. W. Ursinus, and P. W. G. Groot Koerkamp. 2021. Good animal welfare by design: an approach to incorporate animal capacities in engineering design. *Agricultural Systems* 191: 103154. https://doi.org/10.1016/j.agsy.2021.103154.
5. Makowska, I. J., and D. M. Weary. 2019. A good life for laboratory rodents? *ILAR Journal* 60: 373–388. https://doi.org/10.1093/ilar/ilaa001.
6. Kleinberger, R., J. Cunha, M. M. Vemuri, and I. Hirskyj-Douglas. 2023. Birds of a feather video-flock together: design and evaluation of an agency-based parrot-to-parrot video-calling system for interspecies ethical enrichment. In *Proceedings of the 2023 CHI Conference on Human Factors in Computing Systems*, 1–16. CHI '23. New York, NY, USA: Association for Computing Machinery. https://doi.org/10.1145/3544548.3581166.
7. Calapai, A., D. Pfefferle, L. C. Cassidy, A. Nazari, P. Yurt, R. R. Brockhausen, and S. Treue. 2023. A touchscreen-based, multiple-choice approach to cognitive enrichment of captive Rhesus Macaques (Macaca mulatta). *Animals* 13: 2702. https://doi.org/10.3390/ani13172702.
8. Washburn, D. 2015. The four Cs of psychological wellbeing: lessons from three decades of computer-based environmental enrichment. *Animal Behavior and Cognition* 2: 218–232. https://doi.org/10.12966/abc.08.02.2015.
9. Bennett, A., J. Bailoo, M. Dutton, G. Michel, and P. Pierre. 2018. Psychological science applied to improve captive animal care: a model for development of a systematic, evidence-based assessment of environmental enrichment for nonhuman primates. https://doi.org/10.17605/OSF.IO/79XKY.

10. Fagot, J., and E. Bonté. 2010. Automated testing of cognitive performance in monkeys: use of a battery of computerized test systems by a troop of semi-free-ranging baboons (*Papio papio*). *Behavior Research Methods* 42: 507–516. https://doi.org/10.3758/BRM.42.2.507.
11. Wallis, L. J., F. Range, E. Kubinyi, D. Chapagain, J. Serra, and L. Huber. 2017. Utilising dog-computer interactions to provide mental stimulation in dogs especially during ageing. *ACI2017, Improving Relations: Fourth International Conference on Animal-Computer Interaction: Proceedings: 21–23 November 2017, Milton Keynes, United Kingdom. International Conference on Animal-Computer Interaction (4th: 2017: Milt.* 2017: 1. https://doi.org/10.1145/3152130.3152146.
12. Zeagler, C., S. Gilliland, L. Freil, T. Starner, and M. Jackson. 2014. Going to the dogs: towards an interactive touchscreen interface for working dogs. In *Proceedings of the 27th Annual ACM Symposium on User Interface Software and Technology*, 497–507. UIST '14. New York, NY, USA: Association for Computing Machinery. https://doi.org/10.1145/2642918.2647364.
13. Westerlaken, M., and S. Gualeni. 2014. Felino: the philosophical practice of making an interspecies videogame. In *The Philosophy of Computer Games Conference*, Istanbul 2014.
14. Driessen, C., K. Alfrink, M. Copier, H. Lagerweij, and I. van Peer. 2014. What could playing with pigs do to us? Game design as multispecies philosophy. *Antennae*: 79.
15. Santa, A., K. Kanda, T. Kako, M. Miyajima, and I. Adachi. 2024. Matching-to-sample task training of a Killer Whale (Orcinus orca). *Animals* 14: 821. https://doi.org/10.3390/ani14060821.
16. Cunha, J., and C. Rhoads. 2020. Use of a tablet-based communication board and subsequent choice and behavioral correspondences in a Goffin's Cockatoo (*Cacatua goffiana*). In *Proceedings of the Seventh International Conference on Animal-Computer Interaction*, 1–9. ACI '20. New York, NY, USA: Association for Computing Machinery. https://doi.org/10.1145/3446002.3446063.
17. Hopper, L. M., C. L. Egelkamp, M. Fidino, and S. R. Ross. 2018. An assessment of touchscreens for testing primate food preferences and valuations. *Behavior Research Methods* 51: 639–650. https://doi.org/10.3758/s13428-018-1065-0.
18. Vonk, J., J. Truax, and M. C. McGuire. 2023. A food for all seasons: stability of food preferences in Gorillas across testing methods and seasons. *Animals* 12: 685. https://doi.org/10.3390/ani12060685.

19. Cook, R. G., A. I. Geller, G.-R. Zhang, and R. Gowda. 2004. Touchscreen-enhanced visual learning in rats. *Behavior Research Methods, Instruments, & Computers: A Journal of the Psychonomic Society, Inc* 36: 101–106. https://doi.org/10.3758/bf03195555.
20. Harrison, R. A., T. Mohr, and E. van de Waal. 2023. Lab cognition going wild: implementing a new portable touchscreen system in vervet monkeys. *Journal of Animal Ecology* 92: 1545–1559. https://doi.org/10.1111/1365-2656.13857.
21. Huber, L., W. Apfalter, M. Steurer, and H. Prossinger. 2005. A new learning paradigm elicits fast visual discrimination in pigeons. *Journal of Experimental Psychology. Animal Behavior Processes* 31: 237–246. https://doi.org/10.1037/0097-7403.31.2.237.
22. Krakenberg, V., M. Wewer, R. Palme, S. Kaiser, N. Sachser, and S. Richter. 2020. Regular touchscreen training affects faecal corticosterone metabolites and anxiety-like behaviour in mice. *Behavioural Brain Research* 401: 113080. https://doi.org/10.1016/j.bbr.2020.113080.
23. Seitz, B. M., K. McCune, M. MacPherson, L. Bergeron, A. P. Blaisdell, and C. J. Logan. 2021. Using touchscreen equipped operant chambers to study animal cognition. Benefits, limitations, and advice. *PLOS ONE* 16: e0246446. https://doi.org/10.1371/journal.pone.0246446.
24. Johnson-Ulrich, Z., and J. Vonk. 2018. Spatial representation of magnitude in humans (Homo sapiens), Western lowland gorillas (*Gorilla gorilla gorilla*), and American black bears (Ursus americanus). *Animal Cognition* 21: 531–550. https://doi.org/10.1007/s10071-018-1186-y.
25. Aust, U., F. Range, M. Steurer, and L. Huber. 2008. Inferential reasoning by exclusion in pigeons, dogs, and humans. *Animal Cognition* 11: 587–597. https://doi.org/10.1007/s10071-008-0149-0.
26. O'Hara, M., A. M. I. Auersperg, T. Bugnyar, and L. Huber. 2015. Inference by exclusion in Goffin Cockatoos (*Cacatua goffini*). *PLOS ONE* 10: e0134894. https://doi.org/10.1371/journal.pone.0134894.
27. Mikolasch, S., K. Kotrschal, and C. Schloegl. 2011. African grey parrots (*Psittacus erithacus*) use inference by exclusion to find hidden food. *Biology Letters* 7: 875–877. https://doi.org/10.1098/rsbl.2011.0500.
28. Kurz, D. 2011. Touch-screen experiments: common marmosets can discriminate between positive and negative stimuli, but can they also reason by exclusion? Master's, Universität Wien.
29. Marsh, H., A. Vining, E. Levendoski, and P. Judge. 2015. Inference by exclusion in Lion-Tailed Macaques (*Macaca silenus*), a Hamadryas Baboon (*Papio hamadryas*), Capuchins (*Sapajus apella*), and Squirrel Monkeys

(*Saimiri sciureus*). *Journal of Comparative Psychology (Washington, D.C.: 1983)* 129. https://doi.org/10.1037/a0039316.

30. Stephan, C., A. Wilkinson, and L. Huber. 2012. Have we met before? Pigeons recognise familiar human faces. *Avian Biology Research* 5: 75–80. https://doi.org/10.3184/175815512X13350970204867.
31. Bodily, K. D., J. S. Katz, and A. A. Wright. 2008. Matching-to-sample abstract-concept learning by pigeons. *Journal of Experimental Psychology. Animal Behavior Processes* 34: 178–184. https://doi.org/10.1037/0097-7403.34.1.178.
32. Rivas-Blanco, D., I.-M. Pohl, R. Dale, M. T. E. Heberlein, and F. Range. 2020. Wolves and dogs may rely on non-numerical cues in quantity discrimination tasks when given the choice. *Frontiers in Psychology* 11: 573317. https://doi.org/10.3389/fpsyg.2020.573317.
33. Mancini, C. 2017. Towards an animal-centred ethics for animal–computer interaction. *International Journal of Human-Computer Studies* 98: 221–233. https://doi.org/10.1016/j.ijhcs.2016.04.008.
34. Mancini, C., and E. Nannoni. 2022. Relevance, impartiality, welfare and consent: principles of an animal-centered research ethics. *Frontiers in Animal Science* 3. Frontiers. https://doi.org/10.3389/fanim.2022.800186.
35. Mancini, C. 2011. Animal-computer interaction: a manifesto. *Interactions* 18: 69–73. https://doi.org/10.1145/1978822.1978836.
36. Hirskyj-Douglas, I., P. Pons, J. C. Read, and J. Jaen. 2018. Seven years after the manifesto: literature review and research directions for technologies in animal computer interaction. *Multimodal Technologies and Interaction* 2: 30. https://doi.org/10.3390/mti2020030.
37. Lee, S. P., A. D. Cheok, T. K. S. James, G. P. L. Debra, C. W. Jie, W. Chuang, and F. Farbiz. 2006. A mobile pet wearable computer and mixed reality system for human–poultry interaction through the Internet. *Personal and Ubiquitous Computing* 10: 301–317. https://doi.org/10.1007/s00779-005-0051-6.
38. Haladjian, J., Z. Hodaie, S. Nüske, and B. Brügge. 2017. Gait anomaly detection in dairy cattle. In *Proceedings of the Fourth International Conference on Animal-Computer Interaction*, 1–8. ACI '17. New York, NY, USA: Association for Computing Machinery. https://doi.org/10.1145/3152130.3152135.
39. Hirskyj-Douglas, I., and J. C. Read. 2018. DoggyVision: examining how dogs (*Canis familiaris*) interact with media using a dog-driven proximity tracker device. *Animal Behavior and Cognition* 5: 388–405. https://doi.org/10.26451/abc.05.04.06.2018.

40. Pons, P., J. Jaen, and A. Catala. 2017. Towards future interactive intelligent systems for animals: study and recognition of embodied interactions. In *Proceedings of the 22nd International Conference on Intelligent User Interfaces*, 389–400. IUI '17. New York, NY, USA: Association for Computing Machinery. https://doi.org/10.1145/3025171.3025175.
41. French, F., and M. Kaltenbrunner. 2021. User Experience for Elephants: Researching Interactive Enrichment through Design and Craft. PhD thesis United Kingdom: Open University. https://doi.org/10.13140/RG.2.2.23671.60326.
42. Kobayashi, H., K. Muramatsu, J. Okuno, K. Nakamura, A. Fujiwara, and K. Saito. 2015. Playful rocksalt system: animal-computer interaction design in wild environments. In *Proceedings of the 12th International Conference on Advances in Computer Entertainment Technology*, 1–4. ACE '15. New York, NY, USA: Association for Computing Machinery. https://doi.org/10.1145/2832932.2837012.
43. Wolf, J. E., C. M. Urbano, C. M. Ruprecht, and K. J. Leising. 2014. Need to train your rat? There is an App for that: a touchscreen behavioral evaluation system. *Behavior Research Methods* 46: 206–214. https://doi.org/10.3758/s13428-013-0366-6.
44. Scarf, D. 2022. Getting out of the lab: The development of a free-range learning apparatus for pigeons (FLAP). In *Conference: Australia and New Zealand Council for the Care of Animals in Research and Teaching*.
45. Ritvo, S. E., and S. E. MacDonald. 2020. Preference for free or forced choice in Sumatran orangutans (*Pongo abelii*). *Journal of the Experimental Analysis of Behavior* 113: 419–434. https://doi.org/10.1002/jeab.584.
46. Mueller-Paul, J., A. Wilkinson, U. Aust, M. Steurer, G. Hall, and L. Huber. 2014. Touchscreen performance and knowledge transfer in the red-footed tortoise (*Chelonoidis carbonaria*). *Behavioural Processes* 106: 187–192. https://doi.org/10.1016/j.beproc.2014.06.003.
47. Drucker, C. B., T. Baghdoyan, and E. M. Brannon. 2016. Implicit sequence learning in ring-tailed lemurs (*Lemur catta*). *Journal of the Experimental Analysis of Behavior* 105: 123–132. https://doi.org/10.1002/jeab.180.
48. Brent, L., and D. Weaver. 1996. The physiological and behavioral effects of radio music on singly housed baboons. *Journal of Medical Primatology* 25: 370–374. https://doi.org/10.1111/j.1600-0684.1996.tb00031.x.
49. Honess, P. E., and C. M. Marin. 2006. Enrichment and aggression in primates. *Neuroscience and Biobehavioral Reviews* 30: 413–436. https://doi.org/10.1016/j.neubiorev.2005.05.002.

50. Line, S. W., A. S. Clarke, H. Markowitz, and G. Ellman. 1990. Responses of female rhesus macaques to an environmental enrichment apparatus. *Laboratory Animals* 24: 213–220. https://doi.org/10.1258/002367790780866245.
51. Claidière, N., J. Gullstrand, A. Latouche, and J. Fagot. 2017. Using automated learning devices for monkeys (ALDM) to study social networks. *Behavior Research Methods* 49: 24–34. https://doi.org/10.3758/s13428-015-0686-9.
52. Hirskyj-Douglas, I., and V. Kankaanpää. 2021. Exploring how white-faced sakis control digital visual enrichment systems. *Animals* 11: 557. https://doi.org/10.3390/ani11020557.
53. Mejdell, C. M., T. Buvik, G. H. M. Jørgensen, and K. E. Bøe. 2016. Horses can learn to use symbols to communicate their preferences. *Applied Animal Behaviour Science* 184: 66–73. https://doi.org/10.1016/j.applanim.2016.07.014.
54. McEwen, B. S., and R. M. Sapolsky. 1995. Stress and cognitive function. *Current Opinion in Neurobiology* 5: 205–216. https://doi.org/10.1016/0959-4388(95)80028-x.
55. Väätäjä, H., P. Majaranta, H. Törnqvist, M. Ainasoja, V. Surakka, O. Juhlin, and C. Mancini. 2017. Technology for bonding in human-animal interaction. In *Proceedings of the Fourth International Conference on Animal-Computer Interaction*, 1–5. ACI '17. New York, NY, USA: Association for Computing Machinery. https://doi.org/10.1145/3152130.3152153.
56. Horner, A. E., C. J. Heath, M. Hvoslef-Eide, B. A. Kent, C. H. Kim, S. R. O. Nilsson, J. Alsiö, et al. 2013. The touchscreen operant platform for testing learning and memory in rats and mice. *Nature Protocols* 8: 1961–1984. https://doi.org/10.1038/nprot.2013.122.
57. Byrne, C., C. Zeagler, L. Freil, A. Rapoport, and M. M. Jackson. 2018. Dogs using touchscreens in the home: a case study for assistance dogs operating emergency notification systems. In *Proceedings of the Fifth International Conference on Animal-Computer Interaction*, 1–10. ACI '18. New York, NY, USA: Association for Computing Machinery. https://doi.org/10.1145/3295598.3295610.
58. Jackson, M., S. Sun, D. Peet, A. Lau, X. Han, J. Alcaidinho, T. Starner, et al. 2018. Technology for working dogs, 1–5. https://doi.org/10.1145/3295598.3295615.
59. Huber, L., N. Heise, C. Zeman, and C. Palmers. 2015. The ALDB box: automatic testing of cognitive performance in groups of aviary-housed pigeons. *Behavior Research Methods* 47: 162–171. https://doi.org/10.3758/s13428-014-0462-2.

60. Laméris, D. W., J. Verspeek, M. Eens, and J. M. G. Stevens. 2022. Social and nonsocial stimuli alter the performance of bonobos during a pictorial emotional Stroop task. *American Journal of Primatology* 84: e23356. https://doi.org/10.1002/ajp.23356.

61. Tomonaga, M., K. Kumazaki, F. Camus, S. Nicod, C. Pereira, and T. Matsuzawa. 2015. A horse's eye view: size and shape discrimination compared with other mammals. *Biology Letters* 11: 20150701. https://doi.org/10.1098/rsbl.2015.0701.

62. Friedman, S. 2016. Tsk, No, Eh-eh: Clearing the Path to Reinforcement with an Errorless Learning Mindset. A paper present at the ABMA Conference, 2016.

63. Clark, F. 2017. Cognitive enrichment and welfare: current approaches and future directions. *Animal Behavior and Cognition* 4: 52–71. https://doi.org/10.12966/abc.05.02.2017.

64. Pepperberg, I. M. 2000. *The Alex Studies: Cognitive and Communicative Abilities of Grey Parrots*. Harvard University Press. https://doi.org/10.2307/j.ctvk12qc1.

65. Pepperberg, I. M. 2017. Animal language studies: What happened? *Psychonomic Bulletin & Review* 24: 181–185. https://doi.org/10.3758/s13423-016-1101-y.

66. Ford, C. A., L. Bellward, C. J. C. Phillips, and K. Descovich. 2021. Use of interactive technology in captive great ape management. *Journal of Zoological and Botanical Gardens* 2: 300–315. https://doi.org/10.3390/jzbg2020021.

67. Perdue, B. M. 2016. The effect of computerized testing on sun bear behavior and enrichment preferences. *Behavioral Sciences* 6: 19. https://doi.org/10.3390/bs6040019.

68. Hliňák, Z., and E. Rozmarová. 1986. The locomotor-exploratory behaviour of laboratory male rats tested under the "red" and "white" light conditions. *Activitas Nervosa Superior* 28: 202–203.

69. Lockard, R. B. 1963. Some effects of light upon the behavior of rodents. *Psychological Bulletin* 60: 509–529. https://doi.org/10.1037/h0046113.

70. Freil, L., C. Byrne, G. Valentin, C. Zeagler, D. Roberts, T. Starner, and M. Jackson. 2017. Canine-centered computing. *Foundations and Trends® in Human-Computer Interaction* 10: 87–164. https://doi.org/10.1561/1100000064.

71. Zeagler, C., J. Zuerndorfer, A. Lau, L. Freil, S. Gilliland, T. Starner, and M. M. Jackson. 2016. Canine computer interaction: towards designing

a touchscreen interface for working dogs. In *Proceedings of the Third International Conference on Animal-Computer Interaction*, 1–5. ACI '16. New York, NY, USA: Association for Computing Machinery. https://doi.org/10.1145/2995257.2995384.

72. Waal, F. de. 2016. *Are We Smart Enough to Know How Smart Animals Are?* W. W. Norton & Company.
73. McKinley, J., H. M. Buchanan-Smith, L. Bassett, and K. Morris. 2003. Training common marmosets (Callithrix jacchus) to cooperate during routine laboratory procedures: ease of training and time investment. *Journal of Applied Animal Welfare Science* 6: 209–220. https://doi.org/10.1207/S15327604JAWS0603_06.
74. Pryor, K., and K. Ramirez. 2014. Modern animal training. In *The Wiley Blackwell Handbook of Operant and Classical Conditioning*, 453–482. John Wiley & Sons, Ltd. https://doi.org/10.1002/9781118468135.ch18.

Notes

(1) Note that the current authors do not promote or endorse weight control or main dietary food restriction. These practices were repeatedly found in the literature, but we align practices with Mancini's [33, 35] ethical practices and find that animals respond better without restricting food or controlling weight.

(2) Note, however, that luring has been less effective in complex, non-discrimination, multi-step shaping behaviors such as line drawing (e.g., [71]).

Chapter 13

An Investigation into the Encounter between Social Robots and Animals

Oliver Bendel
FHNW School of Business,
University of Applied Sciences and Arts Northwestern Switzerland,
Windisch, Switzerland
oliver.bendel@fhnw.ch

Increasingly, social robots and certain service robots encounter, whether this is planned or not, domestic, farm, or wild animals. They react differently, some interested, some disinterested, some lethargic, some panicked. Research needs to turn more to animal-robot relationships, and to work with engineers to design these relationships in ways that promote animal welfare and reduce animal suffering. This chapter is about social robots that are designed for animals, but also those that—for different, rather unpredictable reasons—meet, interact, and communicate with animals. It also considers animal-friendly machines that have emerged in the context of machine ethics. In the discussion section, the author explores the question of which of the presented robots are to be understood as social robots and what their differences are in their purpose and in their relationship to animals. In addition, social and ethical aspects are addressed.

Animals, Ethics, and Engineering: Intersections and Implications
Edited by Rosalyn W. Berne

ISBN 978-981-5129-56-4 (Hardcover), 978-1-003-65231-1 (eBook)
www.jennystanford.com

13.1 Introduction

This chapter is an expanded, updated, and translated version of a contribution that appeared in a German-language book on social robots in 2021 [1]. It offers examples of hypothetical and actual social robots that relate to animals in some way and provides an overview of activities and innovations in this area, seeking to generate interest in the topic among engineers and students of engineering, scientists, policymakers, and the public. It also seeks to strengthen the definition of social robots, i.e., to show that the openness to all living beings is justified.

In the disciplines of robotics and social robotics, animals have so far been given little consideration. Since the turn of the new millennium, there have been significant attempts to establish animal–computer interaction and animal–machine interaction [2, 3]. Machine ethics has also contributed several scholarly artifacts directed at animals [4, 5] but remains largely focused solely on humans. The closest related contributions have come from cognitive science and behavioral research and from the entertainment and toy industry, which have only a loose connection to the academic and technical communities researching social robotics [6–8].

Social robots are most familiar to the general population from science fiction movies, such as R2-D2 and C-3PO from "Star Wars" or WALL-E from "WALL-E". These are prototypes and role models many are aware of. In real life, there are actual social robots like Pepper, NAO, or Cozmo, designed to interact with humans, such as for the purpose of companionship. Humans maintain social relationships and are social beings. The same can be said for animals [9], and controversially, even some plants. In fact, there are social robots explicitly designed to handle animals, and a growing market for them based on current supply and demand. For example, many people acquired dogs or other pets during the COVID-19 pandemic, which needed to be fed, maintained, and exercised; jobs that could be undertaken by a form of social robot. Indeed, as social robots become more widespread for human use, they will almost inevitably encounter not just humans, but our animal companions, too. However, social robot interactions with animals are not limited to our personal pets; they may also occur with livestock and even wild animals, when these living beings venture into our gardens or agricultural domains and encounter a working robot.

Elsewhere, I have defined social robots as sensorimotor machines created for dealing with humans and animals [10]. This incorporates a definition by Christaller, who defined robots as sensorimotor machines for the purpose of expanding human agency [11]. Furthermore, social robots could be determined by five dimensions, namely: interaction with living beings, communication with living beings, proximity to living beings, representation of (aspects of) living beings, and, vitally, benefit or serve a utility for living beings [10]. If four or five of these elements are present in a robot, then it could be classified as a social robot. Admittedly, it is not always easy to decide whether some of the examples provided here are actual social robots, whereby, e.g., the five dimensions are not completely or convincingly fulfilled. This will be further discussed and problematized in this chapter.

13.2 Basic Concepts in the Context of Humans, Animals, and Machines

This section clarifies some basic concepts about the relationship between humans, animals, and machines.[1] Social robotics together with its subject matter has already been explained and is therefore omitted. Ethics is a millennia-old, fundamental discipline of philosophy and applies scientific methods for reasoning and description. Its subject is morality, i.e., the setting of general rules of action and value standards or personal convictions regarding what is good and evil or just and unjust [12]. Machine ethics is a design discipline with proximity to artificial intelligence (AI) and robotics [5, 13]. It explores machine morality and produces moral or immoral machines by extending (partially) autonomous machines. Both "machine morality" ("artificial morality") and "moral machine" are

[1] The proceedings of the annual International Conference on Social Robotics (ICSR) or the biennial Robophilosophy and edited volumes such as "Social Robotics" are suitable for further reference. In addition, the ACI conference, which has been taking place since 2016, provides a deeper exploration of this concept.

technical terms. Machine morality simulates human morality. It can apply to humans as well as to other living beings [5].

Animal ethics deals with the duties of humans towards animals, the rights of animals [14], and, in rare cases, with the relationship between animals and (semi-)autonomous intelligent systems, e.g., software agents and robots. Most recently, Peter Singer, one of the world's best-known ethicists, has turned to this area of intersection [15]. Animal ethics is gaining new impetus from animal rights movements and vegetarian and vegan lifestyles, which are increasingly widespread. At the same time, it must maintain its independence without sinking into arbitrariness [16]. Environmental protection consists of measures to preserve the natural environment, not least of which is the satisfaction and health of humans and animals. Nature conservation primarily takes aim at the value, dignity, and rights of nature. Animal welfare is the totality of personal, institutional, and legal measures to protect animals from cruelty, abuse, and killing. Animal observation is the observation of animals for scientific, medical, or media purposes.

Animal–computer interaction (ACI) according to Mancini [2] aims to understand the interactions between animals and computer technologies in terms of the contexts in which they usually live, are active, and form social relationships with living beings of all kinds. Mancini emphasizes the interplay of animal, technology, and context, distinguishing between individuals and species, and between types such as wild, domestic, working, farm, and laboratory animals. Animal–machine interaction (AMI), which is still barely developed, is concerned with the design, evaluation, and implementation of more sophisticated or complex machines and computer systems with which animals interact and that, in turn, interact and communicate with animals [17]. Human–machine interaction (HMI), on the other hand, is an established discipline that deals with the design, evaluation, and implementation of machines with which humans interact.

Animal enhancement is the extension of the animal, primarily for its apparent or actual improvement in relation to its own interests or those of humans, for example, in economic or scientific terms [18]. In view are, among others, performance enhancement, quality of life improvement, and optimization of utilization. For the most part, it is a matter of further subjugation of the animal as an addition to classical animal husbandry and use. This is initially different with human enhancement [19]. In this, the aim is an improvement that serves the individual and, in a certain way, liberation from the apparent constraints of nature. However, it may well end up putting electronic chains on people, and government and business may have an interest in this direction. Robot enhancement is the extension and improvement of robots, especially social robots, with, for example, the help of attachments, clothing, accessories, and appropriate programming or AI augmentation [20, 21].

13.3 Technology in Animals

Devices and machines of all kinds care for, use, and kill animals under the operation of or on behalf of humans. These include automatic feeders, milking robots, and gassing, mushing, and slaughtering apparatus. Thus, the areas of application are business and agriculture. In the latter, one finds scarecrows, which in rare cases are robots. Automatic feeders are sometimes used in private households, for example, to feed a cat in the absence of its owners. In the house and garden, service robots are on the move to meet animals, such as vacuum robots, lawn-mowing robots, and pool robots. This is by no means intentional, but almost inevitable due to the omnipresence of insects, small wildlife such as hedgehogs and martens, and free-roaming pets. Public spaces are teeming with semi- or fully automated vehicles, and if they get too close to animals (with the human or the system making the decision), it is often fatal for the animal. Finally, in the wild or wilderness, machines are rare, but as drones they can be used for observation and reconnaissance, detecting forest fires, monitoring bases, etc., and as aircraft for transporting goods and passengers—again, collisions with animals, especially insects, birds, and bats, are inevitable. Overall, questions arise from animal ethics and animal–machine interaction.

While one could say that the deadly encounters described above are collateral damage (which could be avoided to some extent), this is a little different in the case of animal enhancement. The extension or even improvement by technical means happens, as already indicated, rarely in the sense of the animal. To be sure, this is certainly possible when one thinks of intelligent prostheses, medical implants, or social robots for care (such as feeding). But in reality, it's more likely that an invention like RoboRoach is a role model—an extremely negative one in the eyes of animal rights activists. An American company markets an electronic component that you attach to the back of a cockroach; after a few simple steps, you can control it remotely with an app on your smartphone [22] against its own interests. Other experiments aim to attach tiny microphones and cameras to insects. The military and police are likely to be very interested in such animal cyborgs. The development of a mini-robot is time-consuming and cost-intensive—in an insect, on the other hand, the basic motor and some sensory capabilities are already fully developed and only need to be forcibly conquered and purposefully supplemented. No collateral damage, then, but targeted interventions and damage in the human interest. Another approach is virtual fences: cows are fitted with a collar or a structure on their heads, are given more freedom of movement, and are prevented from moving on at a defined point via light electric shocks.

Social robots have hardly been mentioned in this context so far, or perhaps we have considered them more than we thought. It would be possible, for example, that some of the machines mentioned could become social robots through minor or major modifications, or at least exhibit social characteristics. This is well known in certain fields. For example, Relay by Savioke from California is at first glance an ordinary transport robot. It has an abstract shape, is the size of a child, and rolls through the corridors of a hospital or hotel, for example, to bring something from A to B [23]. At closer inspection, you can see its slanted display in the upper area, where eyes can appear. It can produce sounds, thus immediately gaining benefits of social sympathy—similar to R2–D2 or BB-8 from "Star Wars"—and it is helped into and out of the elevator, for example, and one is less inclined to damage it. Even household robots like vacuum robots sometimes have such extensions into the social by interacting and communicating via a display or a voice. Some of them can also affect animals in the process of their work—this will be discussed further below.

Social robots are best known, when it comes to animals, in the functions of entertainment (or play) and physical exercise. Not all devices that exist on this market are actual robots. Some are merely cleverly marketed with that name—"robot" sounds more significant and valuable than "device". But a few certainly deserve the designation. Pets chase after artificial animals or tempting objects and indulge their play instincts to be active. Observation robots, familiar to a wide audience from animal films, serve a completely different purpose. The majority of them are amazingly similar to the mammals, reptiles, and birds they observe. This makes sense, because the better they fit in in terms of appearance and behavior, the more likely they are to be accepted by the herd or flock and by the individuals—which can then be observed at close range. Admittedly, some of this is behavior that does not normally occur in the wild because robots do not normally exist there—but you can get very close to the animals and watch their usual goings-on with minimal disruption.

Social robots are often found in the home and sometimes in the shopping mall—or at train stations and airports. There they meet cats, dogs, and other pets. Animal reactions to such meetings vary and can sometimes be unclear. Among other things, the reaction has to do with whether the social robots are perceived as enemies or friends, i.e., whether the animal should fight, familiarize, or copulate with the robots. This can be dependent on the animal's preexisting experiences with robots. Robotic dogs and cats (often cartoon- and comic-book-like in design) fall into this realm, but even the humanoid (also cartoon-like in design) Pepper and NAO or Cozmo and Vector, which look like little caterpillar vehicles with a face, can pique interest for a while. In science laboratories, such arrangements are readily made and explored. Furthermore, in hospitals and nursing homes, social robots encounter real animals. For example, when a cat is in a nursing or old people's home where Paro, a therapy robot modeled on a baby harp seal, or a robotic therapy dog is being used. Even on streets and in squares, social robots can be in close proximity to animals—for example, when security robots resembling police officers and security guards, such as the uniformed model REEM from PAL Robotics in Spain used in Dubai, encounter dogs, crows, and seagulls [24].

Animal-friendly machines are machines that—often in contrast to those mentioned at the beginning—do something that indirectly or directly helps animals, protects and prolongs their lives, protects them from injury, or lets them do something they otherwise could not do. These include, for example, artifacts of machine ethics, such as those conceived by the author since 2014 and created under his supervision at his university [4]. The focus was on household robots, specifically vacuuming robots and lawn-mowing robots, although the principle can be extended to other service robots (not only in private households). Whether these are social robots remains to be determined. They do not include other wildlife-friendly components such as supplements for wind turbines in the manner of DTBird, which ensure that the rotors stop when flocks of birds or bats approach [25], or specially equipped drones that fly ahead of mowers in the field and report fawns, as in DLR's Flying Wildlife Rescuer [26]. Neither is the prototypical Wildlife Vehicle Collision Avoidance System a social robot [27]—but it is an interesting solution for warning drivers of deer and getting them to brake or take evasive action.

13.4 Non-Human Living Beings and Social Robots

The first part of this section belongs to robotic models that are designed to care for, entertain, observe, drive away, etc., animals. The second part deals with social robots that come into contact with animals accidentally or that—for experimental purposes, for example—are temporarily oriented towards them. The third part deals with animal-friendly machines and systems that the author has conceptualized or prototyped in the context of machine ethics. The aim is not to be exhaustive but to find suitable and typical examples for each area.

13.4.1 Robotic or Robot-Like Models for Animals

There is no shortage of simple toys for cats and dogs. One example is the SlowTon remote cat feather toy with remote control. According to the manufacturer, the highlight of the simple, mouse-like machine is the fur tail, which allows you to attract the attention of the cat. The animal is enabled to "experience the thrill of running and jumping around from the safety of their home" [1]. Overall, it is a kind of toy car and not a robot.

The so-called DADYPET cat toy from the company DADYPET goes one step further. According to them, it simulates the escape behavior of mice, which attracts the cat's attention and awakens its hunting instinct. "The feather also swings left and right, your cat will have incredible fun with it." [28] (own translation) Moreover, "after 5 min of long playing, the toy stops moving and makes a sound. The sound is emitted once every 7 sec and the LED lights up." [28] (own translation). In this state, the toy remains for 2 min, during which the cat can freely play with the feather. After another 2 min, the game continues. A built-in sensor can avoid obstacles to help the mouse escape—"in a clever way" (own translation). "You don't have to worry about the toy getting stuck anywhere" [28] (own translation). In this case, one can definitely speak of a robot—at least the sensor-motor is present in a simple form.

For dogs, there are significantly fewer intelligent toys and hardly any social robots. One reason for this could be that some of them have enormous strength and would destroy the devices with a bite. This is bad for the devices and possibly for the dogs if they get electrocuted or otherwise hurt. Overall, there are probably more differences between dogs themselves than with cats with regard to how they handle toys. In size and strength, at least, dogs differ enormously from each other, and the spectrum ranges from the Chihuahua to the St. Bernard or German Shepherd. What is striking is that, on the other hand, there are a great many social robots on the market that are modeled after dogs (or the cartoon version of dogs), far more than cats, and are suitable for children and curious adults. They are available with smooth, washable plastic surfaces and covered with fur, from the low-price models such as CaCaCook, GoolRC by DIY, and Goolsky LE Neng by Goolsky, to the luxury models such as AIBO by Sony, an ever-reinvented classic that costs several thousand dollars, or like CyberDog by Xiaomi for about $1,400. They are in the tradition of the first robot dog, Sparko, shown at the World's Fair in New York in 1940, alongside the already existing humanoid robot Elektro.

Play and physical exercise are one thing; care and nurturing are another. Social robots could supervise, protect, wash, and clean pets. An obvious thought is walking dogs—the robot holds the leash and supervises its charge as it romps and poops. In fact, service robots have been unveiled that could, in principle, perform this task. Gita, from Piaggio, is a self-balancing two-wheeled cargo robot that strictly follows its owner or goes its own way, i.e., a self-rolling suitcase or messenger on behalf of its master. It is 66 cm tall, can load up to 18 kg, and can reach up to 35 km/h [29, 30]. Schäfer writes in her article, "This robot should walk your dog" (own translation): "Blue, round and rolls—may we introduce: Gita. What looks like a large suitcase from the outside can do much more on the inside, because Gita is a robot that can do your shopping and even wants to take your dog for a walk" ([31], own translation). In a video integrated into her article, this is shown—but it remains unclear whether the robot can interact or communicate with the dog, or even consider its needs, such as sniffing, marking, and relieving itself.

Filmmaker John Downer has created artificial monkeys, wolves, hippos, turtles, alligators, etc., in order to observe corresponding wild animals and obtain spectacular images. His robots are very intricately designed and resemble their subject in every detail. They can often move their limbs and move forward on four legs. Behind their eyes are cameras for observation. The majority will be remote-controlled robots but there is nothing to prevent autonomous ones from being used. The important thing here is that the artificial animal always aligns itself with the object of interest in order to fulfill its purpose. Regarding the BBC series "Spy in the Wild", the website states: "In one of the most innovative natural history series ever presented, Spy in the Wild deploys over 30 ultra-realistic animatronic Spy Creatures to go undercover in the animal world. [...] These robotic look-alikes make all the right moves to not only be accepted by animals but also interact with them, providing revelatory insights into their world." [32] The animals approach the robots curiously, touch them, poke them, try to involve them, and—in the case of a turtle—mate with them. They are not obviously foreign bodies and are seen as conspecifics—which raises the question of deception and trickery, which is otherwise dealt with mainly with regard to humans [33].

The Super Monster Wolf was developed in 2017 by JA Kisarazushi (a Japanese agricultural association) and the University of Tokyo. It uses infrared sensors to detect when animals approach the rice field. Golem magazine wrote on August 28, 2017: "It drives away deer, birds, wild boar, and even bears with loud noises: at up to 95 decibels, it can howl, hiss, speak like a human, imitate gunshots—in total, the machine can make 18 different sounds. The sounds are supposed to vary so that the animals to be driven away do not get used to them." [34] (own translation) However, according to the editor, the wolf cannot run after other animals. It stands on fixed metal legs and can only turn its head back and forth with its red eyes illuminated by LEDs. The power for sensors, LEDs, motors, and sound generator is supplied by a solar module. Obviously, the robot is not particularly nice to animals, but that is no reason not to see it as a social robot in the general sense.

13.4.2 Social Robots with Animals

Social robots of all kinds can be found in shopping malls, train stations, airport buildings, and in streets and squares. They automatically come into contact not only with humans but also with wild animals (in Europe, for example, birds, as well as martens, foxes, and wild boar) and domestic animals (in Europe, mainly cats and dogs). Interestingly, reports of animals attacking social robots rarely circulate. There is little evidence of other problems such as collisions or accidents due to falls. One explanation is that such incidents are not reported or that media, science, nor even animal welfare organizations are not interested in them so far. In addition, one must consider that social robots are not very common in public spaces. Collisions between service robots and animals have been documented [35]. Some of the service robots may become social robots, as mentioned. Incidentally, assaults by humans are not at all rare. For example, Pepper was attacked in a store [36], and the Hitchbot—a hitchhiking robot that was picked up by drivers—was the victim of (real or feigned) vandalism [37].

An early set of experiments—the first of which took place in 2002 [6]—involved the encounter between a dog and an AIBO, a Sony dog-like robot. According to the authors, animal-like autonomous robots that are recognized by animals as social partners could offer new possibilities in the study of interactions. A 2004 paper written by a team from the Department of Ethology at Eötvös Loránd University states that 24 adult and 16 four-to-five-month-old dogs were tested in two situations in which they encountered one of four different partners: a remote-controlled car, an AIBO robot, an AIBO with a puppy-scented fur cover (i.e., a form of robot enhancement), and a two-month-old puppy [38]. In the neutral situation, the dog was allowed to interact freely with one of the named partners for 1 min in a closed arena and in the presence of its owner. In the specific situation, the encounters were started while the dog was eating food [38].

> Our results show that age and context influence the social behaviour of dogs. Further, we have found that although both age groups differentiated the living and non-living test-partners for some extent, the furry AIBO evoked significantly increased responses in comparison to the car. These experiments show the first steps towards the application of robots in behavioural studies, notwithstanding that at present AIBO's limited ability to move constrains its effectiveness as social partner for dogs [38].

In the neutral situation, the dogs preferred the furry AIBO to the normal AIBO and the car. This also seems unsurprising, especially since the enhanced social robot not only looks more animal-like, but also smells more dog-like. There could be potential here for robot enhancement.

Tufts University confronted dogs with an NAO robot over a decade later. According to researchers' perceptions, animals, and dogs in particular, are everywhere in our social spaces, which in turn are increasingly infused with technology [8]. Previous research has primarily examined human-robot interactions and has not considered potential impacts of this technology on animals in the home and, in particular, the potential benefits of using social robots or service robots for animal care. However, for dog-robot interactions to be successful and effective, dogs would need to accept such robots and exhibit positive behaviors toward them [8]:

> Thus, research must determine possible robot characteristics such as particular movements or vocalizations that might be able to facilitate the dog's trust in and acceptance of the robot. The goal of the present exploratory study was to investigate the reaction of dogs to a small humanoid robot under different conditions of vocalization and movement. Our main finding from these dog-robot interaction experiments is that dogs unacquainted with the robot prefer robot vocalizations to robot movement. [8]

Meiying Qin from Yale University and her co-authors brought a dog together with NAO around 2020. In the process, they used its synthetic voice. In two experiments, they studied whether dogs respond to a social robot after it calls their name and whether they follow its "Sit!" commands. A loudspeaker was used alongside the NAO. When the dogs were called by their name, they responded to the social robot rather than the loudspeaker, and they did so by looking at it. They were also more likely to obey the "Sit!" commands given by the NAO than those given by the loudspeaker.

> The contribution of this study is that it is the first study to provide preliminary evidence that (1) dogs showed positive behaviors to social robots and that (2) social robots could influence dog's behaviors. This study enhance [sic] the understanding of the nature of the social interactions between humans and social robots from the evolutionary approach. Possible explanations for the observed behavior might point toward dogs perceiving robots as agents, the embodiment of the robot creating pressure for socialized responses, or the multimodal (i.e., verbal and visual) cues provided by the robot being more attractive than our control condition. [39]

It is striking and obvious that the social robots that were particularly well known in their time were preferred, whether they had an animaloid or humanoid form, and that the focus was on dogs, which can be found in many households and are useful for certain experiments.

13.4.3 Animal-Friendly Machines and Systems

This part is subdivided with headings to better distinguish the different projects. They are all projects that—including the preparatory work—the author himself carried out at the FHNW School of Business between 2014 and 2022.

13.4.3.1 The Animal-friendly Drone

One of the author's machine ethics projects, which he carried out at his university in 2015, involved the design of an animal-friendly photographic drone. This was to be at least partially autonomous and take pictures of fauna and flora. Humans, it was assumed, were not relevant as subjects and should be protected in their privacy and intimacy or informational autonomy [4]. For the drone, as in similar projects, an annotated decision tree was developed. At its core, this is a classic decision tree, such as those known in business informatics, but with moral assumptions and justifications added at the nodes. Such a model was developed also for animal-friendly robot cars, among other things, and presented at Stanford University in 2016 [40].

The modeling starts with the activity of flying. From above, the drone checks whether there is an object on the ground. If this is the case and it turns out to be a human, the recording function is not activated. If the object is an animal, the drone proceeds according to the animal species. Collisions with birds such as eagles or storks should be avoided, and shy animals should be left alone. The latter should only be photographed or filmed from a great distance. To give an example, the comment at "Is it a bird?" (to be checked by cameras and sensors) is "Birds must not be harmed by the drone", referring to both animal welfare and operational safety [40]. Rare species are captured from various heights depending on the endangerment effect the device may have on a specific animal species. If the machine does not detect an animal, it considers several other possible circumstances.

13.4.3.2 LADYBIRD

LADYBIRD is a prototype of an animal-friendly—more precisely ladybug-friendly—vacuum robot [41]. As early as 2014, the design study, which provided rough information about the desired appearance and planned functions of the device, was created and published via the website maschinenethik.net. The idea was repeatedly mentioned at lectures, in publications, and interviews. It was met with goodwill among listeners and readers, while garnering media and scientific interest, because the sense and purpose of a simple moral machine was made visible and comprehensible, and the concern of machine ethics was made understandable. Years later, a foundational work on machine ethics would enter with this very example of a moral machine [42]. In 2015, an annotated decision tree for LADYBIRD was created.

In the modeling, the activity of vacuuming is assumed [41]. The suction robot checks whether something lies in its path and, if this is the case and it is an animal, it clarifies what size it is. A cat is not problematic given the size of the suction device, but a ladybug is. If the vacuum cleaner robot encounters a ladybug, it will stop working. The moral assumptions are crude and simple. They do not have to be shared by everyone. Nor is it necessary, because different devices can be offered, the customers can be made aware of the extensions and limitations at the time of purchase, via product information, labels, and certificates, and they can be offered to modify the machine if they have divergent needs. For example, some people get out the vacuum cleaner to kill spiders, woodlice, or flies. It would not help them that LADYBIRD makes an exception for these animals. Admittedly, this contradicts the approach of being kind to animals. If no living creature is involved, other possible facts are included in the modeling.

In 2017, LADYBIRD was implemented as a prototype at the aforementioned university as part of a practice project [5]. The three-member team used the annotated decision tree described in this work. For recognition, it built a color sensor into the machine. Other desired and useful components—such as motion sensors or systems with image and pattern recognition—were not included because there was too little experience in these areas among the business information specialists and too little time. The result was a primitive robot, which was at least able to illustrate the concern and implementation possibilities. It drives through the area, recognizes an abstracted ladybug, stops for it, and emits a signal tone to let the user know. The paper on LADYBIRD was presented at the AAAI 2017 Spring Symposium "AI for Social Good (AISOC)" and published in the proceedings volume [41].

13.4.3.3 HAPPY HEDGEHOG (HHH)

HAPPY HEDGEHOG (HHH) is the prototype of an animal-friendly—or more precisely, hedgehog-friendly—robotic lawnmower. This was a revival of the LADYBIRD idea at the same university in 2019. However, this one was mainly about showing that moral rules can be implanted into simple machines, turning them into simple moral machines. In reality, ladybugs on the floor of apartments and houses are not a pressing problem, even though they themselves might think otherwise (their thinking capacity and their ability to feel and suffer are limited, however). Hedgehogs killed by robotic lawnmowers are definitely a major problem, and their sentience and capacity for suffering is undisputed. Probably tens of thousands are injured and thousands killed outright each year worldwide, predominantly young individuals blindsided by the mobile machine and unable or unwilling to move. This may be few compared to the casualties of the same species on the roads, but it is suffering that can be avoided without major effort. An annotated decision tree did not emerge in this case. However, the team of four who took on the challenge as part of a practice project had studied the one from LADYBIRD.

HHH is technically much more advanced than its predecessor LADYBIRD. Like the latter, it drives around autonomously. It is equipped with a thermal imaging camera. This enables it to detect living beings and warm objects in its path. When it comes to it, it pauses and, while pointing its high-positioned camera at the unknown object, applies its second method. Using machine learning—the team had fed it numerous hedgehog images—it was able to identify hedgehogs. When it does, it stops working immediately. It would make sense at this point for it to send a message to the owner. A signal tone, as with LADYBIRD, is only of limited use, because when operating a lawn-mowing robot—especially in larger areas—there is not always someone around.

13.4.3.4 ANIFACE

ANIFACE, a project at the university in 2021, was dedicated to facial recognition in animals [43] and is part of a larger context in which systems and machines of all kinds play a role [3]. Facial recognition in the sense of identifying individuals is controversial from a legal, social, and ethical perspective. In particular, its use in public spaces for mass surveillance purposes is rejected. Facial recognition in animals, on the other hand, seems to be uncontroversial from a social and ethical point of view—if one does not assume that they enjoy privacy and intimacy in the strict sense [44]—and might even have a potential for animal welfare. The idea was to design a system to identify and determine the position of individual brown bears in the Alps. The purpose was to avoid encounters between animals and humans and to protect both sides by informing citizens and tourists. The concept should be applicable not only to bears but also to wolves. Other animals returning to the Alps are less problematic.

The authors propose a system consisting of several components, namely wildlife cameras, ground robots, and camera drones. At the current stage of development, wildlife cameras should be used primarily in conjunction with facial capture and recognition. The wildlife cameras should be state of the art and automatically transmit photos. Once ground robots and camera drones reach a suitable stage of development, they will be added. Ground robots include models, such as the four-legged Spot from Boston Dynamics or the four-legged Go2 from Unitree, which can handle rough terrain. All systems are connected to each other and to a control center that houses technical systems and human forces or uses cloud computing. In principle, the three purely technical components combined with a computer solution or cloud computing application can perform the object and facial recognition capabilities described above. All systems should disturb the animals as little as possible and be active in the background. The concept was presented at ACI

2022 in Newcastle upon Tyne and published in the proceedings [43].

13.5 Social and Ethical Discussion of the Examples

This contribution is dedicated to the encounter between social robots and animals. The question is, of course, whether all the machines discussed in the last section are social robots. To answer this question, the definition cited at the beginning of this article should be recalled [10]. Social robots are sensorimotor machines, created to interact with humans and animals. They can be recognized through five dimensions: interaction and communication with living beings, proximity to living beings, representation of aspects or features of living beings, and their utility for living beings.

- The robots of the first part, which serve to entertain and exercise pets, seem to more or less fulfill this definition. Of course, they do not have distinctive communication capabilities. The robot that could walk dogs is probably more of a classic service robot. The observation robots can offer more here, depending on the design. It is also interesting that they are in the middle of the social structure of the representatives of an animal species. The animal scarecrow is an exception, its function a special case, but not an explicit special case in the social. Animals and humans also try to drive away beings that want to use up their resources.
- The social robots of the second part are predominantly typical representatives of their kind. AIBO and NAO can be regarded as highly developed, early classics. What is interesting here is that they are intended for humans but then also fulfill functions for animals or can be used in experiments that provide information about animal behavior. In a technologized world, this seems to be important and the experiment with NAO impressively showed that dogs react quite differently to different technologies. In

the case of AIBO, robot enhancement was crucial—i.e., equipping it with a puppy-scented coat. Research on social and humanoid robots has always revealed much about humans. Now it is also yielding insights about animals.

- The animal-friendly machines of the third part are the least obvious. Their benefit is clear. Proximity is present and crucial (although it is precisely excessive proximity with the risk of injury that is to be avoided). The representation of animal or human characteristics is variously present, in the design of LADYBIRD, in the imitation of eyes by cameras, in the emission of sounds. Communication and interaction with the animals, however, are very limited. LADYBIRD detects the ladybug and reacts to it—the other way round this does not have to be the case in any way. This is not much different from HHH, though the hedgehogs can theoretically curiously explore it when it is stationary, it is not exactly what one would think of as interaction. Perhaps the benefit, to return to this, of sparing the animals, is precisely what might lead one to conceive of the two artifacts as social robots. In this case, one would interpret "social" as various other definitions do, namely as "compatible", "serving", and "protecting". This is quite possible, although not necessary in the given context. The ANIFACE concept bridges the gap between observation robots and animal-friendly machines. The robots and drones used are also observing but stay in the background as much as possible—and ultimately try to help animals and humans.

It is worthwhile at this point to consider the benefits in another way. For example, some will doubt—from an animal ethics point of view—that social robots should be created to interact with and confront animals, suggesting that other approaches should be taken instead of animal-friendly machines. The social robots of the first section partly replace humans, partly complement them. They perform tasks that humans cannot perform in the same way for various reasons. The latter may be absent for a while or unable to take care of the animals—this may be objected to, but obviously the machine can help to solve a bottleneck here. Moreover, it is necessary to consider whether the pets are dependent on continuous human attention. This may vary from species to species and from individual to individual. The social robots of the second part serve experimental purposes. There is probably little to be said against testing them on cats and dogs if they are not unsettled or disturbed. One can even argue that these are some of the few animal experiments that are ethically justifiable. Ultimately, it's just a chance to learn more about the living creatures at our side—and about social robots' impact through their appearance and speech. Finally, the animal-friendly machines have emerged by "moralizing" ordinary autonomous machines in the context of machine ethics. When one uses autonomous machines in certain contexts, they should abide by certain rules, not least moral ones. This principle can be applied to quite a few systems, even those that are not social robots, such as automatic or autonomous cars. These should not quantify and qualify against humans (i.e., count and evaluate possible victims of a traffic accident)—if they do so with animals, there is little to object to [5].

Furthermore, it is necessary to examine which areas of applied ethics come into play and how they are linked. In the case of the toy and entertainment robots of the first part, questions from technology ethics and information ethics arise. In addition, animal ethics is relevant—for example, one can ask about the welfare of pets or how welfare is endangered through things like injury. The social robots of the second part also call for the same domain ethics. Here, moreover, the ethics of science must be included. The fact that such experiments are ethically acceptable and useful has already been mentioned. The animal-friendly machines of the third part come from a completely different area of applied ethics, namely machine ethics. It is about implanting moral rules into autonomous systems. There is a recurring narrative in the literature that information ethics and other areas of applied ethics can help machine ethics find the right moral rules, especially when machine ethics is conceived of as more of a technical discipline [5]. In addition, information ethics can also help solve problems that machine ethics has caused. Interdisciplinarity in this area can mean that philosophical ethicists with different orientations work together. Animal-friendly machines, of course, also raise issues of animal ethics and animal welfare. Thus, one can discuss which animals should be protected at all and in what way.

A final question is whether similarities and differences can be discerned in relation to the animals to which the robots depicted relate and whether they are immutable. The toy robots are intended for pets, especially cats. They could certainly entertain and exercise other smaller animals kept at home, such as hamsters and rats. They may even be helpful for some farm animals while being annoying and disruptive for others. The observation robots are all aimed at wild animals. They could also be used to observe domestic animals and livestock. The Super Monster Wolf should keep wild animals like wild boar off the field. It must be taken into account that the social function here is quite different, namely, to scare them away. The social robots in the experimental context are all geared towards dogs. Presumably, these are well suited to such settings, they are well researched themselves, are familiar to their owners, obey and follow them, and are available. The ACI conference is also heavily dominated by dog researchers, as the proceedings show [43]. However, it would be interesting to include other animals to learn more about them and the effect of social robots. The animal-friendly robots and systems discussed are focused on wildlife. This does not necessarily have to be the case—one can also think of some that help domestic animals or farm animals when they are in danger or become a danger.

13.6 Summary and Outlook

This chapter has considered social robots meant for animals and those that meet, interact, and communicate with them. It also presented animal-friendly machines created in the context of machine ethics, elucidating the possibilities and challenges around social robots and their interactions with animals. Without a doubt, given the plethora of engineering projects happening in this area, it would be beneficial to develop a comprehensive database containing product names, descriptions, and categorizations.

It is not entirely clear whether all the examples provided are best understood as social robots. In the case of entertainment and training robots, there was some evidence in support of doing so, also with the robots used in animal films. The case of the animal scarecrow is more dubious. With the social robots of the experiments, the classification was beyond discussion. The question here is which other models are suitable for handling. Most of them have wheels, not legs. The social robots involved in the experiments have two or four extremities and are thus more like living beings in this respect. Alpha Mini has legs and arms but is problematic with cats and dogs because of its small size. For the animal-friendly machines, the question could not be answered conclusively. However, this is also not essential for their development or use.

Regardless of how social robots designed to interact with animals develop, there is no doubt that those designed to interact with humans will become increasingly common, with new models such as Unitree's H1 and G1 or Figure 01 and Figure 02 from Figure. But as people surround themselves with animals, whether pets or livestock, the design must adapt accordingly to account for animal interaction. First, the exterior must be subjected to closer scrutiny. Perhaps an Uncanny Valley effect exists for animals? Then the voice must be inspected. Is there a voice available or imaginable that is equally pleasant or unpleasant (if deterrence is desired) for humans and animals? Last but not least, expressions of artificial intelligence must stand up to critical scrutiny, especially regarding decision-making behavior. The approach of a "nature-centered AI" could help here—however, such an approach would first have to be developed. In the future, large language models and large vision models could play a role. Such models are also used for robots, which can use them to better perceive and control their environment. Incidentally, it is likely that all-purpose robots such as G1 and Figure 02 will not only support humans, but also animals—looking after them, feeding them, and keeping them moving.

Animal ethics and animal protection agencies must also take part in the solution of this issue and discover what helps the living creatures in our environment and in the wild, and what does not. Finally, they should be willing to discuss scenarios and utopias. What would it be like if wildlife didn't live in reserves and parks, but we did? And what would it be like if they were watched and cared for by automatons and robots, including social robots that come into their immediate vicinity? Such possibilities are already discussed in the literature [3]. What if animal enhancement were implemented in the literal sense, for the benefit of animals, to enable them to live comfortably even under adverse circumstances [45]? Perhaps a highly engineered world is not necessarily a worse one, and social robots or service robots of other kinds could take over some important tasks, at least for wild animals.

Moreover, information ethics must address these issues. It is the key discipline in moral issues that arise in the use of information and communication technologies and information systems, as well as AI systems and robots. Together with robot ethics and technology ethics, information ethics can illuminate the relationship between technology and animals. This must involve collaboration with machine ethics, especially as it is often understood and operated in technical terms. All this leads again to the already raised demand: ethicists of all fields have to cooperate more across disciplines and bring in their respective expertise. This is necessary in many areas of application that have technical aspects—almost all areas of applied ethics cannot do without information ethics [46]. This is particularly necessary regarding animals. Here, we are not simply dealing with a special area of application within human activities, but with a huge, morally relevant group of living beings. Namely, living beings that encounter robots and AI systems for the first time in their history.

References

1. Bendel, O., ed. 2021. Soziale Roboter: Technikwissenschaftliche, wirtschaftswissenschaftliche, philosophische, psychologische und soziologische Grundlagen. Wiesbaden: Springer Gabler. https://doi.org/10.1007/978-3-658-31114-8.
2. Mancini, C. 2011. Animal-computer interaction: a manifesto. *Interactions* 18: 69–73. https://doi.org/10.1145/1978822.1978836.
3. Bendel, O. 2022. Passive, Active, and Proactive Systems and Machines for the Protection and Preservation of Animals and Animal Species. *Frontiers in Animal Science* 3. https://www.frontiersin.org/articles/10.3389/fanim.2022.834634/full.

4. Bendel, O. 2018. Towards Animal-Friendly Machines. *Paladyn, Journal of Behavioral Robotics* 9: 204–213. https://www.degruyter.com/view/journals/pjbr/9/1/article-p204.xml.
5. Bendel, O., ed. 2019. Handbuch Maschinenethik. Wiesbaden: Springer VS. https://doi.org/10.1007/978-3-658-17483-5.
6. Kubinyi, E., Á. Miklósi, F. Kaplan, M. Gácsi, J. Topál, and V. Csányi. 2002. Can a dog tell the difference? dogs encounter AIBO, an animal-like robot in two social situations. In *From Animals to Animats 7: Proceedings of the Seventh International Conference on Simulation of Adaptive Behavior*. MIT Press, Cambridge, 403–404.
7. Lakatos, G., M. Janiak, L. Malek, R. Muszynski, V. Konok, K. Tchon, and Á. Miklósi. 2014. Sensing sociality in dogs: what may make an interactive robot social? *Animal Cognition* 17: 387–397. https://doi.org/10.1007/s10071-013-0670-7.
8. Morovitz, M., M. Mueller, and M. Scheutz. 2017. Animal–Robot Interaction: The Role of Human Likeness on the Success of Dog–Robot Interactions. In *1st International Workshop on Vocal Interactivity in-and-between Humans, Animals and Robots*, Skövde, 25–26 August 2017.
9. Wild, M. 2018. Tiere als soziale Wesen. In *Handbuch Tierethik: Grundlagen—Kontexte—Perspektiven*, ed. J. S. Ach, and D. Borchers, 70–74. Stuttgart: J. B. Metzler. https://doi.org/10.1007/978-3-476-05402-9_12.
10. Bendel, O. 2020. Soziale Roboter. Gabler Wirtschaftslexikon. Wiesbaden: Springer Gabler.
11. Christaller, T., M. Decker, M. J. Gilsbach, G. Hirzinger, K. W. Lauterbach, E. Schweighofer, G. Schweitzer, and D. Sturma. 2001. Robotik. Berlin, Heidelberg: Springer. https://doi.org/10.1007/978-3-642-56422-2.
12. Höffe, O. 2008. Lexikon der Ethik. München: C.H. Beck.
13. Anderson, M., and S. L. Anderson, ed. 2011. *Machine Ethics*. Cambridge: Cambridge University Press. https://doi.org/10.1017/CBO9780511978036.
14. Wolf, U. 2012. Ethik der Mensch-Tier-Beziehung. Frankfurt am Main: Vittorio Klostermann. https://doi.org/10.5771/9783465141617.
15. Singer, P., and Y. F. Tse. 2023. AI ethics: the case for including animals. *AI and Ethics* 3: 539–551. https://doi.org/10.1007/s43681-022-00187-z.
16. Bendel, O. 2014. Tierethik. Gabler Wirtschaftslexikon. https://wirtschaftslexikon.gabler.de/definition/tierethik-54000. Wiesbaden: Springer Gabler.

17. Bendel, O. 2015. Überlegungen zur Disziplin der Tier-Maschine-Interaktion. In *GBS Schweiz*, 14 February 2015. https://gbs-schweiz.org/blog/ueberlegungen-zur-disziplin-der-tier-maschine-interaktion/.
18. Bendel, O. 2016. Animal Enhancement. Gabler Wirtschaftslexikon. Wiesbaden: Springer Gabler.
19. Bendel, O. 2015. Human Enhancement: Die informationstechnische Erweiterung und ihre Folgen. TATuP, Vol. 2: 82–89.
20. Bendel, O. 2021. Robot Enhancement. Gabler Wirtschaftslexikon. Wiesbaden: Springer Gabler.
21. Bendel, O. 2022. Möglichkeiten und Herausforderungen des Robot Enhancement. In *Mensch-Maschine-Interaktion—Konzeptionelle, soziale und ethische Implikationen neuer Mensch-Technik-Verhältnisse*, 267–283. Münster: Mentis.
22. Bendel, O. 2020. Überlegungen zu Bio- und Bodyhacking. HMD Praxis der Wirtschaftsinformatik 57: 480–492. https://doi.org/10.1365/s40702-020-00605-y.
23. Bendel, O. 2020. Der Einsatz von Servicerobotern bei Epidemien und Pandemien. HMD Praxis der Wirtschaftsinformatik 57: 1286–1301. https://doi.org/10.1365/s40702-020-00669-w.
24. Bendel, O. 2020. Sicherheitsroboter. Gabler Wirtschaftslexikon. Wiesbaden: Springer Gabler.
25. May, R., Ø. Hamre, R. Vang, and T. Nygård. 2012. Evaluation of the DTBird video-system at the Smøla wind-power plant. *Detection Capabilities for Capturing Near-Turbine Avian Behaviour*. https://doi.org/10.13140/RG.2.1.1909.9928.
26. Wimmer, T., M. Israel, P. Haschberger, and A. Weimann. 2013. Rehkitzrettung mit dem Fliegenden Wildretter: Erfahrungen der ersten Feldeinsätze. In *Bornimer Agrartechnische Berichte*, 85–95.
27. Kurain, N. S., S. Poojasree, and S. Priyadharrshini. 2018. Wildlife Vehicle Collision Avoidance System. *International Journal of Electronics and Communication Engineering* Volume 5. SSRG International Journal. https://doi.org/10.14445/23488549/IJECE-V5I3P104.
28. Katzenspielzeug—Roboter Maus [sic] für die Katze. 2020. https://www.spielzeug-katze.de/dadypet-katzenspielzeug-roboter-maus-fuer-die-katze/. May 7.
29. Bendel, O. 2020. Serviceroboter aus Sicht der Ethik. In *Schöne neue Welt? Drei thematische Einblicke*, ed. Mathias Lindenau, Marcel Meier Kressig. Vadian Lectures Volume 6. Bielefeld: transcript, 57–76. https://doi.org/10.1515/9783839452516-003.

30. Coxworth, B. 2017. "The Vespa of Robots" follows you home. New Atlas. January 3. http://newatlas.com/piaggio-gita/47646/.
31. Schäfer, M. 2017. Dieser Roboter soll mit deinem Hund Gassi gehen. TECHBOOK. February 16. https://www.techbook.de/easylife/life-hacks/dieser-roboter-soll-mit-ihrem-hund-gassi-gehen.
32. Spy in the Wild. 2024. John Downer Productions. http://jdp.co.uk/programmes/spy-in-the-wild.
33. Schulze, H., O. Bendel, M. Schubert, M. Binswanger, M. Simmler, R. T. D. Reimer, A. Tanner, et al. 2021. Soziale Roboter, Empathie und Emotionen—Eine Untersuchung aus interdisziplinärer Perspektive. Zenodo. https://doi.org/10.5281/zenodo.5554564.
34. Pluta, W. 2017. Roboterwolf vertreibt Wildschweine. Golem. https://www.golem.de/news/landwirtschaft-roboterwolf-vertreibt-wildschweine-1708-129719.html.
35. LaVallee, A. 2008. When Dogs and Robots Collide, Somebody Needs a Talking To. Wall Street Journal, June 11, sec. News. https://www.wsj.com/articles/SB121314664909963011.
36. Collins, K. 2015. Drunk man attacks adorable robot that was just trying to help. Wired UK. September 8. https://www.wired.co.uk/article/drunk-man-attacks-pepper-robot-in-japan.
37. Sokolov, D. A. J. 2015. Trampender Roboter HitchBOT zerstört: Wunderliches Überwachungsvideo. Heise Online. August 4. https://www.heise.de/newsticker/meldung/Trampender-Roboter-HitchBOT-zerstoert-Wunderliches-Ueberwachungsvideo-2767051.html.
38. Kubinyi, E., Á. Miklósi, F. Kaplan, M. Gácsi, J. Topál, and V. Csányi. 2004. Social behaviour of dogs encountering AIBO, an animal-like robot in a neutral and in a feeding situation. *Behavioural Processes* 65: 231–239. https://doi.org/10.1016/j.beproc.2003.10.003.
39. Qin, M., Y. Huang, E. Stumph, L. Santos, and B. Scassellati. 2020. Dog Sit! Domestic Dogs (Canis familiaris) Follow a Robot's Sit Commands. In *Companion of the 2020 ACM/IEEE International Conference on Human-Robot Interaction*, 16–24. HRI '20. New York, NY, USA: Association for Computing Machinery. https://doi.org/10.1145/3371382.3380734.
40. Bendel, O. 2016. Annotated Decision Trees for Simple Moral Machines. In *The 2016 AAAI Spring Symposium Series*. AAAI Press, 195–201. Palo Alto.
41. Bendel, O. 2017. LADYBIRD: the Animal-Friendly Robot Vacuum Cleaner. In *The 2017 AAAI Spring Symposium Series*, 2–6. Palo Alto.

42. Misselhorn, C. 2018. *Grundfragen der Maschinenethik*. 4. ed. Ditzingen: Philipp Reclam jun..
43. Bendel, O., and A. Yurekkirmaz. 2022. A Face Recognition System for Bears. In *9th International Conference on Animal-Computer Interaction: Defining Tomorrow, ACI 2022, December 5, 2022 — December 8, 2022. ACM International Conference Proceeding Series*. Newcastle upon Tyne, United Kingdom: Association for Computing Machinery. https://doi.org/10.1145/3565995.3566030.
44. Paci, P., C. Mancini, and B. Nuseibeh. 2022. The Case for Animal Privacy in the Design of Technologically Supported Environments. *Frontiers in Veterinary Science* 8. https://www.frontiersin.org/articles/10.3389/fvets.2021.784794/full.
45. Bendel, O. 2021. Chips, Devices, and Machines within Humans: Bodyhacking as Movement, Enhancement, and Adaptation. In *Mensch. Maschine. Kommunikation.*, ed. Christa Dürscheid, and Sarah Brommer. Beiträge zur Medienlinguistik. Tübingen: Narr Francke Attempto.
46. Bendel, O. 2013. Die Medizinethik in der Informationsgesellschaft. *Informatik-Spektrum* 36: 530–535. https://doi.org/10.1007/s00287-012-0661-0.

The False Moon

Lightbulbs: the technology that revolutionized the way people work through the day by increasing productivity and lengthening the amount of time we can spend together. How can such a transformative tool negatively impact the natural world?

Many tropical coasts are the hatching grounds of endangered sea turtle species like the loggerhead, leatherback, and green sea turtles [1]. In recent years, turtle hatchlings have been found in environments far from the sea. Turtles instinctually follow the brightest light: the sunlight bouncing off the water during the day, and the moon reflecting off the waves in the night. It is an instinct meant to help hatchlings differentiate the dark vegetation of the coast from the bright ocean waves and lead them towards the sea. With the rise of tourism, bright lights, and humans inhabiting every mile of coastline, turtle hatchlings become disoriented when they emerge from the sand. They follow the false moon, the lights of coastal cities. Baby turtles wander into territory they are unprepared for, far from the waves they are supposed to meet, and become trapped in sewers and caught under cars.

There are methods which have been implemented to prevent these animals from being lost to the light [2]. Red lights and low-pressure sodium-vapor (LPS) lights have been found to be less disruptive to turtle nests, and tinted windows or light blocking curtains can decrease the glow coming from areas inhabited by humans. Locals can be informed of turtle nesting and hatching season to stay conscious about the amount of time they leave lights on during the night. The disruption of unnatural light serves as an example of the unexpected ways technology can interfere with the lives of non-human animals and alerts prospective engineers to be proactive in their implementation of new technologies by investigating all potential side effects, including those of the natural environment.

Citation

1. Homestead, Mailing Address: 40001 SR-9336, and FL 33034 Phone: 305 242-7700 Contact Us. 2024. Sea Turtles and Light - Dry Tortugas National Park (U.S. National Park Service). https://www.nps.gov/drto/learn/nature/seaturtleslight.htm. Accessed May 9.
2. Information About Sea Turtles: Threats from Artificial Lighting - Sea Turtle Conservancy. 2024. https://conserveturtles.org/information-sea-turtles-threats-artificial-lighting/. Accessed May 9.

Chapter 14

Lead a Horse to Water: Beach Ponies in the American South and the Cultivation of Ferality

Abraham Gibson
Department of History, University of Nebraska, Lincoln, Nebraska, USA
agibson12@unl.edu

14.1 Introduction

The Spanish mustangs who roam the American West are one of the most famous animal populations on Earth, and yet few people realize that free-roaming horses can also be found in the American South. There are currently nine distinct bands in the region, and most of these animals live on remote barrier islands along the southeastern coast of the United States, from southern Maryland to northern Florida. For untold generations, these animals have grazed in the marsh, frolicked in the surf, and reproduced beyond humanity's meddlesome dominion. Like their more famous cousins out West, these "beach ponies" have played a critically important role in local history. Unlike their western cousins, however, these beach ponies are not federally protected. As a result, each population

Animals, Ethics, and Engineering: Intersections and Implications
Edited by Rosalyn W. Berne

ISBN 978-981-5129-56-4 (Hardcover), 978-1-003-65231-1 (eBook)
www.jennystanford.com

finds its fate tied to an ever-evolving constellation of locally contingent social, cultural, and ecological factors.

This chapter examines the history of free-roaming horses in the American South, including the extant populations that currently live on barrier islands. In doing so, it gives scholars an opportunity to reflect on the significance of *engineering* in animal history. Ostensibly "wild," beach ponies are actually "feral," which means that they are descended from once-domesticated ancestors, but that they now live in the wild. As many scholars have now shown, feral animals are uniquely instructive precisely because they are neither wild nor domestic, and yet they are somehow both. Far from their reputation as icons of raw liberty, beach ponies belong to highly engineered populations that are in some ways more tightly controlled than their "domesticated" cousins. They undermine the dichotomy that distinguishes wild from domestic, and they invite us to reexamine traditional assumptions about both engineering and animals [1–5].

Studying the history of beach ponies also allows one to ask whether and how *ethics* may have contributed to their survival. Up and down the southeastern seaboard, people have gone to extraordinary lengths to preserve and support these animals, devising ingenious methods to manipulate their numbers, their distribution, and their well-being. To understand what motivates these people, one must understand the unique relationship between horses and people in each area. Beach ponies have played many different roles over the years, and each one has influenced their fate in dramatic fashion. Nowadays, they are most frequently cast as ancient relics from a bygone era, final vestiges of the open range, or romantic icons of unbridled liberty. Because of this long history and uncertain future, many people feel an ethical obligation to help preserve these animals. Beach ponies thus provide scholars with an excellent opportunity to reexamine the complex relationship between engineering and ethics, but from a unique perspective.

14.2 Colonial Husbandry

Horses accompanied the Spanish conquistadors when they first explored the American South in the early sixteenth century, but there is no evidence that any of them established feral populations apart from humans. The first horses in the English colony of Virginia were unceremoniously eaten during the "Starving Time" that befell Jamestown in the winter of 1609–1610, but subsequent arrivals fared much better. Crucially, Virginia's colonists chose to build fences around their crops, rather than their animals, and horses were therefore left to fend for themselves. They managed surprisingly well. By the 1670s, thousands of horses stalked the woods of eastern Virginia. Some colonists treated these horses like game, hunting them for sport. Others treated them like a sort of colonial mass-transit system, drafting the animals into service on an as-needed basis. One visiting Frenchman noted that colonists rode horses everywhere, even a mere hundred yards down the road, and that they would usually just surrender the animals back into the woods after they reached their destination [6]. Still others viewed the horses as means to political ends. For example, Nathaniel Bacon conscripted hundreds of horses out of the woods when he incited a rebellion against the colonial government [7].

The other southern colonies followed Virginia's lead, stipulating that colonists should fence their crops and not their animals, thereby giving horses the run of the land. Even so, the open range looked different in different places. In the 1740s, John Brickell reported that droves of horses roamed the savannas of South Carolina, more than 50 miles from the closest person [8]. To harvest these feral bands, planters enlisted their enslaved laborers, many of whom had worked with horses in their ancestral West Africa. It is no exaggeration to say that America's earliest cowboys were enslaved Africans [9]. Meanwhile, this overabundance of unsupervised horses allowed both the Creek and Cherokee confederacies to obtain horses by mid-century. When the first proto-Seminoles began moving into northern Florida around 1763, they did so on horseback [10].

14.3 Antebellum Novelties

Feral horses persisted in sparsely populated Florida for several more decades, but they grew increasingly scarce in long-settled parts of the South. Other species continued to roam free, but horses were deemed too large, too conspicuous, and too valuable. In the Revolutionary War, every available horse had been pressed into service. By the early 1800s, feral horses had been absent from the range for several generations. The last scattered bands were confined to the most remote, inaccessible margins of the South: barrier islands off the coast of Virginia, North Carolina, and Georgia.

The earliest published reference to free-roaming horses on barrier islands of the American South appeared in *American Turf Register* in September 1829. The article noted that "small horses, known by the name of Beach Ponies," roamed various islands along the southeastern seaboard, from Delaware to Georgia, but otherwise did not elaborate [11]. Six years later, Thompson Holmes, a physician who lived on the Delmarva Peninsula near Chincoteague Bay, described the Virginia horses at length for the *Farmer's Register.* He explained that horses roamed free on nearby Assateague Island for most of the year, but that they were rounded up during an annual "penning" event that attracted thousands of visitors one weekend every summer. The whole frenzied scene was marked by "unrivalled noise, uproar and excitement, which few can imagine who had not witnessed it, and none can adequately describe" [12].

The annual penning event grew even more popular in the years following the Civil War. Every summer, thousands of city-dwellers from the urban Northeast would travel down to the eastern shore of Virginia, a land that time had seemingly forgotten, and marvel that horses still roamed free across the land and in the surf. Significantly, neither locals nor tourists had any clue when or how the horses first arrived on the island. Even so, every newspaper or magazine article describing the animals mentioned the popular legend that the animals owed their origin to a mythical shipwreck. "A vessel loaded with horses, sailing to one of the Elizabethan settlements of Virginia, was wrecked upon the southern point of the island, where the horses escaped," *Scribner's Monthly* explained in 1877, adding that the horses were already roaming the islands when European settlers first arrived [13].

The feral horses who roamed the Outer Banks of North Carolina first attracted notice on the eve of the Civil War. In the 1850s, noted agriculturalist Edmund Ruffin, the editor of the aforementioned *Farmer's Register,* confirmed that hundreds of feral horses also roamed the Outer Banks. He explained that locals held an annual penning festival every summer, like the one in Chincoteague, and that those lucky enough to attend a penning on the banks would not forget it. Locals did not use lassos to subdue the animals, but instead used their bare hands and their steel courage. Some of the horses were loaded onto small barges and taken to the mainland to be sold [14]. It was hard to beat the investment. The market for horses was stronger than ever, and those herds on the Outer Banks cost literally nothing. The horses who were not ferried across to the mainland were turned loose, left to fend for themselves until the next go-round. Yet another similarity that North Carolina's horses shared with the Virginia herd was that nobody knew how they got to the islands. Not surprisingly, a similar origin myth emerged. Articles sometimes cited a long-forgotten shipwreck to explain the presence of horses in the remote Outer Banks, but they more frequently cited Walter Raleigh and his doomed Roanoke colony [15–17].

14.4 Divergent Fates

Demand for horses had increased throughout the nineteenth century, but the invention of the horseless carriage changed everything. The first automobile arrived in Chincoteague by ferry in 1909 and a causeway connecting the island to the mainland was finally completed in 1922. One might think the arrival of automobiles would spell disaster for Virginia's feral horses, but the opposite was true. By the 1920s, the newly formed Chincoteague Fire Department had assumed control of the annual penning event. The firefighters, who did not actually own the horses, were charged with rounding up the ponies on Assateague and ushering them to Chincoteague. Since there were more horses than boats, they directed the horses to swim across the channel. The horses were rushed ashore, herded through the streets of Chincoteague, and enclosed in a holding pen near the center of town where they could be auctioned off. The first "pony swim" conducted by the Fire Department was a huge success, attracting more than 15,000 people [18].

It is significant that the origin myth surrounding these horses intensified during the automobile era. Every article that was written about the animals mentioned, without evidence, that they had survived a shipwreck at some point in the distant past. Eventually, the myth became part of the official record. Virginia artist Jack Woodson painted a dramatic mural in the Hotel Russell on Chincoteague that showed the skeleton of a shipwreck just offshore, dramatizing the horses' noble (if unfounded) pedigree. His mural was later printed on postcards and mailed around the world. In 1947, Marguerite Henry published *Misty of Chincoteague*, a beloved children's book, which repeated the fable that horses first arrived on Assateague when a Spanish galleon wrecked against the reef centuries earlier [19].

Automobiles had a much different impact on beach ponies in the Outer Banks. Melville Chater, a writer for *National Geographic*, observed immense droves of feral horses when he visited the Outer Banks in 1926. He estimated that somewhere between 5,000 and 6,000 beach ponies roamed the island, and, like everyone else, he wondered about their mysterious origins. "Current tradition has it that they are descended from Barbary ponies which were brought over by Sir Walter Raleigh's colonists," Chater wrote. He watched as a chosen few were auctioned off for $6 per head. He learned that the same horses would have sold for more than $100 each a few years earlier. When he inquired about the change, one local explained: "too much gasoline about nowadays!" [20].

By the 1930s, the number of horses living on the Outer Banks had grown to such an extent that authorities decided to destroy them. Beginning in 1935, an extermination campaign was initiated, and thousands of horses were shot down. The extermination campaign was executed by county deputies, who were assisted by members the United States Forestry Service and the Federal Bureau of Fisheries. Although estimates had placed the number of ponies as high as 6,000 as recently as 1935, fewer than 50 feral horses roamed the entire Outer Banks just three years later. On June 14, 1938, newspapers reported that eastern North Carolina's most expert marksmen had set out that morning armed with high-powered rifles and dum-dum bullets, in pursuit of the "final extinction of the banker ponies" [21–23]. Alas, while their numbers were drastically reduced, the horses were never completely exterminated.

Meanwhile, the feral horses of Georgia faced similar challenges. They too had been released on coastal Sea Islands during the colonial period. James Oglethorpe had personally placed horses on Cumberland Island and Amelia Island [24]. Horses were still roaming Georgia's various Sea Islands in 1886, when a cohort of millionaires, including James Pierpont Morgan, William Vanderbilt, and Joseph Pulitzer, purchased Jekyll Island for $125,000. The island contained more than 200 feral horses at the time of purchases. Finding little use for the animals, the new owners "found it necessary to wage a war of extermination on the wild horses" [25]. Not all of the animals were killed, however. Some of them were trapped and transported to nearby Cumberland Island, where a different set of millionaires, the Carnegies, had recently purchased land. They allowed the refugee horses from Jekyll Island to join the "half-wild horses" who were already roaming Cumberland Island prior to their arrival [26].

Rather than spelling doom, the arrival of the automobile reaffirmed the Carnegies' commitment to feral horses. In 1921, aspiring archaeologist Oliver Ricketson (25-year-old son of Margaret Carnegie Ricketson) observed immense droves of feral horses during his first professional expedition to Arizona. Impressed with their sprightly vigor, he loaded a boxcar full of "unbroken mares" and sent them back home to be released on Cumberland Island [24]. Additional horses were released throughout the twentieth century, most recently in 1992. These frequent, well-documented introductions on Cumberland Island highlight an important point. Namely, these horses may evoke romantic stirrings of unbroken spirits and unbridled passions, but they are nevertheless intensively managed, and they probably would not exist if people did not replenish their numbers.

14.5 Modern Period

There are currently nine bands of beach ponies, and they are each subject to unique pressures. Take the fabled Assateague horses, for example. When a hurricane killed half of the herd on Ash Wednesday,

1962, locals released new horses to help supplement their numbers. In 1978, the Chincoteague Fire Department was compelled to euthanize numerous horses who had recently tested positive for equine infectious anemia. To supplement the suddenly smaller herd, they procured 40 mustangs from California, and released them on Assateague. There are now more than 150 horses on the island. Despite these well-documented introductions, the shipwreck myth persists. For example, the Visitor Center at the Chincoteague Wildlife Refuge features a dramatic mural that depicts horses scampering ashore during a storm while their ship founders behind them.

There is an interesting juxtaposition between the feral horses living on the southern half of Assateague Island, and thus in Virginia, and the horses living on the northern half of the island, and thus in Maryland. In the 1960s, the federal government bought the Maryland side of the island from beleaguered investors, established Assateague Island National Seashore, and erected a wire fence that would prevent Virginia's horses from intermingling. At the same time, the Parks Service removed all of the privately owned horses from the Maryland area. In response, local Jaycees donated 21 horses to form the nucleus of a new herd. Park officials adopted a hands-off approach to managing these animals, and, as a result, more than 60 horses roamed the Maryland side of the island by 1970. To prevent them from overrunning the island, biologist Jay Kirkpatrick devised a new method to deliver immuno-contraceptives remotely. In short, he shot the horses with birth-control darts [27]. Assateague Island thus functions like a natural laboratory, with two distinct bands of beach ponies that are engineered in radically different ways.

There are three extant populations in the Outer Banks of North Carolina, and they each face different prospects. The beach ponies in Ocracoke attracted national attention in the 1950s, when *Boy's Life* magazine published an article on the nation's most unique Boy Scout troop. Each boy in the troop was responsible for catching and taming a horse of his choosing. Even so, the number of horses continued to dwindle. By the 1970s, there were just nine horses left on Ocracoke. In response, the National Parks Service made a special effort to save them from extinction. Officials imported an Andalusian stallion named Cubanito to breed with the remaining mares on hand.

The herd gradually increased, and there are now about 25 horses on Ocracoke [28].

Another population lives at the southern end of North Carolina's barrier islands, on Shackleford Banks. There were about 240 horses in the area by the 1990s. In response to this perceived overcrowding, the Park Service rounded up the horses to give them health inspections. More than 70 of the animals tested positive for equine infectious anemia and were subsequently euthanized. The public was outraged, which resulted in federal legislation. In 1998, President Bill Clinton signed a law protecting the horses on Shackleford Banks, provided they number no fewer than 100 and no more than 110 [29].

Yet another population lives near Corolla, in the northern Outer Banks. This area was developed much later than the rest of the Outer Banks (Corolla did not even get electricity until 1955), and so there is less documentation regarding the historical distribution of feral horses in the region. We know there were dozens roaming the northern banks when members of the Spanish Mustang Registry visited Corolla in 1982, and that six horses were killed in a single automobile accident in 1989. In response, several concerned citizens of Currituck County (where the horses were located) established the Corolla Wild Horse Fund. Since then, these individuals have tirelessly sought to protect Corolla's feral horses. The organization assures visitors to its website that the horses are descended from Spanish mustangs, touting endorsements from the Spanish Mustang Registry and the American Livestock Breeds Conservancy [30]. According to the *Official Horse Breeds Standards Guide*, the feral horses of Corolla are among the "purest herds" of Spanish mustangs in existence [31]. This is more than just idle boasting about fabled pedigrees. If the animals are descended from Spanish stock, they may well represent a unique genetic reservoir and gain federal protection accordingly. This could be vitally important to the herd's continued existence [32, 33].

Farther south, a sixth population continues to roam Cumberland Island off the coast of southern Georgia. This population had already been supplemented with several well-documented introductions by 1972, when the federal government established Cumberland Island National Seashore. The National Park Service (NPS) allowed the horses to remain on the island, and they even allowed a nearby resident to introduce several Arabian horses to the herd in 1992. There are currently about 140 horses on the island. For the past decade or so, the foaling rate and the death rate have remained relatively close to one another, meaning park officials have not had to remove any horses during that span. The NPS allows feral horses to roam Cumberland Island, though it should be noted that officials also provide a list of items (dunes, wetlands, water quality, ground stability, historic structures, archaeological sites, native grasses, and wildlife) that the free-roaming horses purportedly threaten.

Finally, three populations trace their founding event to deliberate introductions in the twentieth century. There are roughly twenty "marsh tackies" (the colloquial name for Lowcountry horses) now roaming Little Horse Island off the coast of South Carolina. Citing the animal's lack of cultivation over the years, breed-standard guides refer to the marsh tacky as living time capsules from colonial times. That may well be true, but they are more recently descended from horses who were deliberately released onto the island in the 1950s [34]. Meanwhile, two populations of beach ponies have been transplanted away from barrier islands altogether. Fifty horses from Assateague were released in Grayson Highlands State Park in southwestern Virginia in 1975, and their descendants continue to roam the mountainous terrain decades later. Last but not least, local boosters donated six "cracker horses" (the colloquial name for Florida horses) to Paynes Prairie State Park near Gainesville, and there are now 30 horses in the park [31].

14.6 Final Thoughts

Many scientists believe that Earth has entered a new geological epoch known as the Anthropocene. They explain that human activities have transformed the climate, the continents, and the oceans on a planetary scale, and that humanity's collective influence is now so widespread that it is, literally, a global force of nature. Many insist that there is no place on Earth that is free from anthropogenic influence, and that there are no more truly "wild" places left on the planet. Some hasten to add that humanity's influence on the biosphere is not only pervasive but also negative. They note that humans have destroyed ecosystems around the globe, and that we have now ushered in the planet's sixth great extinction event.

Beach ponies occupy a strange niche in the Anthropocene. Their very existence is noteworthy, especially since most of the planet's large mammals are rapidly disappearing, but that is not the only thing that makes them unique. After all, beach ponies are defined by their purported divorce from humanity, and yet they exist on a planet in which every nook and cranny is subject to humanity's influence. One might conclude that beach ponies are doomed, and yet that is not the case. Instead, beach ponies continue to exist because people cannot stop creating them.

Colonists once feralized horses so that they could focus on their cash crops, but those times are long gone. Nowadays, people feralize horses so that the animals can be romanticized in dramatic settings, and, if future generations want to manufacture the same sublime experience, they will eventually have to feralize additional horses. And yet, strange though it may sound, beach ponies could face a brighter future than their domestic cousins. Their numbers might be limited, but most beach ponies have now been granted state or federal protection in perpetuity. Our compulsion to romanticize these free-roaming horses guarantees that they will survive for generations hence. What is that if not evolutionary success?

Developing a comprehensive strategy to guide humanity's interactions with beach ponies will not be easy, but nor is it impossible. We can start by abandoning the idea that beach ponies live apart from us. Humans are the primary force shaping the distribution and composition of all feral populations, beach ponies included. These animals are subject to a much different kind of anthropogenic selection, but it is anthropogenic selection, nonetheless. In other words, we retain dominion over these animals whether we like it or not. It is a daunting responsibility, but it is also an opportunity to fix what previous generations got wrong. After all, we inherit our forebears' decisions, however noble or shameful they might have been, but that does not mean that we are also beholden to them. We have free will and we can change course. The sooner we accept this fact, the sooner we can finally start cultivating a future that affords dignity to other animals and to one another.

References

1. Gibson, A. H. (2016). *Feral Animals in the American South: An Evolutionary History*. Cambridge University Press.
2. Anderson, V. D. (2006). *Creatures of Empire: How Domestic Animals Transformed Early America*. Oxford University Press.
3. Gibson, A. H. (2021). 4 Harvesting Hogzillas: Feral Pigs and the Engineering Ideal. In *4 Harvesting Hogzillas: Feral Pigs and the Engineering Ideal*, 60–70. University of Chicago Press. https://doi.org/10.7208/chicago/9780226783574-005.
4. O'Connor, T. (2013). *Animals as Neighbors: The Past and Present of Commensal Animals*. MSU Press.
5. Kóvskaya, M. (2021). Feral Atlas: The More-Than-Human Anthropocene. *Asian Anthropology* 20 278–281. https://doi.org/10.1080/1683478X.2021.1966913.
6. Dauphiné, D. (1923). *A Frenchman in Virginia: Being the Memoirs of a Huguenot Refugee in 1686*. [Richmond]: Priv. print.
7. A Narrative of the Rise, Progresse and Cessation of the Late Rebellion in Virginia. (2005). In *Samuel Wiseman's Book of Record*, 205. Plymouth, UK: Lexington Books.
8. Brickell, J. (1737). *The Natural History of North Carolina*. Johnson Publishing Company.
9. Sluyter, A. (2015). How Africans and Their Descendants Participated in Establishing Open-Range Cattle Ranching in the Americas. *Environment and History* 21: 77–101. https://doi.org/10.3197/096734015X14183179969782.
10. Hoffman, P. E. (2002). *Florida's Frontiers*. Indiana University Press.
11. Natural History: Horse. (1829). *American Turf Register and Sporting Magazine*, September.
12. Holmes, T. (2024). Some Account of the Wild Horses of the Sea Islands of Virginia and Maryland | The Countryside Transformed: https://eshore.iath.virginia.edu/node/1957. Accessed January 10.
13. Pyle, H. (1877). Chincoteague: The Island of Ponies. *Scribner's Monthly*, April.
14. Ruffin, E. (1861). *Agricultural, Geological, and Descriptive Sketches of Lower North Carolina, and the Similar Adjacent Lands*. Raleigh: Printed at the Institution for the Deaf & Dumb, & the Blind.
15. Olds, F. (1902). The Wild Horse of the Banks. *Forest and Stream*, November 15.

16. The Wild Horse in America. (1903). *Forest and Stream*, January 3.
17. Wild Horses of North Carolina. (1902). *Springfield Republican*, December 7.
18. Mariner, K. (1996). *Once Upon an Island: The History of Chincoteague*. Miona Publications.
19. Henry, M. (1947). *Misty of Chincoteague*. Rand McNally.
20. Chater, M. (1926). *Motor-Coaching Through North Carolina*. Nat Geographic Mag.
21. Progress May Mean End of Wild Ponies. (1935). *Raleigh News and Observer*, May 5.
22. Dare Sharpshooters Begin Banker Ponies' Extinction. (1938). *Raleigh News and Observer*, June 14.
23. Federal Writers' Project (N.C.). (1939). *North Carolina, a guide to the old north State*. Chapel Hill, The University of North Carolina Press.
24. Bullard, M. (2003). *Cumberland Island: A History*. University of Georgia Press.
25. Van Wormer, J. (1898). An Island on the Georgia Coast. *The Cosmopolitan*, January.
26. Ober, F. (1880). Dungeness, General Greene's Sea-Island Plantation. *Lippincott's Magazine of Popular Literature and Science*, August.
27. Keiper, R. (1985). *The Assateauge Horses*. Tidewater Publishers.
28. Urquhart, B. S. (2002). *Hoofprints In The Sand: Wild Horses of the Atlantic Coast*. Lexington, Ky: Eclipse Press.
29. Prioli, C., and Taylor, S. (2007). *Wild Horses of Shackleford Banks*. Illustrated edition. Blair.
30. McCalpin K. H. (2010). *Saving the Horses of Kings The Wild Horses of Currituck Outer Banks*. Outer Banks Press, Kitty Hawk, NC.
31. Lynghaug, F. (2009). *The Official Horse Breeds Standards Guide: The Complete Guide to the Standards of All North American Equine Breed Associations*. 1st ed. Minneapolis: Voyageur Press.
32. Conant, E. K., Juras, R., and Cothran, E. G. (2012). A Microsatellite Analysis of Five Colonial Spanish Horse Populations of the Southeastern United States. *Animal Genetics* 43: 53–62. https://doi.org/10.1111/j.1365-2052.2011.02210.x.
33. Sponenberg, D. P. (2000). *Deciding Which Feral Horse Populations Qualify as a Genetic Resource*. Resource Notes no. 25. Denver, CO: U.S. Department of the Interior, Bureau of Land Management, National Science & Technology Center.

34. Kidd, J. (2015). Update: Marsh Ponies of Little Horse Island Make it Through Difficult Winter. *Island Packet*, March 9.
35. Russell, A. L., and Vinsel, L. (2018). After Innovation, Turn to Maintenance. *Technology and Culture* 59. Johns Hopkins University Press: 1–25.
36. Pollan, M. (1991). *Second Nature—A Gardener's Education*. 1st ed. New York, NY: Atlantic Monthly Press.

Chapter 15

Mechanical Symbiotes: Robotics in Coral Restoration

Nicole E. Carey
Autodesk Robotics Laboratory, Autodesk Inc. San Francisco, California, USA
nic.carey@autodesk.com

It took over 2 hours by boat to reach our destination; a coral nursery in the Houtman Abrolhos, an island chain off the coast of Western Australia near the town of Geraldton. The winds in that region are almost ceaseless, either blustery south-westerlies which drive the waves high and bend the scraggly trees clinging to the coastline, or hot dry easterlies that sweep in dust and relentless heat from the desert plains. We were heading into a southern swell, plummeting in and out of deep-sea troughs and watching the occasional pod of dolphins ride the bows. It was fun at first, then grimly nauseating.

On-shore the temperature was an oppressive 45°C, but 80 km to the northwest we emerged quite suddenly from the swell into a serene and balmy turquoise lagoon, floating above thickly bristling trails of natural and artificial reef. Wrasse and damselfish thronged, feeding on the thick algae that encroaches on coral during the warm months, and biting our fingers as we detached large artificial reef plugs covered in half-grown, experimentally seeded coral. Equipped with scuba and snorkeling gear, we ferried these one by one to the

Animals, Ethics, and Engineering: Intersections and Implications
Edited by Rosalyn W. Berne

ISBN 978-981-5129-56-4 (Hardcover), 978-1-003-65231-1 (eBook)
www.jennystanford.com

surface, where they were rapidly transferred to massive insulated coolers of seawater, then briefly removed to a lightbox to be photographed from all angles. Afterwards we returned them, one by one, to their original anchorage.

15.1 The Holobiont: Coral as a Unique Symbiotic Entity

It is a truism in conservation that charismatic animals attract more attention, more research funding, and more donations. "Charismatic" crosses genera, but usually it means *Mammalia*: furry, cute, personable. The unsettling metamorphoses of dragonfly larval forms, for example, do them few favors in their role as ambassadors for waterway health. Corals, despite being as deeply weird as an animal can get, are something of an exception to this rule—perhaps precisely because of their weirdness. A close-up view of a reef shelf looks like a minute, alien city. Fractally telescoping structures recede into nooks and crevices, from which emerge monstrous crustaceans, eels, tubeworms, all of which dwarf the tiny fleshy coral polyps pulsating their tentacles from the calcite towers. This complexity and variety emerge from coral's dual nature, plant and animal living in deeply entwined symbiosis. The gastrodermic *Symbiodinium* algae that gives corals their phenomenal color arrays has a morphological resonance in the forest- and moss-like growth of their structural skeletons. In shallow-water reefs, *Symbiodinium* cells supply most of each coral's energy needs, via photosynthate secretions, in exchange for protection against predators and the environment [1]. It is this energetic lift that enables the high calcification rates necessary to build the labyrinth of structures that makes up a reef [2].

The collective organism comprising the coral animal, its symbiotic algae, and the bacteria that inhabit both, is known as a holobiont (a discrete symbiotic ecological unit). Shallow hard-coral reefs are created by the calcification activities of many unique holobionts, resulting in clusters of nutrient- and shelter-rich habitats which support up to a third of the life in the oceans [3]. Despite their keystone role, the workings of reef ecosystems remain opaque. Even obtaining a rough estimate of the number of coral species is non-trivial in that coral hybridization challenges the traditional concept of a biological species (conventionally based on reproductive isolation [4]). Considered holistically, the combinatorial space for coral holobionts comprises atomic permutations of well over a thousand scleractinian coral species, hundreds of *Symbiodinium* species, and thousands of prokaryotic and acellular organisms [2].

The alien complexity of coral extends to their lifecycle and reproduction. Coral can reproduce sexually or asexually [4], their symbiotic algal colonies are often acquired horizontally, but can be passed on vertically [5] and even selectively, distributing different assemblages to different offspring and thus equipping them with a variety of physiologies and tolerances to suit a multitude of habitats [6]. Sexual reproduction, conducted via mass synchronized spawning events [7], produces planktonic larvae, which drifts in ocean currents, and swells until it finds a place to settle. Once established, these newly attached polyps begin the business of reef building—a process of calcification whose precise chemical details and biological mechanism of control, remains, again, mysterious [8]. Over time (and if it can survive predators and a hostile marine environment), a newly incepted tropical reef, usually home to a community of distinct corals [3], will develop in topological complexity and concomitant biodiversity, acting as an oasis of marine life in otherwise sparsely populated shallow waters of the continental shelf [9].

15.2 Dependence and Destruction: The Relationship Between Humans and Reefs

Reefs have risen and ebbed with sea level and global temperature but have been geological and biological mainstays of ocean life in some form for around 400 million years. Enabled by their algal powerhouses, reef growth is phenomenally swift when considered on a geological timescale. The Great Barrier Reef covers an area of

348,700 km^2 and stretches over 2300 km in length [10]. At the end of the last Ice Age, it sat fully exposed, high and dry above shallow plains in the north-eastern Australian continent. Its enormous breadth was re-populated by living coral over the course of a single myria-annum, with its newest reaches built from scratch within the last 8000 years [11]. The drowning of the plains and the emergence of what is now the Great Barrier Reef is recorded in the oral history of the Yidjini people. Much of the world's present shallow-water coral distribution is similarly contemporate with the human inventions of agriculture, writing, and centralized government.

Our dependence on reefs is multilayered, multi-scale, and growing rather than diminishing. The localized buffer reef calcification provides against chemical fluctuations in the water creates a self-supporting biochemical feedback system which ensures the long-term health of dependent fish, sponge, and sea-cucumber populations [12]. Along with their associated predators, these animals are vital caloric resources for many coastal human communities. In American Samoa, reef fish have provided the bulk of people's staple diet for millennia [13], and in-shore reef fisheries still contribute up to 80% of Fiji's annual fish catch [14]. Protective and structural functions also abound. In the Maldives, coral rock has long provided building material for mosques, shrines, and tombstones [15]. Banks of reefs defend Indonesia and the Philippines from hundreds of millions of dollars in flood damage every year [16]. Recent discoveries have led to new avenues for exploitation—*Halameda* algae found on reefs in the Bahamas contain powerful antimicrobials [17]; reef macro-algae from French Polynesia are used in a huge array of cosmetics [18]; Malaysian porite corals provide graft material and substrates for bone tissue engineering [19].

Given this network of reliance, covering everything from sustenance to storm protection to medicine, a naive onlooker might expect our modern, knowledgeable societies to prioritize reef health, adopting practices to nurture and sustain reef infrastructure and communities. A more cynical audience will be unsurprised to discover the ways we cavalierly disrupt, degrade, and destroy these habitats, both deliberately and incidentally.

On 11 June 1770, the HMS Endeavour, captained by James Cook, struck a stretch of coral rock now known as the Endeavour Reef, part of the Great Barrier Reef chain. 60 years before Darwin formulated his initial theories of reef formation aboard the Beagle, coral reefs were regarded by Western navigators chiefly as hazards. Unfamiliar with the ecological significance of such outcroppings and atolls, European sailors saw them as enigmatic and dangerous, as indicated by the Portuguese map note *abro-olhas*, tr. "look out" [11]. This injunction gave the Houtman Abrolhos their name in 1619—aptly so, as a decade later the Dutch trading ship Batavia would end up catastrophically wrecked on one of the Houtman reefs. Historical records detail the damage to both the Batavia and the Endeavour, and in the latter case, the frantic steps that had to be undertaken to render the ship seaworthy [20]. They do not note the corresponding damage to the reefs.

Over 200 years later, the 230 m long freighter Shen Neng 1, owned by Shenzhen Energy and carrying 975 tons of oil and 65,000 tons of coal, ran aground in a nominally protected area of the Great Barrier Reef, immediately disgorging 4 tons of oil into the surrounding ocean [21, 22]. The resulting reef grounding scar was 3 km long and 250 m wide, in some areas completely denuded of life. Oil tarballs continued to wash up into protected turtle nesting colonies and bird rookeries months later [23].

Although Shen Neng 1 was found to be off-course when it ran aground, it had deviated only 12 km outside of an approved shipping lane that daily transports thousands of tons of fossil fuels around and through some of the most fragile marine ecosystem in the world. In 2014, the Australian federal government negotiated a deal with mining magnate Gina Rinehart to allow the export of 60 million tons of coal annually through the Great Barrier Reef Marine Park, a process that would involve clearing 28 hectares of sensitive on-shore wetland and dredging 5 million tons of seabed within the park itself. In return for this licensed destruction undertaken in service of climate-destabilizing fossil fuel transport, the companies responsible agreed to pay $50,000 per year to the marine park authority, an amount that would cover approximately 375 square foot of coral restoration using conventional methods [24]. They were also permitted to refuse a requirement to prevent coal dust entering the Great Barrier Reef environment [25].

In the above cases, damage to the reef is either fully inadvertent, or an unfortunate by-product of "business as usual" commercial practices, where the wheels of commerce must be greased with the blood of nature. But just as frequently, damage to reefs and the associated habitats and animals is more deliberate; a direct, albeit paradoxical, result of our cultural and nutritional dependence. In Malé, in the Maldives, coral rock has long been used as building material for culturally important structures like shrines and tombs. The islanders mine the rock by choosing a site close to its intended destination, and manually remove large branching and massive corals from the shallow reef top. They then break up these living corals into manageable chunks, using iron bars, and transport them to the build site. Here the corals are left to die, stranded on the shoreline, until rain and sun has killed the animals and stripped their skeletons. The resulting rock is then further broken down manually into nodules which are used directly in construction. Over the last 80 years, the use of this prized mineral has expanded, until by the 1980s it had become the main construction material on the capital and surrounding resort islands, used for everything from houses to walls to road surfacing. Houses built with coral nodules are believed to be stronger and more temperate than those constructed with other building materials, leading to escalating demand as the population increases. In 1988 it was estimated that over 20,000 cubic meters of coral rock would need to be mined per annum to meet demand [15]. It is unlikely that this number has since decreased; aside from the cultural significance of coral-built dwellings, importation of conventional cement and concrete remains prohibitively expensive.

Meanwhile, the remaining reef structures in the shallows are left stripped bare and smooth. The loss of living coral has an immediate impact on the diversity of dependent marine life in the area, but it's the structural smoothing which has more long-term consequences—without the shelter of a craggy substrate, newly settled corals are easily dislodged and abraded by water movement, and vulnerable to predators. Even 20 years after such a mining event (an amount of time usually sufficient to allow significant coral regrowth after natural decimation by storms or typhoons), such areas have barely begun to recover. Knock-on effects continue to impact both the reefs and the society that exploits and depends on them—*Gambierdiscus toxicus* algae proliferates, becoming a dominant food for herbivorous reef grazing fish, and resulting in waves of ciguatera fish poisoning for people consuming these [26].

In the absence of cheap, reliable, and immediately available building alternatives to coral nodules and coral lime cement, suggestions for damage mitigation around Malé usually amount only to delaying tactics. Brown et al [15] propose a sacrificial offering of individual large faros or knolls, mined to significant depth, in order to (temporarily) save the more ecologically crucial surface corals. It is not clear how such a solution would be achieved using the traditional tools and methods available on the islands, nor what the impact of this more localized loss might be on the protective qualities of the coral atolls and reefs surrounding the islands.

Half a globe away, Terrence Hughes [27] details the cascade failure of reef communities in northern Jamaica due to over-fishing. Between the early 20th century and 1960, fish biomass on the fringing reefs had already been reduced by 80%. Large predatory species had virtually disappeared; the herbivores remaining were rarely permitted to mature to a reproductive age. For a time, the impact of this biomass loss on the corals themselves appeared minimal—algal growth, which might otherwise smother or kill live corals (and is usually kept at bay by grazing herbivores), was mitigated by an increase in the numbers of the echanoid sea urchin *Diadema antillarum*, whose traditional predators were casualties of the increase in human fishing activity. However, this singular dependency led to a singular point of failure. Between 1982 and 1984, the entire geographic range of *Diadema* was affected by a species-specific pathogen which caused mass die-off, resulting in an urchin density loss of 99%. Without this keystone herbivore, the Jamaican reef system underwent a decades long benthic algal bloom, down to depths of more than 40 m, which killed most adult coral and severely hindered larval recruitment. Today, even the more resilient mound corals cover only 2% of their pre-1980s range. The fabulously diverse reefs of Jamaica, cherished and admired for centuries, no longer exist.

In this they are not unique. Similar fishing pressure-related collapses have been observed in eastern Africa [28] and southern Japan [29]. Often it's not the amount of fishing that is problematic, but the technique. In many parts of southeast Asia, grouper and humphead wrasse are caught for the live-fish restaurant trade using cyanide poisoning, which temporarily stuns large fish but mass kills small fish, invertebrates, and coral [30]. In the Philippines, the adoption of muro-ami fishing (where hundreds of divers frighten fish into wide-strung nets or bags by bouncing rocks or bamboo against the coral substrate) has critically impacted populations of shoaling herbivores and planktivores, which both depend upon and support coral health [31]. Almost everywhere, dynamite or blast fishing destroys marine life indiscriminately, damaging reefs, decimating habitats, breaking natural seawalls and barriers that protect the coastline, killing (and severely injuring) fish of every scale. The noise from blasts is known to harm or confuse marine mammals, and the mixture of explosives (often including fertilizer) causes further damage through pollution and nutrient poisoning [31].

Beyond the immediate, localized impacts of these types of human activity slouch the behemoths of anthropic climate change. Ocean warming, ocean carbonization, and ocean acidification are all byproducts of the same Industrial Age acceleration in carbon levels. 90% of modern anthropic carbon dioxide emissions are absorbed by the great sponge of the ocean, causing a chemical reaction which decreases the ambient pH. As the earth's surface warms, so does the water, and hotter oceans absorb more CO_2—a compounding feedback loop winding the spring on an ecological time bomb [32]. Immediate reductions in carbon emissions can rapidly decelerate surface level warming, but subsurface thermochemical systems operate on a much longer, slower cycle. The deep ocean is presently locking in carbon reservoirs that will continue to acidify the seas for centuries [33].

Coral reefs are particularly vulnerable to ocean acidification. Not only the rate, but the structural output of reef calcification is strongly affected by a decrease in ocean pH [34]. Algal photosynthesis, the energetic process that powers calcification, is enabled by a potential differential between the coral's symbiocyte cells and the (weakly alkaline) surrounding seawater [11]. At lower external pH (a higher acidity environment), there is a corresponding drop in calcification rate. In turn, skeletal density and growth rate decreases, leaving both animal and algal cobionts more vulnerable and fragile. Non-calcifying reef organisms may start to out-compete the calcifying coral species, reducing habitat availability for the reef macrofauna [35]. Gradually, even species less individually sensitive to changes in ocean acidity begin to decline, until a tipping point is reached and the reef undergoes (at minimum) an ecological regime change, if not total collapse [34].

Faced with the scale of the existential threat facing coral reefs in this century, a profusion of conservation and educational programs has sprung up. But the scope of current—and imminent—loss is daunting, particularly since most efforts at rehabilitation are slow and laborious. Using existing methods, an ambitious coral restoration program might succeed at recovering perhaps 1 hectare of reef per year. In 2016 alone, the Great Barrier Reef experienced a post-bleaching mortality rate of 22%, amounting to upwards of 7 million hectares of dead coral in a single summer [36].

Recognizing the need for rapid and widespread action, and the arduous nature of manual restoration, it might seem expedient to attack immediate chemical threats at the source, through geo-engineering approaches. Unfortunately, most climate engineering proposals do not address (or even impact) ocean acidification [37, 38]. Direct alkalinization could feasibly decrease ocean acidity, but falls down in practical terms—the amount of carbon that could be extracted by ocean liming is at least balanced, if not exceeded, by the emissions resulting from mining, processing, and transportation of the raw materials [37]. Even given a hypothetical future technological development that could be applied effectively, the nuances and complexities of the ocean's contribution to the biosphere and its relationship with localized, niche biomes such as reefs are still poorly understood. Visions of "designer ecosystems" bespeak an unwarranted arrogance regarding our understanding of planetary scale dynamic systems. As David Grinspoon has it, "given how much we have yet to learn, we really should not be let anywhere near the controls of planet Earth," [39].

15.3 Automating Intervention: Robotic and AI Initiatives

The tension between the inevitability of collapse given a lack of action, and the possibility of catastrophe given the wrong action, underpins much ecological rehabilitation work. Reef restoration on a global scale could itself be seen as a type of geoengineering, albeit of a somewhat more cautious and subtle variety. To guard against accidental destabilization, enough of the conditions of the original ecosystem must remain to allow for system stabilization within expected bounds, and we must understand enough of its function to accurately capture a representative snapshot of the keystone species that must be restored.

Understanding reefs requires data—at every spatial or temporal scale, describing previous and current reef conditions, cycles, coverage, composition, processes, contribution to ocean chemistry, erosion rates, accumulation rates, biodiversity—and so on. Moreover, the complex interplay between reef, ocean, and climate presents a significant analytical challenge before raw information can be formulated into predictive models. But data gathering is expensive and difficult. Ecological and ethological research is perpetually under-resourced because the timescale of most biomonitoring programs is an order of magnitude greater than most grant funding cycles, and novelty of approach is increasingly valued over longitudinal explorations [40]. A lack of information doesn't necessarily lead to a lack of action, which would be bad enough—poorly understood systems can encourage poor management decisions, with conservation metrics focused myopically on some preconceived model of likely stressors while ignoring more acutely relevant impacts [41, 42]. Without cohesive, consistent data which can be leveraged for evidence-based management practices, it can be impossible to assess whether apparent changes in reef conditions are problematic, what has caused them, and whether intervention is required.

This age of cascading environmental disasters is also one of great mechanical and cybernetic innovation, marked by an explosion in the autonomous capabilities of hardware and software. The introduction of novel technologies into conservation programs has the potential to be both a direct and indirect force multiplier. Robotics and AI, with their high up-front costs but long operational tail and incremental publication norms, lend themselves well to the predominant scientific funding paradigm [43].

People love to remind robotics researchers that the English word "robot" derives from a Czech word for "slave" [44]. This semantic framing has allowed philosophers, ethicists, and auteurs to use robots to explore questions of cognition, hierarchy, responsibility, and sentience, using robot stand-ins. We project all sorts of anxieties and neuroses onto robots, from the loss of our own humanity and empathy through use and abuse, to an apparently inevitable revolt and uprising. This lensing still shapes the public perception of automation.

Sitting in a wire tangle of deconstructed sensors, robots, and ill-tempered routers in Western Australia—an area with its own bloody history of forced labor, differing from slavery principally only in name [45]—the comparison seems clumsy. Perhaps our ongoing obsession with enslaved robots is at least partly a way to displace instances of systemic brutality and violent repression onto inhuman hypotheticals. Robots—insensate, inanimate, and constructed rather than born—should occupy a more fluid and nuanced position in the social imagination. They allow us to do things we cannot do alone, but only with the investment of massive amounts of time, effort, and energy. Robots and their sister technology of machine learning are manifestations of both our priorities and our biases; they may reflect our nature, but they also extend it in new ways, and in so doing, cause us to adapt to them. Rather than slaves, they are symbionts.

Like corals, humans are a composite species—holobionts who have evolved in parallel with our colonies of microbes, bacteria, and algae, on which we now depend utterly, even integrating their DNA through horizontal gene transfer [46]. Robots are simply the latest technology to make the jump from creation to codependent, and they are progressively making inroads into all aspects of human society, including ecological research and protection. On our Geraldton testing site, I'm getting hands-on experience with the manual components of live coral transplanting and seeding, in order to understand more deeply how to apply robotics to problems in coral conservation. We are urgently trying to answer several crucial questions: can we hand off the slowest and most laborious activities to a fleet of robots? Can machine learning help us analyze data and monitor marine environments? Can technology help lift the burden on existing restoration teams, and allow us to scale restoration efforts to match or outpace global reef destruction?

The longest-serving robots in coral reef support do not fit the conventional mental model of robot morphology. Satellite imaging has been a game changer for large-scale reef monitoring, especially now that access to low-earth orbit is no longer solely the domain of governments and military organizations. Private providers of satellite imaging services are increasingly keen to partner with conservation efforts. Planet's global nano-satellite network [47] is currently employed in surveys of remote coral atolls [48]. But other industries they serve include Energy & Infrastructure, and Forestry (a euphemism for forest clearance). The apparent contradiction in purpose is superficial—climate and conservation contracting is a less lucrative business model, but the public relations gloss they provide is considerable. Publicly owned infrastructure must also serve multiple competing interests. Landsat, whose imaging data has been used for shallow-water reef monitoring from the 1970s until today [49] and forms one of the foundational dataset of the Allen Coral Atlas [50], touts its utility for mineral exploration and "natural resource exploration" [51]. Conservation can make use of these private and public facilities and services, but their proliferation is not necessarily a boon to ocean health.

Even the most cutting-edge of satellite imaging cannot yet penetrate far beneath the water's surface, or reliably resolve sub-meter details. This caps its ability to assist within-reef modelling and leaves deep-sea corals effectively invisible [52]. To truly understand life beneath the water, we have to get up close and personal. Historically, this has meant equipping divers with an array of cameras, sensors, and sampling equipment—but relying on human divers inevitably biases the sites we study, since only a tiny fraction of marine ecosystems are regularly accessible. Additionally, human exploration is slow and expensive. Increasingly, scientists are turning to autonomous underwater vehicles (AUVs).

The difficulties inherent in long-range communication and precise navigation underwater drive many reef monitoring projects towards multi-agent systems and robot swarms. Often, these even look like animals—evocative cinematography of schools of fish converging and scattering as though under one mind or pods of whales traversing multi-continental distances year upon year lend narrative credence to the benefits of biomimicry [53]. SoFi, a soft robot designed at MIT to have explicitly biomimetic morphology and dynamic modalities [54] is a pioneer of autonomous robotic reef exploration. Tested to depths of ten fathoms on the reefs of the Somosomo Strait in Fiji, SoFi undulates along shelf gradients mere inches from the reef outcroppings, recording data with a forehead camera reminiscent of the bulbous cranial protrusion of the humphead wrasse. Both SoFi and the similarly shaped (but hard-bodied) BlueSwarm platform [55] are intended to be singular autonomous instances in a distributed homogeneous robotic swarm. Small, adept, able to maneuver precisely around both static but delicate corals and unpredictably motile reef inhabitants without causing damage or alarm, these platforms promise to deliver a trove of data with negligible habitat disturbance. The potential trade-off for too-accurate bimimicry might be ingestion (or at least destruction) by undiscriminating predators, but one of the attractive features of swarm robot is their redundancy; sheer numbers can overcome the stochastic inaccuracies of single-point data gathering and compensate for attrition.

While cheaper per square foot of monitoring than diver-based missions, these miniature robots have limited depth range and battery life, so do not greatly expand the marine geography that can be explored. For deep or truly remote reefs, researchers must invest in much larger and more robust UAVs (unmanned autonomous vehicles) and ROVs (remotely operated vehicles) such as C-Bot [53] and BlueROV2 [56], which are intended for longer-term monitoring missions with a high initial cost outlay but minimal per-deployment expenses. These robots are less vulnerable and more powerful and can be used long-term to map diver-inaccessible transects of reef and ocean floor at regular intervals. For example, C-Bot missions are intended to last six months, with the robot lying in a low-energy dormant state between its regular high-resolution transect scans.

More advanced autonomy can allow these kinds of robots to adjust to conditions and unexpected events, and independently divert their path to target areas of high salience or interest. But even simple autonomous platforms can provide enormous value, at a somewhat lower cost, like Murray Robert's benthic landers which were used to investigate deep cold-water reefs in the early 2000s [57].

As with satellites, the costs of deploying dedicated solutions without a parallel profit generation mechanism means continuity of data and opportunities for iterative technological improvement are difficult to guarantee. So it's not surprising that underwater exploratory missions often serve multiple purposes. Alongside undiscovered species, geological formations, and climate regulation mechanisms, the deep sea also contains oil, gas, and minerals. Combining these two expeditionary branches, as with the MORPH heterogenous mapping system [58]—or WiMUST [59], or RO-BUST [60]—allows scientists to gather much-needed data, at the cost of further enabling the destruction of the systems they are trying so desperately to understand and quantify.

Robots under the employ of the fossil fuel and mining industries will no doubt be surveying the sea floor regardless, so we may as well lean on them to help solve our data acquisition problems. But this influx of information merely compounds the corresponding analytical challenge [61]. The AI revolution may be the nostrum enabling management and interrogation tools fast and accurate enough to interpret this flood of extant and forthcoming data, allowing us to build effective models of the complex, multi-scale feedback systems that comprise coral reefs.

Computational intelligence techniques have already been applied to dozens of problems in coral research [62, 63]. Several of the most effective leverage human input to create a symbiotic analytical gestalt that reduces time and monotony for human researchers without ceding responsibility to unsupervised black box algorithms. For example, CoralNet provides a free, open-source, collaborative framework for coral classifier training [64], where a minimal expertly labelled dataset is used to initialize a network and suggest potential labels to a human collaborator. The system then uses the collaborator's acceptance, rejection, or novel labelling input to update and improve the network. Since its first release in 2011, researchers have uploaded somewhere in the region of 2 million images, with an accompanying 65 million annotations. As learning techniques increase in sophistication, more ambitious capabilities may be realized—for example, deep ensembling provides a way for automated coral classification to cautiously dip its toe in the waters of deep learning, promising trustworthy ab initio labelling of corals in novel environments [65], although due to the difficulties in precise coral species identification from visual data for even subject matter experts, such efforts are necessarily confined to broad categorical sorting [4].

The problem with identifying system interactions and anomalies using machine learning is that most of our training data has been gathered in an ocean already heavily impacted by climate change and a century of intensive commercial fishing operations. It raises questions about what can be considered a "pristine" reef, or "normal" baseline conditions. This is in addition to the known perils of over-reliance on machine learning in poorly understood spaces [66]. Ocean life—reef life—is so diverse that classification systems must be highly specialized to match the local ecology, topological conditions, even lighting. However, this kind of bespoke network diversification may require huge amounts of processing power and water consumption [67]. The percentage of AI resource consumption dedicated to conservation tasks is comparatively tiny when compared to, e.g., a general-use generative language model like ChatGPT. But as large model approaches like LLMs and deep learning are folded into more specific perceptual training pipelines [68, 69], this distinction will become increasingly blurred. Fundamentally, as long as any proportion of restoration work is dependent on high-energy machine learning practices (not to mention plastics and oil derivatives, chemical pollutants, rare earth minerals, and fossil fuel consumption), then we will at some level be burning the village in order to save it [70]. That said, there are usage frameworks and best practice policies that can be applied to highly consumptive processes which might minimize this harm, such as transparent sustainability reporting and the climate geography of data-center locations [71, 72].

Equipped with robot-assisted raw data acquisition and AI-informed models, coral conservation still has to address the biggest bottleneck in mass restoration—the physical labor involved in coral fragmentation, seeding, and securing reef plantings, and in long-term monitoring of reef health. Beyond passive observation, can robots be safely used as active participants in reef preservation? As costs come down and capabilities increase, robots are increasingly cast in more adventurous roles, such as guardian, gardener, doctor, and midwife.

Guardians operate in both air and sea. 45 km off the coast of Belize sits Glover's Reef Atoll, a small island archipelago arcing around 380 km^2 of coral lagoon. Like northern Jamaica, reefs here are under threat from over-fishing—mostly illegal poaching in the Glover's Reef Marine Reserve. Overstretched enforcement agencies cannot hope to adequately patrol huge stretches of open sea, so a partnership with Conservation Drones, an organization explicitly targeting conservation initiatives in developing nations [73], has been enjoined [74]. Equipped with high-resolution camera and video; quiet, low-altitude, with a range of over 50 km and the ability to stay airborne and autonomously monitoring for up to an hour, a single drone might not be able to offer significantly more coverage, but a double-digit fleet radically expands the effective surveillance network guarding the lagoon.

Underwater, guardians can serve dual roles, becoming exterminators or assassins. Off the coast of Florida, escaped—now feral—lionfish threaten the local populations of reef fish [75]. ReefSweeper robots, at a cost of $300,000 each, promise to autonomously identify, target, harpoon, and harvest invasive lionfish, culling the wild population while establishing the fish as a sustainable—even ecologically beneficial—source of fresh seafood for locals and tourists. The RSE Guardian Bot, supported by iRobot, aims to fill a similar niche at a lower price point [76]. In the southern hemisphere, the Great Barrier Reef Foundation has proposed using Queensland University of Technology's COTSbot as a fleet of small, low-cost, fully autonomous lethal injection machines, targeting Crown-of-Thorn starfish, outbreaks of which are deemed a serious threat to the Great Barrier Reef [77].

Energetic though these coalface projects may be, they do not address the upstream problems threatening reefs. Prosecuting individual poachers does nothing to address food insecurity issues faced by subsistence farmers and fishers in Belize. Employing killer robots to target scapegoat species—who are merely responding to environmental factors far beyond their control—could be framed as a form of ethical displacement [78]. Casting a non-human in the role of ecological villain, and then eliminating them, allows us to dodge the question of why these pests or predators are proliferating to such a harmful level, and does not necessarily lead to a systematic effort to prevent similar incursions happening in the future. Moreover, using robots to deal out death to nature further abstracts our fundamental responsibility—both the problem and the solution are rendered less tangible, less visible, by the employment of remote mechanical proxies.

Not all robotic gardeners are intended as killers. RangerBot, an upgraded cousin to COTSbot, can also capture millions of coral larvae and deposit them on damaged reefs in a mass midwifery operation [79]. CoralBot [80] aims to mend broken reefs, returning recently dislodged and still-alive coral fragments to suitable reef outcroppings and affixing them, optimizing their orientation for regrowth potential. A phalanx of CoralBots might travel in the wake of the deep-sea trawl nets which frequently devastate cold-water reefs. My own work in Geraldton is intended to support and prototype what promises to be a step-change in active localized restoration—using cooperative robots for large-scale parallelized automation of coral fragmentation and seeding. For several days we (researchers and robots, together) experiment with manipulating different types and shapes of coral. We take measurements of force, friction, and slippage, examine polyps for damage, then make adjustments to the robots' end-effectors, grasp locations, speed, sensitivity. And we adapt our overall goals, as well, in response to the robots' successes or limitations. The process is collaborative, it is not dictated purely by human wants; both coral and robot establish the bounds of feasibility and humans can, at best, in-fill and optimize within these constraints.

Our efforts here are just part of a larger attempt to bring industrialized manufacturing processes into coral conservation initiatives. The team is also addressing mechanical challenges that can abound when anchoring fragile coral transplants, which can be easily dislodged by tides, storms, and fish, to reefs or ocean beds without adding plastics or detrimental volatile substances to the marine environment. Coral anchoring is a well-explored space, because coral cultivation for decorative, aquarium, and scientific consumption is a serious commercial proposition. Maritechture is one company tackling speed bottlenecks in coral transplanting and manual reef building, taking a motion optimization approach to connectivity and modularity. Rapidity and effectiveness of upscaling is the main focus here—the ecological impact of the substrate material is of secondary concern. In the non-profit sphere, these priorities are often flipped. ReefBalls, although still constructed out of high-emission concrete, are carefully designed to be chemically inert in order to not affect water quality. Well over half a million ReefBalls have been deployed globally [81]—but studies indicate no obvious benefits over constructing reefs in situ from much cheaper and more readily available rubble and piled coral rocks [81].

Such political syllogism plagues conservation work—the urgency of most problems is so immediate that maintaining momentum through active intervention is preferred over the expensive and time-consuming exhaustive research required to identify best practices. Anthropic bias, which assumes designed structures are inevitably better than the happenstance of nature, may also play a role. But even eco-friendly substrates can fail to clear "better than nothing" baseline for active intervention—artificial reefs can create ecological traps, forming superficially attractive environments that nevertheless do not offer the concomitant benefits of natural reef formations, and so negatively impact the mortality rate of reef species [82]. At least the latter are not usually toxic, unlike a 10-year-long Floridian practice of reef building using piles of old car tires which disrupted the local ocean chemical balance so badly it called for a multi-million dollar cleanup operation in the early 2000s [83].

The latter would not be the first, only, or last "artificial reef" that actively degraded the marine environment. Since the term entered the English lexicon in the late 1940s, it has been deployed to deflect from or spin all manner of marine disasters, insurance scams, and lazy waste disposal projects, which have resulted in tons of polluting material being jettisoned into the depths of the ocean, as if what this environment requires is merely topology.

Each rotting hull or concrete jumble squatting on the seabed represents thousands upon thousands of expert hours spent in design, engineering, and construction. We often speak of waste in purely material terms, but the energetic and labor costs of these edifices are often sunk along with them. What of the functional waste of energy, of labor? Our own restoration project has already consumed years of person-hours; if it is not widely adopted—or if it is adopted but quickly abandoned—its legacy might only be in an unrecoverable cost of both materials and labor. How do we design resilient uptake, as well as resilient technology?

15.4 Scaling Solutions: Aligning Humans, Nature, and Technology

To keep pace with the projected annual loss of coral biomass, we need to multiply the collective results of all current restoration efforts a thousand-fold. If we want to restore what has been lost over the last century, we must tack on an additional order of magnitude—perhaps two. This will require hundreds, if not thousands, of seeding operations, nurseries, and planting operations in critical sites around the globe. The confluence of history and geology means the nations and peoples most in need of large-scale reef rehabilitation generally lack the infrastructure, financial resources, and political power to act alone. But Western models of conservation are poorly understood in non-Western regions, and often viewed with suspicion. Top-down programs initiated and driven by external actors without cultural buy-in from residents and local communities can often become mired down by competing interests and perverse incentives, or discarded altogether.

Just as we adjust and adapt our robotic workflows to meet the requirements of the coral we support, the larger intervention models we adopt need to extend and mesh with existing social and cultural practices. A critical danger with generalist, pre-packaged conservation solutions developed in the global north to be rolled out in the global south is that too often they end up retreading colonial pathways of displacement and dispossession.

On a small coral island in the Spermonde archipelago of Indonesia, a coral restoration project sponsored and managed by an American-owned multinational corporation seeks to "conserve and restore the biodiversity of local coral reefs, while improving the food security of the local community" [84]. Roughly once a month, islanders from the village closest to the restoration site gather and, under the direction of marine experts, create coral "spiders"—sand-coated steel skeletons with live fragmentary corals attached—that will be placed on barren, dead, or damaged coral beds in the hopes of generating regrowth. Villagers are paid for their work, and some are also employed as reef guards, to limit poaching and illegal fishing around the restoration area.

Elsewhere in Indonesia, the same corporation is engaged in multiple large natural resource extraction activities, including forest clearance and commercial-scale fishing operations. Their coral rehabilitation is part of a market-based approach to environmental restoration, in which carbon and biodiversity credits are offered as indulgences to gain moral absolution for the destruction of forests, poisoning of waterways, and massive CO_2 generation [85].

The human archipelago of Spermonde can be viewed as a long-standing network of interdependent communities. Reciprocal social agreements and mutual obligations have evolved over centuries to ensure each network node can maintain a peaceful, equitable, and sustainable existence. The spider coral project, however, did not engage with this broad web of interconnected groups, but liaised only with the village close to the chosen restoration site, skewing any economic benefits towards those most adjacent. Coral guards drawn from this village, untrained, often young, aggressively enforce what they perceive to be rigid categorical bans against fishing and anchoring against boats from neighboring islands. In response to these aggressions, *pungawaa* from other parts of the islands remove the restoration village's rights of access to other traditional fishing grounds [86]. The social adhesive that should give common cause to a collective of resource-poor people, all facing the same existential threats from climate change and Western exploitation, is becoming unstuck. Meanwhile, long-standing community-driven agreements intended to address resource scarcity and buffer the effects of climate change—such as a ban on sand removal from around the islands, meant to mitigate sea level rise and limit coastal erosion—are ignored by the restoration initiative, who see island sand as a plunderable resource for use in its coral support structures.

This kind of privately funded program is often framed as an alternate model to state-level environmental protection, which can be heavily constrained by lack of resources or political will [87]. But how can such alliances be brokered equitably, given the massive power imbalances between those involved? Linking up with customary law and traditional governance hierarchies is a way to extend public awareness of both environmental problems and proposed relief efforts and establish effective cooperative frameworks.

As with many coastal communities, traditional ocean harvesting in Vanuatu is primarily undertaken on nearshore reefs. Marine resource management was historically just one facet of a complex underlying cosmology linking both land and sea, where taboo areas act as a mosaic of refugia or sanctuaries for motile marine life [88]. Temporal separation of activity in marine and land environments to avoid nutrient and chemical contamination is encoded in traditional taboos, as are seasonal closures and intermittent species-specific taboos which help to alleviate pressure on fish and other marine animals. Areas of great spiritual significance, encompassing both land and sea, may be subject to more formal and permanent traditional protections—many of these are also zones of high biodiversity [88].

Understanding local cosmology, how it shapes community practices, and reinvigorating its acceptance in colonized regions may create a far more flexible, adaptable, ecosystem-relevant approach to conservation and restoration than a top-down Western model of protection. But global trade, emigration, immigration, population dispersal, and other cultural disruptions can weaken the ability of traditional practices to obviate destructive exploitation. What constitutes effective deterrence for breaches of community-agreed conservation practices? In Kubulau, Fiji, the council of chiefs (Bose Vanua), recognizing the links between unsustainable land use and the health of the marine ecosystem, established the Kubulau Resource Management Committee to create distributed, horizontally integrated management structures for protection of marine, freshwater, and terrestrial resources. This in turn led to the establishment of a network of village-managed protected areas, and a holistic, scientifically informed "ridge to reef" management plan. Synergies between community decision making and traditional sanctioning practices which could be used to shape culturally acceptable enforcement methods were identified [87]. This kind of holistic approach is particularly important in low-income and subsistence regions, where immediate financial gains may well outweigh community sanctions. The ability to remove upstream incentives for protection violations may be crucial, hence it is necessary to give community managers the rights, flexibility, and legal space to make decisions that reflect localized unique circumstances and priorities [88, 89].

There is no magic bullet for long-term conservation program retention, and perhaps that is no bad thing—as scientific understanding, applicable technology, and local conditions change, so too should best practices. Debate and disagreement about the extent and degree of protected areas will always exist. But more powerful private entities and NGOs seeking to enter into conservation partnership with indigenous and impoverished communities can take a more equitable and ethical approach by expending political capital and using their economic leverage to give such communities a more equal footing and a stronger voice in local decisions. This is especially the case in regions where the dominant political and economic power is unrepresentative and externally imposed, as on the occupied island of Guam.

More than half of Guam is built on limestone. Even its jungles—complex supersystems of plants, mosses, fungi, animals—grow on a thin topsoil layer overlaying a limestone bed, in an arrangement known as a limestone forest. The bedrocks of these forests were once submerged tropical reefs. Now, the ancient coral skeletons provide the porous substrate that allows the trees purchase and anchors the topsoil. In the waters around Guam sit around 100 km^2 of extant, living coral, which protect the whole Mariana island chain from storms, waves, and typhoons. Pre-colonization, the Chamorro people—the traditional owners and inhabitants of the islands—nurtured a relationship with nearshore reefs and their adjacent fauna for at least 3000 years [90].

Guam's Apra Harbor is home to one of the larger and more accessible Mariana reef systems, one that until the 20th century formed the heart of a rich shoreline fishing culture. In 1944, US forces detonated 7000 tons of explosive in the harbor in order to destroy naval vessel-impeding reefs, and forcibly removed the occupants of the Harbor villages. Shortly afterwards, the 1946 US Congress passed Public Law 594, allowing the US Navy to acquire "any and all lands it deems necessary" on Guam [91], thus beginning a near century-long occupation and dispossession in the name of strategic territorial value and paternalistic protectionism. The US military presence in the south Pacific is set only to increase in coming years, as is the scale of its major destructive and constructive operations on Guam's land and sea areas, under the guise of "resource security." These include continued large-scale reef dredging operations in Apra Harbor. In 2009, the US Navy released an Environmental Impact Statement regarding these plans, giving local inhabitants just 90 days to respond to a 10,000-page document. Undaunted, the community launched a guerilla reading campaign, distributing the task of document analysis across its members and pooling their findings [92]. The ensuing multi-front legal challenge forced the Department of the Navy to reconsider the scope of some of their proposed developments—but not their plans for Apra Harbour [91].

15.5 Ancestral Debt: Environmental Costs across Deep Time

On Invasion Day [93] in 2024, I listened to a woman named Katherine Councillor give a Welcome to Country in the Geraldton QEII Centre. "Our ancestors took care of the waterways here for 60,000 years," she said. "It has only taken 200 for settlers to destroy them." She asked the city to consult with the local community and tribal elders when making decisions about resource allocation and land use, to consider both future environmental impacts and the deep past of Aboriginal Australians' relationship with the land. Neither the Mayor nor the Master of Ceremonies responded to the substance of the speech. They seemed like people who sat lightly with history.

A few days later, I walked the 5 km from our field station down the river to Boolungu Springs, the location of a mass killing of Aboriginal people 170 years ago. The site of the massacre is denoted now by a memorial. The informational board with the map of the site is far more visible and ostentatious than the memorial itself—a series of eleven plaques set into concrete discs just a few inches high, scattered along the dirt trail and river bank, telling stories of the place and people in a series of quotes from oral and written histories, by both historians and Elders. Without the gaudy, cheerful map, the plaques are easy to miss—obscured by trees and overgrown by grasses. Worn by winds and sands, they are quite literally self-effacing, which made it all the stranger that several had been actively effaced. Some of the words and phrases describing the killing of Yamatji people were deeply scratched out with a knife or blade. Was someone trying to erase that narrative, or change it?

Inspecting the plaques near the head of the walking trail and making an offering of sand to Bimurra, the rainbow serpent, can feel stilted and dutiful. The continent-encircling highway a kilometer to the east, connecting the populous south with the isolated north, is still audible in the distance. But walking deeper into the trees and scrub, bird calls gradually replacing engine noises, provokes a sense of eerie isolation, as if you move amongst watching ghosts. I felt a growing awareness of the many, many generations of peoples who had walked this land before me; the air was pregnant with a sense of expectation and obligation. History no longer sat lightly with me.

Aboriginal Australian law holds that land is held in trust for future generations. The people who occupy it in the present are mere temporal custodians, able to live off the bounty of rivers, forests, lakes and grasslands, but primarily responsible for passing all of these forwards to the next generation [94]. The notable feature of this cosmology is not that it is unique, but that it echoes a precept found in cultures the world over. Disparate environments lead to different manifestation of generational obligation, but the underlying philosophy is remarkably universal. Most people understand and connect with this simple, foundational idea, even though short-term thinking, limitations in understanding, and the perverse incentives of global capitalism may override its inherent injunctions.

Similarly, a common core of underlying values can be used to guide site-specific practices in long-term coral restoration. A (non-exhaustive) list of tenets might be as follows:

> Restoration efforts need to reflect local knowledge and priorities. They should be transparent—local communities should have full access to the information behind strategic restoration decisions; and approached with a degree of humility—publicized project goals should avoid setting unrealistic expectations or presenting an unrealistically rosy view of future conditions.

These types of commitments are key to building trust and establishing common goals—to ensure resilience over decades, collaborative programs need to promote community self-monitoring and reporting; to engage with local enforcement protocols grounded in traditional law, while clearly articulating the relationship between local and state or federal regulation, and the boundaries of each.

Turning this lens on our own high-volume coral seeding goals, we must acknowledge that however slow they might be, manual data collection, coral fragmentation, seeding, planting, and so on, must still have a place in future reef management strategies. Manual fragmenting and planting can be integrated into existing ocean use practices and provide immediate avenues for community engagement. Inclusive, hands-on collaborations encourage a symbiosis between externally led research or conservation efforts and local environmental programs. Finding ways to scale up coral restoration without displacing people with deep cultural and emotional (let alone material) ties to reefs and their animals gives us a far greater change at long-term success. A question conservation managers in all fields will need to address in the coming decade is: how to harmonize the benefits of machine learning and robotics with human-centered programs? [95]

It's pointless to pour hundreds of thousands of dollars into reef restoration while also spending billions expanding the industrial processes and agricultural expansions that kill them. To quote Christian Voolstra: "If something got lost, it got lost for a reason.

If you don't change the factors, and just restore, it will have the same destiny or fate. . . . Restoration implies you can fix it. I don't think that's possible" [78]. While massive scaling of coral rescue operations and perhaps even mass coral migration is imperative, it is also wildly insufficient.

No intervention can survive without a broad base of commitment from those most affected by its presence or absence. Conservation models that disregard the social relationships embedded in the landscape of operational areas forget that we are, foremost, social animals [96] and that all our endeavors ultimately rest on social bonds and contracts. Solutions are always upper bounded by their uptake, and ingenuity must be subordinate to acceptance. The loss of coral reefs is not solely a secondary effect of climate change and ocean warming, it is also driven by highly localized material needs. We cannot hope to halt or reverse damage from agricultural run-off, local chemical pollution, over-fishing, and coral mining, without directly addressing upstream causes. Hence it behooves reef conservation organizations to partner or liaise with other socially and environmentally focused engineering efforts. These might include improving local access to renewable energy sources; finding low-cost, eco-friendly, passively insulated building material for reef-rock-dependent societies; increasing the availability of alternative and sustainable protein sources to eliminate both localized over-fishing and commercial deep-sea trawling practices; adapting agricultural and irrigation practices to minimize reef-damaging run-off; undertaking education and advocacy to reduce the use of dynamite and cyanide fishing; even restructuring the economic levers of cultural groups with strong ties to reef ecosystems to allow them to protect these biomes without impacting their livelihoods. There is also an imperative to engage with the legal and political frameworks governing ocean use and abuse, to more accurately reflect the demands of coral restoration and delineate responsibility [97].

An honest accounting of environmental costs surrounding automation—whether it is used to cut costs for a corporate entity, make life easier for a wealthy urbanite, or rapidly restore a fragile ocean ecosystem—includes the debts of chemical decay and emissions from all sources over each component lifespan. The expected duty life of an industrial manipulator such as the ones we are unpacking in our Western Australian test site is perhaps 10 years under heavy operation, but the pollution half-life of its non-recyclable components is measured in anywhere from tens to hundreds of years. This kind of imbalance should weigh heavily on the spreadsheets of tech corporations, instead of being virtually invisible. The techno-utopic narrative that infests entrepreneurial spaces tends to imagine we can fix the world's problems with no sacrifices; an inherently childish notion. But coupling technologically informed imagination with a deeper sense of accountability to the planet allows us to ask questions like: are the things we might need to sacrifice (cheap processors, same-day delivery, military firing ranges) adding much to our quality of life? Can we be innovative about enforcing a tighter coupling between environmental exploitation and climate restoration [52]? Disrupting the perverse incentives underlying the ecological destruction of the ocean will not magically reverse what has already been lost, but it can give us the space and time to develop a new conservation landscape, one which prioritizes net-zero power consumption, eco-beneficial materials, social integration, and long-horizon planning.

References

1. Schlesinger, A., E. Zlotkin, E. Kramarsky-Winter, and Y. Loya. 2008. Cnidarian Internal stinging Mechanism. *Proceedings of the Royal Society B: Biological Sciences* 276: 1063–1067. https://doi.org/10.1098/rspb.2008.1586.
2. Blackall, L. L., B. Wilson, and M. J. H. van Oppen. 2015. Coral-the World's Most Diverse Symbiotic Ecosystem. *Molecular Ecology* 24: 5330–5347. https://doi.org/10.1111/mec.13400.
3. Knowlton, N., R. E. Brainard, R. Fisher, M. Moews, L. Plaisance, and M. J. Caley. 2010. Coral reef biodiversity. In *Life in the World's Oceans: Diversity, Distribution, and Abundance*, ed. A. D. McIntyre, 1st ed., 65–78. Wiley. https://doi.org/10.1002/9781444325508.ch4.
4. Harrison, P. L. 2011. Sexual Reproduction of Scleractinian Corals. In *Coral Reefs: An Ecosystem in Transition*, ed. Z. Dubinsky and N. Stambler, 59–85. Dordrecht: Springer Netherlands. https://doi.org/10.1007/978-94-007-0114-4_6.

5. Kirk, N. L., and V. M. Weis. 2016. Animal–Symbiodinium Symbioses: Foundations of Coral Reef Ecosystems. In *The Mechanistic Benefits of Microbial Symbionts*, ed. C. J. Hurst, 269–294. Cham: Springer International Publishing. https://doi.org/10.1007/978-3-319-28068-4_10.
6. Padilla-Gamiño, J. L., X. Pochon, C. Bird, G. T. Concepcion, and R. D. Gates. 2012. From Parent to Gamete: Vertical Transmission of Symbiodinium (Dinophyceae) ITS2 Sequence Assemblages in the Reef Building Coral Montipora capitata. *PLoS ONE* 7: e38440. https://doi.org/10.1371/journal.pone.0038440.
7. Sorek, M., E. M. Díaz-Almeyda, M. Medina, and O. Levy. 2014. Circadian Clocks in Symbiotic Corals: The Duet between *Symbiodinium* Algae and Their Coral Host. *Marine Genomics* 14: 47–57. https://doi.org/10.1016/j.margen.2014.01.003.
8. Allemand, D., É. Tambutté, D. Zoccola, and S. Tambutté. 2011. Coral Calcification, Cells to Reefs. In *Coral Reefs: An Ecosystem in Transition*, ed. Z. Dubinsky and N. Stambler, 119–150. Dordrecht: Springer Netherlands. https://doi.org/10.1007/978-94-007-0114-4_9.
9. Velásquez, J., and J. A. Sánchez. 2015. Octocoral Species Assembly and Coexistence in Caribbean Coral Reefs. *PLOS ONE* 10: e0129609. https://doi.org/10.1371/journal.pone.0129609.
10. Facts. 2024. *Great Barrier Reef Foundation*. https://www.barrierreef.org/the-reef/facts. Accessed March 29.
11. Sale, P. F. 2021. *Coral Reefs: Majestic Realms under the Sea*. New Haven: Yale University Press.
12. Anthony, K. R. N., J. A. Kleypas, and J.-P. Gattuso. 2011. Coral Reefs Modify Their Seawater Carbon Chemistry – Implications for Impacts of Ocean Acidification. *Global Change Biology* 17: 3655–3666. https://doi.org/10.1111/j.1365-2486.2011.02510.x.
13. Valentin, F., E. Herrscher, F. Petchey, and D. J. Addison. 2011. An Analysis of the Last 1000 Years Human Diet on Tutuila (American Samoa) Using Carbon and Nitrogen Stable Isotope Data. *American Antiquity* 76: 473–486. https://doi.org/10.7183/0002-7316.76.3.473.
14. Tuqiri, N. K., H. S. Valtisson, and H. Ármannsson. 2009. Reef Fish Assessment in Fiji's "I Qoliqoli." *UNESCO*.
15. Brown, B. E., and R. P. Dunne. 1988. The Environmental Impact of Coral Mining on Coral Reefs in the Maldives. *Environmental Conservation* 15: 159–165. https://doi.org/10.1017/S0376892900028976.
16. Beck, M. W., I. J. Losada, P. Menéndez, B. G. Reguero, P. Díaz-Simal, and F. Fernández. 2018. The Global Flood Protection Sav-

ings Provided by Coral Reefs. *Nature Communications* 9: 2186. https://doi.org/10.1038/s41467-018-04568-z.

17. Paul, V. J., and W. Fenical. 1983. Isolation of Halimedatrial: Chemical Defense Adaptation in the Calcareous Reef-Building Alga Halimeda. *Science* 221: 747–749. https://doi.org/10.1126/science.221.4612.747.
18. Stiger-Pouvreau, V., and M. Zubia. 2020. Macroalgal Diversity for Sustainable Biotechnological Development in French Tropical Overseas Territories. *Botanica Marina* 63: 17–41. https://doi.org/10.1515/bot-2019-0032.
19. Manassero, M., A. Decambron, N. Guillemin, H. Petite, R. Bizios, and V. Viateau. 2016. Coral Scaffolds in Bone Tissue Engineering and Bone Regeneration. In *The Cnidaria, Past, Present and Future: The World of Medusa and Her Sisters*, ed. S. Goffredo and Z. Dubinsky, 691–714. Cham: Springer International Publishing. https://doi.org/10.1007/978- 3-319-31305-4_43.
20. Sharman, R. 1971. Guns from the 'Endeavor. *Queensland Heritage* 2: 691–714.
21. Connolly, E., and J. Henley. 2010. The Great Barrier Reef Scandal. *The Guardian*, April 6, sec. Australia news.
22. Shen Neng 1, 3 April 2010 |Australian Maritime Safety Authority. 2024. https://www.amsa.gov.au/marine-environment/incidents-and-exercises/shen-neng-1-3-april-2010. Accessed April 23.
23. Harris, T., 2010. China Ship 'Gouged Two-Mile Scar' in Great Barrier Reef, *Phys.org*, https://phys.org/news/2010-04-china-ship-gouged-two-mile-scar.html#:~:text=The%20Chinese%20coal%20carrier%20Shen,a%20top%20expert%20has%20said. Accessed April 23, 2024.
24. Won, D. 2023. A Review of Coral Reef Restoration Methods. *Journal of Student Research* 12. https://doi.org/10.47611/jsrhs.v12i2.4394.
25. Milman, O. 2014. Criticism Over Great Barrier Reef Deals for Gina Rinehart's Mining Company. *The Guardian*, August 20, sec. Environment.
26. Bagnis, R., S. Chanteau, E. Chungue, J. Hurtel, T. Yasumoto, and A. Inoue. 1987. Ciguatera Fish Poisoning: An Objective Witness of the Coral Ree Stress. In *Human Impacts on Coral Reefs: Facts and Recommendations*, ed. B. Salvat, 241–253. Frnch Polynesia: Antenne Museum.
27. Hughes, T. P. 1994. Catastrophes, Phase Shifts, and Large-Scale Degradation of a Caribbean Coral Reef. *Science* 265: 1547–1551. https://doi.org/10.1126/science.265.5178.1547.
28. McClanahan, T. R., A. T. Kamukuru, N. A. Muthiga, M. Gilagabher Yebio, and D. Obura. 1996. Effect of Sea Urchin Reductions on

Algae, Coral, and Fish Populations. *Conservation Biology* 10: 136–154. https://doi.org/10.1046/j.1523-1739.1996.10010136.x.

29. Wilkinson, C. R. 1999. Global and Local Threats to Coral Reef Functioning and Existence: Review and Predictions. *Marine and Freshwater Research*. https://doi.org/10.1071/MF99121.
30. Warren-Rhodes, K., Y. Sadovy, and H. Cesar. 2003. Marine Ecosystem Appropriation in the Indo-Pacific: A Case Study of the Live Reef Fish Food Trade. *Ambio* 32: 481–488. https://doi.org/10.1579/0044-7447-32.7.481.
31. Mindanao State University-Tawi-Tawi College of Technology and Oceanography, Albaris Tahiluddin, Mi?ndanao State Uni?versi?ty-Tawi?-Tawi? College of Technology and Oceanography, and Jurmi?n Sarri?. 2022. An Overview of Destructive Fishing in the Philippines. *Acta Natura et Scientia* 3: 116–125. https://doi.org/10.29329/actanatsci.2022.352.04.
32. Buck, H. 2020. "Climate Engineering Doesn't Stop Ocean Acidification": Addressing Harms to Ocean Life in Geoengineering Imaginaries. In *Blue Legalities: The Law and Life of the Sea*, Irus Braverman and Elizabeth Johnson, eds., Duke University Press, 2020.
33. Mathesius, S., M. Hofmann, K. Caldeira, and H. J. Schellnhuber. 2015. Long-Term Response of Oceans to CO_2 Removal from the Atmosphere. *Nature Climate Change* 5: 1107–1113. https://doi.org/10.1038/nclimate2729.
34. Kleypas, J., and K. Yates. 2009. Coral Reefs and Ocean Acidification. *Oceanography* 22: 108–117. https://doi.org/10.5670/oceanog.2009.101.
35. Smith, S. V., and R. W. Buddemeier. 1992. Global Change and Coral Reef Ecosystems. *Annual Review of Ecology, Evolution, and Systematics* 23: 89–118. https://doi.org/10.1146/annurev.es.23.110192.000513.
36. Anderson-King, K. D., C. Wayman, S. Stephenson, et al. 2023. Branching Coral Growth and Visual Health During Bleaching and Recovery on the Central Great Barrier Reef. *Coral Reefs* 42: 1113–1129. https://doi.org/10.1007/s00338-023-02403-6.
37. Morton, O. 2015. The Planet Remade: How Geoengineering Could Change the World. *Princeton University Press*.
38. Tjiputra, J. F., A. Grini, and H. Lee. 2016. Impact of Idealized Future Stratospheric Aerosol Injection on the Large-Scale Ocean and Land Carbon Cycles. *Journal of Geophysical Research: Biogeosciences* 121: 2–27. https://doi.org/10.1002/2015JG003045.

39. Moskowitz, C., and L. Billings. 2017. Earth in Human Hands: Shaping Our Planet's Future. *Scientific American*.
40. Hughes, B. B., R. Beas-Luna, A. K. Barner, K. Brewitt, D. R. Brumbaugh, E. B. Cerny-Chipman, S. L. Close, et al. 2017. Long-Term Studies Contribute Disproportionately to Ecology and Policy. *BioScience* 67: 271–281. https://doi.org/10.1093/biosci/biw185.
41. Brodie, J., and J. Waterhouse. 2012. A Critical Review of Environmental Management of the 'Not So Great' Barrier Reef. *Estuarine, Coastal and Shelf Science* 104–105: 1–22. https://doi.org/10.1016/j.ecss.2012.03.012.
42. Flower, J., J. C. Ortiz, I. Chollett, S. Abdullah, C. Castro-Sanguino, K. Hock, V. Lam, and P. J. Mumby. 2017. Interpreting Coral Reef Monitoring Data: A Guide for Improved Management Decisions. *Ecological Indicators* 72: 848–869. https://doi.org/10.1016/j.ecolind.2016.09.003.
43. González-Rivero, M., O. Beijbom, A. Rodriguez-Ramirez, D. E. P. Bryant, A. Ganase, Y. Gonzalez-Marrero, A. Herrera-Reveles, et al. 2020. Monitoring of Coral Reefs Using Artificial Intelligence: A Feasible and Cost-Effective Approach. *Remote Sensing* 12: 489. https://doi.org/10.3390/rs12030489.
44. Capek, K. 2024. R.U.R. https://standardebooks.org/ebooks/karel-capek/r-u-r/paul-selver_nigel-playfair/text/single-page. Accessed April 23.
45. Martens, J. 2023. Forced Labour, Indenture and Convict Transportation: A Case Study of the Western Australian Pastoral Industry, 1830–1850. *Labour History* 125: 31–54. https://doi.org/10.3828/labourhistory.2023.19.
46. Gilbert, S. F., J. Sapp, and A. I. Tauber. 2012. A Symbiotic View of Life: We Have Never Been Individuals. *The Quarterly Review of Biology* 87: 325–341. https://doi.org/10.1086/668166.
47. Daily Earth Data to See Change and Make Better Decisions. 2024. *Planet*. https://www.planet.com/. Accessed February 21.
48. Asner, G. P., R. E. Martin, and J. Mascaro. 2017. Coral Reef Atoll Assessment in the South China Sea Using Planet Dove Satellites. *Remote Sensing in Ecology and Conservation* 3: 57–65. https://doi.org/10.1002/rse2.42.
49. Kutser, T., J. Hedley, C. Giardino, C. Roelfsema, and V. E. Brando. 2020. Remote Sensing of Shallow Waters – A 50 Year Retrospective and Future Directions. *Remote Sensing of Environment* 240: 111619. https://doi.org/10.1016/j.rse.2019.111619.

50. Kennedy, E. V., C. M. Roelfsema, M. B. Lyons, E. M. Kovacs, R. Borrego-Acevedo, M. Roe, S. R. Phinn, et al. 2021. Reef Cover, a Coral Reef Classification for Global Habitat Mapping from Remote Sensing. *Scientific Data* 8: 196. https://doi.org/10.1038/s41597-021-00958-z.
51. NASA. 2021. Landsat: Practical Uses. November 18.
52. Barbier, E. B., D. Moreno-Mateos, A. D. Rogers, J. Aronson, L. Pendleton, R. Danovaro, L.-A. Henry, T. Morato, J. Ardron, and C. L. Van Dover. 2014. Ecology: Protect the deep sea. *Nature* 505: 475–477. https://doi.org/10.1038/505475a.
53. Zereik, E., M. Bibuli, N. Mišković, P. Ridao, and A. Pascoal. 2018. Challenges and Future Trends in Marine Robotics. *Annual Reviews in Control* 46: 350–368. https://doi.org/10.1016/j.arcontrol.2018.10.002.
54. Katzschmann, R. K., J. DelPreto, R. MacCurdy, and D. Rus. 2018. Exploration of Underwater Life with an Acoustically Controlled Soft Robotic Fish. *Science Robotics* 3: eaar3449. https://doi.org/10.1126/scirobotics.aar3449.
55. Berlinger, F., M. Gauci, and R. Nagpal. 2021. Implicit Coordination for 3D Underwater Collective Behaviors in a Fish-Inspired Robot Swarm. *Science Robotics* 6: eabd8668. https://doi.org/10.1126/scirobotics.abd8668.
56. Wilby, A., and E. Lo. 2020. Low-Cost, Open-Source Hovering Autonomous Underwater Vehicle (HAUV) for Marine Robotics Research based on the BlueROV2. In *2020 IEEE/OES Autonomous Underwater Vehicles Symposium (AUV)*, 1–5. https://doi.org/10.1109/AUV50043.2020.9267913.
57. Roberts, J. M., O. C. Peppe, L. A. Dodds, D. J. Mercer, W. T. Thomson, J. D. Gage, and D. T. Meldrum. 2005. Monitoring Environmental Variability Around Cold-Water Coral Reefs: The Use of a Benthic Photolander and the Potential of Seafloor Observatories. In *Cold-Water Corals and Ecosystems*, ed. A. Freiwald and J. M. Roberts, 483–502. Berlin, Heidelberg: Springer. https://doi.org/10.1007/3-540-27673-4_24.
58. Kalwa, J., A. Pascoal, P. Ridao, A. Birk, M. Eichhorn, L. Brignone, M. Caccia, J. Alvez, and R. S. Santos. 2012. The European R&D-Project MORPH: Marine Robotic Systems of Self-Organizing, Logically Linked Physical Nodes. *IFAC Proceedings Volumes* 45. 3rd IFAC Workshop on Navigation, Guidance and Control of Underwater Vehicles: 349–354. https://doi.org/10.3182/20120410-3-PT-4028.00058.
59. Widely Scalable Mobile Underwater Sonar Technology |WiMUST Project |Fact Sheet |H2020. 2024. *CORDIS | European Commission*. https://cordis.europa.eu/project/id/645141. Accessed April 23.

60. GRAAL LAB. 2018. ROBUST. September 1.
61. Ditria, E. M., C. A. Buelow, M. Gonzalez-Rivero, and R. M. Connolly. 2022. Artificial Intelligence and Automated Monitoring for Assisting Conservation of Marine Ecosystems: A Perspective. In *Frontiers in Marine Science*, 9:918104. https://doi.org/10.3389/fmars.2022.918104.
62. Salcedo-Sanz, S., L. Cuadra, and M. J. A. Vermeij. 2016. A Review of Computational Intelligence Techniques in Coral Reef-Related Applications. *Ecological Informatics* 32: 107–123. https://doi.org/10.1016/j.ecoinf.2016.01.008.
63. Mehta, A., E. Ribeiro, J. Gilner, and R. van Woesik. 2007. Coral Reef Texture Classification Using Support Vector Machines: In *Proceedings of the Second International Conference on Computer Vision Theory and Applications*, 302–305. Barcelona, Spain: SciTePress—Science and Technology Publications. https://doi.org/10.5220/0002062303020305.
64. Chen, Q., O. Beijbom, S. Chan, J. Bouwmeester, and D. Kriegman. 2021. A New Deep Learning Engine for CoralNet. *2021 IEEE/CVF International Conference on Computer Vision Workshops (ICCVW)*. Montreal, BC, Canada: IEEE: 3686–3695. https://doi.org/10.1109/ICCVW54120.2021.00412.
65. Wyatt, M., B. Radford, N. Callow, M. Bennamoun, and S. Hickey. 2022. Using Ensemble Methods to Improve the Robustness of Deep Learning for Image Classification in Marine Environments. *Methods in Ecology and Evolution* 13: 1317–1328. https://doi.org/10.1111/2041-210X.13841.
66. Saunders, M. I., C. Doropoulos, E. Bayraktarov, R. C. Babcock, D. Gorman, A. M. Eger, M. L. Vozzo, et al. 2020. Bright Spots in Coastal Marine Ecosystem Restoration. *Current Biology* 30: R1500–R1510. https://doi.org/10.1016/j.cub.2020.10.056.
67. Critics Furious Microsoft Is Training AI by Sucking Up Water During Drought. 2023. *Futurism*. September 26.
68. Giaccaglia, P., C. A. Bono, and B. Pernici. 2024. Enhancing Emergency Post Classification through Image Information Amplification via Large Language Models. In *Proceeding Conference on Information Systems for Crisis Response and Management*, 1–14. ISCRAM.
69. Saleh, A. 2023. Novel Deep Learning Architectures for Marine and Aquaculture Applications. PhD, James Cook University. https://doi.org/10.25903/4k34-xm39.
70. Inclezan, D., and L. I. Prádanos. 2023. AI for Sustainability: A Dangerous Fantasy or an Unfulfilled Potential. In *Impact of Artificial Intelligence in Business and Society*. Routledge.

71. Falk, S., and A. van Wynsberghe. 2023. Challenging AI for Sustainability: What Ought It Mean? *AI and Ethics*. https://doi.org/10.1007/s43681-023-00323-3.
72. Bliek, L. 2022. A Survey on Sustainable Surrogate-Based Optimisation. *Sustainability* 14: 3867. https://doi.org/10.3390/su14073867.
73. The Story of ConservationDrones.org |. 2024. https://conservationdrones.org/our-story/. Accessed April 23.
74. Society, W. C. 2024. Safeguarding Belize's Barrier Reef with Conservation Drones. https://phys.org/news/ 2014-07-safeguarding-belize-barrier-reef-drones.html. Accessed March 13.
75. Fisheries, NOAA. 2022. Impacts of Invasive Lionfish | NOAA Fisheries. *NOAA*. Southeast. July 29.
76. Antaya, A., E. Peterson, T. Ralph, and K. Conroy. 2019. Lionfish bot 2.0. *Worcester Polytechnic Institute*.
77. Milne, R., M. Anand, and C. T. Bauch. 2023. Preparing for and Managing Crown-of-Thorns Starfish Outbreaks on Reefs Under Threat from Interacting Anthropogenic Stressors. *Ecological Modelling* 484: 110443. https://doi.org/10.1016/j.ecolmodel.2023.110443.
78. Braverman, I. 2020. Robotic Life in the Deep Sea. SSRN Scholarly Paper. Rochester, NY.
79. Mou, S., D. Tsai, and M. Dunbabin. 2022. Reconfigurable Robots for Scaling Reef Restoration. arXiv. https://doi.org/10.48550/arXiv.2205.04612.
80. Braverman, I. 2018. *Coral Whisperers: Scientists on the Brink*. 1st ed. Vol. 3. University of California Press. https://doi.org/10.2307/j.ctvq4c11g.
81. Hylkema, A. 2022. Active Interventions to Rehabilitate Caribbean Coral Reefs: The Use of Artificial Substrates to Increase the Abundance of Herbivorous Fish and Sea Urchins. https://doi.org/10.18174/569699.
82. Komyakova, V., D. Chamberlain, and S. E. Swearer. 2021. A Multi-Species Assessment of Artificial Reefs as Ecological Traps. *Ecological Engineering* 171: 106394. https://doi.org/10.1016/j.ecoleng.2021.106394.
83. Sherman, R. L., and R. E. Spieler. 2006. Tires: Unstable Materials for Artificial Reef Construction. In *Environmental Problems in Coastal Regions VI*, 1: 215–223. Rhodes, Greece: WIT Press. https://doi.org/10.2495/CENV060211.
84. Vandenberg, J. 2020. The Human Dimensions of Coral Reef Restoration: A Case Study in the Spermonde Islands of Indonesia. University of Rhode Island.
85. Rinne, J., and E. Salo. From Carbon to Nature: What the Biodiversity Markets Can Learn from the Voluntary Carbon Market?

86. Vandenberg, J. 2020. The Risk of Dispossesion in the Aquapelago: A Coral Reef Restoration Case Study in the Spermonde Islands. *Shima: The International Journal of Research into Island Cultures* 14. https://doi.org/10.21463/shima.14.2.08.
87. Clarke, P., and S. D. Jupiter. 2010. Law, Custom and Community-Based Natural Resource Management in Kubulau District (Fiji). *Environmental Conservation* 37: 98–106.
88. Hickey, F. 2006. Traditional Marine Resource Management in Vanuatu: Acknowledging, Supporting and Strengthening Indigenous Management Systems. *SPC Traditional Marine Resource Management and Knowledge Information Bulletin* 20: 11–23.
89. Lindsay, J. 2000. Creating a Legal Framework for Community-Based Management: Principles and Dilemmas. *Decentralization and Devolution of Forest Management in Asia and the Pacific. Food and Agriculture Organization of the United Nations and RECOFTC*: 23–38.
90. Hunter-Anderson, R. 2005. An Anthropological Perspective on Marianas Prehistory, Including Guam. *Guam History: Perspectives.*
91. Hornung, J. W., and M. Z. Boradello. 2017. The US Military Laydown on Guam: Progress amid Challenges. Sasakawa USA.
92. Quintanilla, O. 2020. Inafa' Maolek Restoring Balance through Resilience, Resistance, and Coral Reefs: A Study of Pacific Island Climate Justice and the Right to Nature. UC San Diego.
93. Selvanathan, H. P., J. Jetten, and C. R. Crimston. 2023. Australia Day or Invasion Day? Perspectives on the Continuing Impact of Colonialism Underlies Public Contestations Around Australia's National Day. *Political Psychology* 44: 61–77. https://doi.org/10.1111/pops.12818.
94. Pascoe, B. 2018. *Dark Emu: Aboriginal Australia and the Birth of Agriculture, New Edition*. Magabala Books.
95. McClure, E. C., M. Sievers, C. J. Brown, C. A. Buelow, E. M. Ditria, M. A. Hayes, R. M. Pearson, V. J. D. Tulloch, R. K. F. Unsworth, and R. M. Connolly. 2020. Artificial Intelligence Meets Citizen Science to Supercharge Ecological Monitoring. *Patterns* 1: 100109. https://doi.org/10.1016/j.patter.2020.100109.
96. Wilson, E. O., and B. Hölldobler. 2005. Eusociality: Origin and Consequences. *Proceedings of the National Academy of Sciences of the United States of America* 102: 13367–13371. https://doi.org/10.1073/pnas.0505858102.
97. Johnson, E. R., and I. Braverman. 2019. Blue Legalities: Governing More-Than-Human Oceans. https://doi.org/10.1215/9781478007289-001.

Chapter 16

To See the Bees: Reflections of a Beekeeper on Her Journey of Healing

Diane Hellwich-Wellons
Association of Southern Maryland Beekeepers,
Maryland State Beekeepers Association, Leonardtown, Maryland, USA
deezltownbeez@gmail.com

16.1 Introduction

In my little corner of the United States, a rural county in Southern Maryland, we have our own microcosm. Here in the countryside, it is peaceful and serene. I call it my sanctuary for a reason. I am my best self (I believe) when I am in the yard tending to my chickens and playing with the bees. I am really in my element when outside teaching a beekeeping class or helping a new beekeeper see whatever new fascination I see in the hive. It's a rare pleasure to find connection with the natural world in our bustling, electronic and linked-in lives. So, I have come to appreciate solitude, and peace, if not necessarily silence. Rural life means dealing with the pests and problems that are inherent to the living conditions we surround ourselves with.

Animals, Ethics, and Engineering: Intersections and Implications
Edited by Rosalyn W. Berne

ISBN 978-981-5129-56-4 (Hardcover), 978-1-003-65231-1 (eBook)
www.jennystanford.com

I meet people routinely who have never left this county. They haven't traveled further north than the county line and have no desire to. They want to live simply, within the confines of their small surroundings. They are not interested in modern industry coming to their neck of the woods and they prefer to keep things as they have been for the last hundred years. There is the rub, progress waits for no one. It continues to happen at an alarming rate. I am old enough to remember a time before cell phones and computers. I'm also old enough to remember a time when plants were used as medicine. Once upon a time, lavender and other herbs and tree bark could help a headache, or eating certain spices or weeds steeped in water like tea could help lower blood pressure and help a body feel better or heal. There are a whole host of herbs that can still be used as medicine to reduce blood pressure and open pathways for healing. Honey is used for everything from burn injuries to reducing seasonal allergies and to soothe a cough or scratchy throat. My mother made us a "hot totty" when we were kids using lemon, honey, and some small amount of brandy. It really did help soothe and calm the cough from common colds [1–4].

16.2 Through the Eyes of a Beekeeper

As a beekeeper, I have come to appreciate the hidden medicines that can be found in my own backyard. My apiary has become a source of wonder and healing for me. I benefit almost daily, albeit accidentally, from the venom of the very creatures I care for. People with Lyme disease or Rheumatoid arthritis have found relief from pain and swelling using bee venom. It is counterintuitive, but somehow manages to work, what many consider real treatments with real benefits.

If you have ever been into a beehive, you understand that honey bees are anything but silent. From the moment you walk up to the hive, they hear you and they respond. When you crack the lid off, the natural hum of the hive increases in volume. When they register that I am coming in for a peek, not to harm, things generally calm back down. But every frame I pull elicits a response. If the queen is working on the frame I remove, the tempo increases again. If the hive is queen-less, things can get raucous. Bees communicate using vibrations/sounds and pheromones. Most folks think they communicate by stinging, and they do, just not the kind of conversation most of us want to have with the business end of a winged stinging machine. Beekeepers do not like to get stung, but it is a part of the job we learn to respect and appreciate. For me, the sting means a couple of things. I need to slow down and pay more attention. Slow is smooth and smooth is fast. If I slow down, I will be less likely to make mistakes or smash bees. No mistakes to fix means things go faster in the long run. Plus, they give off an alarm pheromone when they sting or get crushed. It also means a kind of healing therapy. The late Dr. Eugene Odum noted that "in nature there are a lot of answers about what we should be doing in society. Nature has been here longer than humans and has survived a lot of catastrophes."

16.3 Bees as Healers

Having suffered from long-term chronic Lyme Syndrome since 2008, I have even learned to look forward to the sting versus fearing it. It took a while to make the connection; the relationship between my health and my honey bees was not obvious at first. In fact, the original Lyme infection took a full year to figure out. Doctors did not exactly know what was wrong with me when I got sick. I thought I was going crazy. Odd intermittent symptoms that progressively became worse. Diagnosing a Lyme infection at the time was not an easy task. New and not always effective testing in 2008 terms meant that they really were looking for a needle in a haystack. To see the infection, they needed to see the spirochete on the slide of the microscope on which they were focused. They needed to be able to pinpoint the parasite and call me the host to name me as sick with Lyme disease versus looking at me as a complete organism that was failing and searching for and naming the cause of that failure. They could hear my raspy voice and see my physical and mental decline. They treated me in the hospital for the recurring intractable migraines that were so debilitating. I was forced to a dark room and bedrest for days at a time. They could see the symptoms of a problem but could not name it. The frustration I felt was palpable. I needed more information; I needed a cure.

A full year after my initial infection and after months of enlisting the help of a whole team of doctors, from voice specialists to neurologists and otolaryngologists all the way to Johns Hopkins, I got a second tick bite. This one I saw, I was able to capture in some tape for my doctor to see and by the next day I had the classic bullseye rash, the one and only time in the last 15 years that I have had the obvious sign of Lyme infection. A full year of going bananas trying to find a cause for the malaise and boom, there it was. It seems like that would be the end of the story– right? I wish with all my heart I could say that was the end. The diagnosis solved one problem but opened a Pandora's box of others that I continue to deal with to this day. New allergies and some creative experimentation with herbals and antibiotics to find a combination that worked took up much of the next few years as I was plagued with intermittent illness. Lyme flares that were intense and debilitating. I would lose my voice due to vocal cord paralysis or be crushed under the weight of the migraines. It was a rough ride. Which takes me back to the wonder of the mighty honey bee and her potent sting. Many will say that there is no medicine in her venom. I will tell you I know otherwise, and many others will too. Science has not tested to the full extent venom therapy benefits, so we lack scientifically tested data. But for me it is a tangible thing. I feel first-hand the benefits of every sting I take.

I began reading about beekeeping and learning what I could to get started in 2015. One of my physicians recommended that I start eating some local honey daily to help boost my immune system. Just a small amount he said. Honey is loaded with enzymes that heal the gut, pollen grains that help reduce seasonal allergies, antimicrobial and antibacterial properties that give the system a boost when ingested regularly. That first jar of local honey was expensive. I paid a dollar an ounce for this liquid gold from a local farmer and that is what originally triggered the idea that I could become a beekeeper. We have land and space to put them, and I thought why not? I figured it would be a fun hobby to replace the horses we had recently sold because I simply wasn't able to take care of them anymore because of my illness. I thought beekeeping would be easier. This is the part where anyone and everyone who has ever kept bees rolls their eyes and outright guffaws. Beekeeping is anything but easy. It is fun and stimulating and constant and new and alive and thought provoking but rarely is it "easy."

For the first two years of beekeeping, I kept a diary of the daily goings on. Every new frame of comb, every sting. (I got stung nine times the first season.) I wore gloves every time I went out but still managed to get stung while taking off my veil or picking up equipment I had used during inspections. The second year saw 26 stings. I had more colonies and was in the bees more often, so I attributed the extra stings to that. It hurt and I was swollen and red and itchy for days after each one. By the third year I realized I had not had another Lyme flair since I started keeping the bees. I shrugged it off as pure coincidence. By 6 years in, I still had not had another flare up and I started researching why that could be. By that point in my beekeeping journey, I had been managing twenty colonies through the seasons and keeping them alive over winter (It is a passion or an addiction). I stopped keeping track of stings and such the third year as I had too many hives and too many stings to document them all. Most notably absent was my illness. No major Lyme flares and almost none of the symptoms I had been living with for years. I wasn't sure if it was the venom or the honey or likely a combination of the two, but I had been relatively well and overall symptom free for several years.

I began researching to try to find the reason. I came across articles about people all over the country who had been stung by honey bees and found their illness gone after a bit of recovery (for more on venom, see [6]). There were naturalists using bee venom therapy to kill off Lyme or reduce the symptoms of the disease by localizing the sting locations to the "hot spots" on the body that were most painful and affected. It was working for so many others that I began to realize it is not likely a psychosomatic response. It is more than likely true data that others have documented. I was not the only person finding relief.

16.4 Foray into the Hive

When I got my first Nucleus (small colony unit of) and Packages (3lbs. of bees in a cage with a queen) of honey bees the following year, I was so excited. After a year of researching and reading everything I could about beekeeping, it was finally happening (for more on bees, see [5]). I thought I was prepared for the journey. I should have taken at the bare minimum a beginner's class. I simply did not have any clue what I was getting into. I had read a whole bunch of books, watched a ton of YouTube videos and I still went out and made a bunch of mistakes. I learned from every single one and to this day I continue to learn and grow in the hobby with my bees, more than any other hobby I have ever had. It's a perpetual mind-bending, learning game and I love it. I love the way it smells inside the hive as I pull the cover off: sweet nectar curing and stored capped honey sealed inside the wax confines of the cells created by the bees one tiny flake of wax at a time. The smell of fresh pollen stored for later use to feed and care for the next generation of that colony and its queen. I enjoy the piney resinous fragrance of the propolis seal surrounding every corner, nook, and cranny. I love to see the activity of the day. Cohesive, cooperative, purposeful, and intentional are about the only ways I can think that accurately describe the inner workings of a hive. The colony works to care for future generations and the queen is a slave to the colony doing what they dictate and when. She lays fifteen hundred—two thousand eggs a day, as the hardest-working member of the colony. It is no holiday to be the queen bee. Her pheromones are what suppress the laying ability of every other member of the colony.

Each foray inside the hive is a remarkably humbling and grounding experience. These tiny insects teach us about resilience and cohesion, about teamwork and selflessness. They teach us about the purest meaning of altruism. Each of them lives and dies for the betterment of the colony. Each would sacrifice her/his life for future generations of honey bees they will never know, to survive in our world with no proof or promise of success. They fly themselves to death to forage for nectar and work tirelessly to create honey and amazing comb structures for a safe and sound home, all for the future generation to enjoy. They work without appearing to harbor even the tiniest bit of jealousy or resentment for those who will inherit the fruits of her and her sisters' labors of love.

There are stories throughout human history of the use of bee venom as medicine, called apitherapy [7–10]. Humans around the globe have been using bees and their sweet honey since the earliest men walked on earth. Man of Bicorp Cave paintings from Bicorp, Spain dating between 6000 and 8000 BC depict beekeeping at its earliest and our fascination with these tiny creatures from the beginning of human history. Additionally, there is much research that indicates the of bee propolis, the antibacterial material they use to fortify the inside of their hives [11–13]. Today we are still just as curious and learn more each year about the inner workings of the hive, the interactions the bees have in nature and how we can help (or hurt in many cases) the future of the honey bees in our world.

This is where the ethical dilemma comes in. Just penning this essay has given me pause for I wish dearly to see the world as a better place. I hope for cures to ailments and solutions to illnesses that plague us. But there are some real concerns for the human use of honey bees. Not just for their honey but, the medicinal benefits of their stings, the continued use of their wax, the potential study and use of them for venom therapies and unknown future potential uses. Will we be causing more harm than good? Are we doing this creature a disservice by taking part of its honey harvest for ourselves? Are the bees struggling because of human machinations in their hive designs and "keeping" of the bees? Is it possible (or rather likely) that we are the cause of the trouble with the bees in the first place?

16.5 Environmental Stressors

In my mind, it seems a requisite for engineers to study non-human animals and learn to integrate lessons from the natural world in the next generation of engineering design and practices, or we just might engineer ourselves out of existence. We infringe on their foraging grounds by cutting down everything in sight to build shopping malls and grocery stores and cineplex theaters. Even farming has become so industrialized that chemicals are used to increase crop yields to feed more people. The problem lies in the fact that those same chemicals are harmful to the very creatures that humans need (and depend on) to perform pollination to produce the crop in the first place or increase the yields even further. Where hedgerows of wildflowers once stood, there are culverts or green grass, the equivalent of a nutritional wasteland for the honey bee and the four thousand other pollinator species that live in the wilds just in the United States. There are approximately 20,000 species the world over according to the US Geologic Survey [14].

We imbed the new crops seed with pesticides, herbicides, or fungicides to help them grow free of pest pressures. Those same chemicals can be found in the wax made by my honey bees that have never left the hive. Nature designed them with special glands on their abdomens that produce tiny flakes of wax. Each flake is a marvel and the beeswax honeycomb an amazing piece of engineering. I could write a whole chapter about the process by which it is made and its uses throughout human history. From sealing and waterproofing leather to burning it for light and religious ceremony. What I wonder is how are those chemicals showing up in newly created bees' wax? The same wax in which they are storing their food resources, laying eggs, feeding larvae, and growing pupating honey bees. We hypothesize that the generation caring for and raising young larvae in the hive are being fed by bees who have foraged the land for nectar and pollen resources. They bring in the chemicals with the food. Or they don't wipe their feet well before entering the hive. The bees are resilient and can overcome and fair amount of pressure from these chemicals. But can they overcome multiple chemicals at one time?

It is not just one thing causing bees to struggle, it is a combination of pressures. In addition to loss of foraging habitat and increased chemical pressure, there are also pest pressures that wreak havoc for every beekeeper. A quick internet search of Varroa mites vs Tropilaelaps mites will send you down a hair raising, skin crawling rabbit hole that you will wish you had not dared to venture. There are new pests like the Asian born spotted lantern fly (*Lycorma delicatula*) which came to the United States much like the honey bees themselves did on cargo ships (and airplanes.) The point is, there is not just one pressure bearing down on our ecosystem, there are many.

16.6 Conclusion

The world is changing and evolving, for better or worse. The creatures that live here need to evolve as well; we need to be better if we want to see this world and the creatures that inhabit it survive, ourselves included. Humans seem to think that the "circle of life" is something we are standing outside of and watching, controlling. We are beginning to understand that we are part of that circle, spinning along with all the other creatures that are fighting for survival alongside us. Just because we are the top of the food chain and the one creature in the chain that can affect our planet by our actions, does not mean that our own extinction is not on the horizon. We should take heed of the natural world around us, begin to look at things from the perspective of each creature. Find a way to coexist with them as sentient beings that deserve life and happiness, just like ourselves. It seems far-fetched to think of insects as sentient. But when I pinch a honeybee between two frames in the hive, she makes noise to tell her family (or me) that she is stuck. She makes noise that sounds like screaming, sympathy effectively forcing me move the frame to release her. She does seem to communicate that she feels something and reacts to that sensation. I cannot be naïve enough to believe her life does not matter to her colony. For it certainly does. I cannot feel superior to the point of pinning her in with no regard to her life. I am learning to take care, to recognize her as a unique independent individual that adds benefit to her colony. To see in myself her uniqueness and her value.

Myriad species and organisms reside on this planet, or at least they did. From one mass extinction event to another, the strongest have survived and we find ourselves coexisting here with creatures big and small. We humans have used many of these creatures to our benefit and some to near or outright extinction. We are not the only life forms here, but we have managed to infuse ourselves into the ecosystems and create spaces for ourselves without any real regard for the native populations or ecological impact of our infiltration. Little thought of sustainability has gone into the engineering advances we call progress. The native ecology does not see it through the same lenses we do. We aim for survival. At what cost?

We really must begin to see ourselves as part of the circle of life and respect the other living entities on the planet, to look at the tiniest of creatures and see infinite possibilities. Maybe then we can begin to learn from them in the way Odum thought us capable.

References

1. Eteraf-Oskouei, T., and M. Najafi. 2013. Traditional and Modern Uses of Natural Honey in Human Diseases: A Review. *Iranian Journal of Basic Medical Sciences* 16: 731–742.
2. Tashkandi, H. 2021. Honey in Wound Healing: An Updated Review. *Open Life Sciences* 16: 1091–1100. https://doi.org/10.1515/biol-2021-0084.
3. Honey. 2024. *Mayo Clinic.* https://www.mayoclinic.org/drugs-supplements-honey/art-20363819. Accessed March 9.
4. Honey: Benefits, Uses, and Properties. 2018. February 14.
5. Dadant. 2015. *The Hive and the Honey Bee.* Hamilton, Illinois: Dadant.
6. Wilcox, C. 2016. *Venomous: How Earth's Deadliest Creatures Mastered Biochemistry.* Farrar, Straus and Giroux.
7. El-Wahed, A. A. A., S. A. M. Khalifa, M. H. Elashal, S. G. Musharraf, A. Saeed, A. Khatib, H. E. Tahir, et al. 2021. Cosmetic Applications of Bee Venom. *Toxins* 13: 810. https://doi.org/10.3390/toxins13110810.
8. Wehbe, R., J. Frangieh, M. Rima, D. El Obeid, J.-M. Sabatier, and Z. Fajloun. 2019. Bee Venom: Overview of Main Compounds and Bioactivities for Therapeutic Interests. *Molecules* 24: 2997. https://doi.org/10.3390/molecules24162997.
9. Bee Venom Therapy: Bee Venom, Its Nature, and Its Effect on Arthritic and Rheumatoid Conditions. 1935. *Journal of the American Medical Association* 104: 2120. https://doi.org/10.1001/jama.1935.02760230068028.
10. Medicinal Uses of Bee Venom. 2024. *Department of Surgery.* https://froemkelab.med.nyu.edu/surgery/content?ChunkIID=13504. Accessed March 9.
11. Pasupuleti, V. R., L. Sammugam, N. Ramesh, and S. H. Gan. 2017. Honey, Propolis, and Royal Jelly: A Comprehensive Review of Their Biological Actions and Health Benefits. *Oxidative Medicine and Cellular Longevity* 2017: 1259510. https://doi.org/10.1155/2017/1259510.
12. Braakhuis, A. 2019. Evidence on the Health Benefits of Supplemental Propolis. *Nutrients* 11: 2705. https://doi.org/10.3390/nu11112705.
13. Przybyłek, I., and T. M. Karpiński. 2019. Antibacterial Properties of Propolis. *Molecules* 24: 2047. https://doi.org/10.3390/molecules24112047.
14. How Many Species of Native Bees Are in the United States? | U.S. Geological Survey. 2024. https://www.usgs.gov/faqs/how-many-species-native-bees-are-united-states. Accessed March 9.

Chapter 17

Designing and Crafting Systems for Non-Human Animals: Inspirations from Elephants

Fiona French
Department and Institutional Affiliation,
School of Computing and Digital Media, London Metropolitan University, London, UK
f.french@londonmet.ac.uk

It is ten years since I first stepped into a sunny meadow in Wales to meet Valli, who was to be my first consistent and reliable prototype tester in the quest to design a novel enrichment device for a sanctuary-housed elephant. Valli destroyed anything that wasn't sufficiently robust or firmly bolted in place, refused to touch objects while I was observing, and demanded an endless supply of apples as her dressing room rider. She was mischievous and stubborn, with diva-like charm. She is a middle-aged Asian female elephant, but in truth, she reminds me strongly of our terrier, Skomer.

As a researcher in the field of Animal-Computer Interaction (ACI), my interests lie in the application of technology to support animal well-being. The prototype elephant enrichment devices were themed around offering control and choice to Valli, in the context

Animals, Ethics, and Engineering: Intersections and Implications
Edited by Rosalyn W. Berne

ISBN 978-981-5129-56-4 (Hardcover), 978-1-003-65231-1 (eBook)
www.jennystanford.com

of acoustic and cognitive stimulation. That involved offering Valli an interface she could use to play synthesized sounds or trigger other outputs.

This chapter draws on several years of trying out different designs for elephant control systems and of facilitating regular multidisciplinary design workshops ("zoo jams") that consider ways we can deploy technology to address animal welfare challenges [1]. The knowledge and insights gained are applied to a discussion of some interaction design features relevant to non-humans, and how we used craft as part of a process to support the design of animal-centered computing systems. Much of the research has been done in the context of managed non-human animals (henceforth, animals), which means there are always human stakeholders to consider. These might include animal stewards or stock hands, management teams, animal health and welfare specialists, other researchers, and in some cases, visitors. As a technologist, having the opportunity to collaborate with animal specialists continues to be one of the highlights of undertaking research that focuses on other species.

We are already familiar with human cognitive, sensory and physical capabilities and emotional responses in different contexts. Moreover, we can rely on our personal intuition to guess what a particular experience might be like for another human being. Designers work with this knowledge each time they consider a new artifact or tool or happening, aiming to please the senses and provide interactive features that are suitable for hands with opposable thumbs and eyes with bifocal vision. Humans are adept at creating designs that are fit for purpose *for a human*. New challenges arise when we try to design systems for other animals, partly because we don't know enough about their perceptual or cognitive abilities, and partly because we often have no idea what they want or what gives them pleasure [2, 3]. We can't conduct user surveys or hold focus groups, therefore, on the surface, there appear to be no shared communication strategies.

Nonetheless, a basic tenet of interaction design is to "know your user." This places responsibility on design teams to learn about their potential animal users, which can be achieved through research and collaborations with species experts. Ideally, as well as undertaking background research, designers should strive to conduct investigations with the target animals, aiming for a multispecies take on participatory design.

How is it possible to achieve that? Environmental enrichment that offers control and choice to managed animals is endorsed by welfare experts, as it gives the animal an opportunity to respond to a stimulus, which in turn requires the exertion of evolved mental and physical skills, and probably involves learning new behaviors [4–6]. As defined by UFAW (University Federation of Animal Welfare): "*The goal of environmental enrichment is to make the environment more complex and dynamic, to provide more stimulating, cognitive challenges, and opportunities to make choices, control social interactions and behave in more diverse species-specific ways.*" [7]. Thus, providing an animal with control and choice around any novel intervention designed with them in mind allows them to make meaningful decisions and express some of their preferences, and this is a helpful form of feedback for designers [8].

17.1 Designing Animal Interfaces

There are many great examples of low-tech systems and devices that offer enrichment for zoo-housed species [9]. However, it's also possible to embed technology into enrichment systems to help create novel experiences, as well as support research and management goals through automation and data capture. Humans have come to rely on tech to provide us with much of our modern stimulation, including social interactions, information gathering, geolocation of assets and a variety of entertainment systems. We didn't evolve to live this way, but we have adapted swiftly to a changing infrastructure, and other lifeforms are having to adapt as well, to cope with human population expansion. In acknowledgement of this, and with welfare at the top of the agenda, the ACI mission is "*to advance the art and science of designing animal-centered interactive systems fit for a participatory multi species society.*" [10]

Technology can be used covertly, for example, as a surveillance technique, or it can be an explicit aspect of an animal-centered system, with an interface that allows humans or animals to interact with it, or both of these uses simultaneously. In an interactive system, technology provides the link between inputs (from users) and outputs (from a system), facilitating a conversation between user and computer. A significant part of our work with elephants involved investigating which interface elements were engaging for the animals, through testing different types of sensors that could detect trunk tip inputs and trying to design appropriate outputs. The key features of an animal-centered system interface were identified as: *consistency, multiplicity, affordance, specificity, differentiation and graduation.*

Consistency refers to the robust functionality that is critical for all prototypes, so that users can discover or gradually learn how to manipulate the controls without the device failing. This feature is important for enabling animals to gain competence [6].

Multiplicity references the multitude of possible ways to create interfaces with technological systems, and is a reminder that other species' sensory, physical, emotional and cognitive characteristics can be explored and used to create relevant designs. Perhaps the most well-known control (for humans) is a simple push-to-make button, such that the user can make a decision whether or not to trigger an output (like a shower, or a sound, or a dispensed treat). Yet, in the context of animal users, this raises some tricky questions. How would an animal without fingers even push a button? With their nose? Designers need to consider species-specific *affordance*, such that the perceivable features of the interface reveal to animal users how they can engage with the controls. Signifiers provide initial cues to interaction possibilities [11], while feedforward is an understanding of what will happen when the interaction happens [12]. Together these features give the user a mental model so they can use the controls effectively. Since most animals have no prior experience of technology, we would not expect them to immediately understand a new system, therefore designing for *learned affordance* is a more achievable goal. For this to be successful, the learning aspect could be facilitated using training techniques, or the animals could learn independently through trial and error. In either case, offering a system with controls that can be activated using a familiar mechanism supports cognitive capability, such as tugging a rope for an elephant or landing on a perch for a parrot.

We could replace a push button, essentially a mechanical device, with a simple touch control, using a different kind of sensor, such as one responsive to heat or light or proximity. But would an animal understand how to use it effectively? Regarding system status, does the touch control remain "ON" after triggering, or does it revert to "OFF" as soon as the interaction stops? If it remains ON, how does the animal change the status? Using the same control? How does the animal know whether the control is currently ON or OFF? Does the control itself offer some indication of its status, in the same way that a light switch changes position? Would this not require the interaction design of the device to offer feedback from the touch control that was distinct from the output that it triggered? Conflating outputs would be difficult to interpret, since humans would not understand which signal was responsible for the animal's response to the device. Thus, controls also require *specificity*, meaning that each one is only responsible for changing one sensory experience.

Research on using touchscreens with parrots [13] showcases some successful applications of touch controls, triggered by parrots tapping the screen with their beaks to obtain food rewards and make choices about different playful activities. In this case, the touchscreen displayed a range of options to the birds. *Differentiation* enables animals to distinguish between different controls, but designing for this also comes with challenges. Humans, for example, understand vertical hierarchies, but we don't know if the location of a control imbues it with any particular meaning for an animal. It is possible that any kind of differentiation, such as color, shape, texture or smell, might impact on an animal's choice and so confound an investigation into output preferences.

For Valli in Wales, interaction design solutions proposed by animal experts included using beam-break technology, which means that moving an object (trunk or ear or stick) across the controller infrared beam will trigger an input. But if the controller automatically reset after the object was removed, this could be very frustrating: imagine a touch tap that you can activate with your hand and then as soon as you try to pick up the soap, it switches off again; or imagine a music player that only plays while you keep your finger on the button. This problem suggests that the control should remain in an ON state, yet that could be stressful if the user didn't know how to switch it off. By contrast, a moving interface element requires the animal to deliberately act upon it, thereby adjusting the associated output in the manner of an analog controller. This could reduce any motivational ambiguity and provide very useful feedback to the design team if the movement was precisely mapped to one specific changeable quality of the design. This feature of the interface we describe as *graduation*.

In our work, we found that moving objects held a lot more interest for our target users than static buttons, which meant we had to invent new ways to capture the elephants' inputs. This was because although objects that move may be the most interesting ones to design and subsequently play with, they are often also the most challenging to construct and make functional. For me, designing with elephants over several years has highlighted the value of *craft* as a contributary method for investigating prototype designs and bridging the gap both between and within different species.

But what exactly is "craft"? Several researchers make a strong connection between *craft* and *design research* [14–17]. Craft has been described as *"a careful form of making"* [18], as a form of *embodied cognition* [19], and importantly, as being *inclusive of computing*, in the context of tangible interaction design [20, 21]. Within the definition of craft, we include the practice of *tinkering* with electronics, which can be more fruitful for developing an appreciation of sensors and actuators than using off-the-shelf solutions. There is some debate as to whether digital making that involves using embedded technology constitutes craft in the "traditional" sense. It has been claimed that the computer then becomes the mediating tool, rather than the action of making [20], but we argue that tinkering with the functionality of a system and recombining the interactive components can be considered a form of crafting—using sensors and wires instead of needles and thread. Within the making community, appreciating the sensitivity and potential of electronic equipment over time and through experimentation can be akin to the experience of working with wood or rope or clay. Although outputs generated through the use of embedded technology are not objects, they are still concrete, perceivable experiences, such as sounds, vibrations, lights and movements. They have aesthetic dimensions that designers are able to discern and control through their practice.

Craft formed part of our method for mediating between humans, objects and animals through sensory and intellectual practice.

17.2 Craft for Sensory Appreciation

Although humans rely extensively on vision, for many species, haptic, kinesthetic and olfactory information is much more important for interpreting the world. Animals typically experience the world at a different scale, and certainly have a different set of common sense principles (e.g., if in doubt, climb a tree). Animals lack the exposure humans have had to computer systems and interactions with technology, even if their abilities transcend our own in areas such as pheromone identification or balance. Additionally, their physical capabilities such as strength and speed, and psychological motivations such as hunting and foraging, may make a significant difference to how an animal perceives and interacts with their environment.

When people work closely with domesticated animals, such as canine companions, a reciprocal understanding emerges, because of the proximity between the human and the dog, and the requirement to observe/listen to/smell/feel each other in order to communicate. Sharing that world space and having mutual awareness gives each species some insight into the other [22]. However, this is unlikely to happen with undomesticated zoo-housed species, who are typically maintained in environments without human social contact—the majority of zoos in the UK now aim for Protected Contact only, in order to carry out routine tasks such as medical checks.

For a developer not in close daily contact with their animal user, it can be difficult to fully appreciate the qualities of the "other" (user) that will help define the most appropriate way of designing an interface or system or experience. While this is true even of humans, who have variable characteristics and requirements within the same species, the dilemma becomes more critical when the user is a different species. Humans have no mental model of what life is truly like for a different species and the *physicality* of an animal is one aspect of their experience that we inevitably find hard to conceive.

If designers conceptualize an interactive product, then hand a blueprint over to someone else to manufacture, they miss an important opportunity to work directly with physical materials, an activity that can support the development of animal-centered designs. The value of craft as a method for understanding material quality and connecting our human senses to the physical world should not be underestimated. It can help people to focus on practical and structural aspects of systems such as the dimensions, location and feasibility of their devices, as well as the form, weight, texture and smell. Indeed, the aesthetic dimensions of devices are highly relevant in the context of designing systems for animal usage. Technical details, manufacturing constraints and electrical hazards can also be considered in relation to the overall design, and this is made easier because of the three-dimensionality of a hand-crafted prototype and the fact that it can be manipulated and examined in space.

17.3 Co-Crafting

Crafting models is a mode of expression that is often ignored in favor of sketches, which require fewer resources. However, not only does 2D visual representation put the final design in the hands of a confident artist, but "*a mark on paper becomes a kind of signature for its author—it can be erased or written over... that is a deliberate and destructive act*" [23]. Although collaborative drawing can be fun and productive, as long as participants remain respectful of each other's contributions, co-crafting can be even more inclusive, offering a flexible, unassuming editing process, comparable with co-writing documents or code on a shared platform. Physical pieces can be moved around freely until a decision is reached; paper and card can easily be lengthened or shortened; co-creation is such a *fluid* process that it is easy for everyone to become involved [23–25].

In addition, a series of rough physical "sketches," evolving over time, has more potential for engaging stakeholder collaboration than a high-fidelity version of a solution, ready to be tested [23, 26]. The less finished the piece of work, the more opportunities there are for others to participate in the design by contributing their own ideas and having some ownership of the process. This flexibility can be extended to the animal users, so that they also have the opportunity to make choices regarding the characteristics of the systems that is designed for them. It can be said that this constitutes a kind of participatory design with the animal. Since humans currently lack the ability to explain to the target user what they have in mind as a final product, the animal is necessarily missing some context and must react "on the fly" to whatever novelty is introduced. Yet, perhaps this results in more honest feedback than if the designer had been able to say (for example): *"Look, it's going to be an exciting musical toy, but what do you think of this button so far?"*

17.4 Interspecies Mediation

We would like to draw attention to two parallel events—the choices made by designers that influence the final experience offered to the animals, and the choices made by animals if they are offered a way to express their preferences during the process of development. Designers' decisions should be heavily influenced by the choices made by their animal users, which strongly suggests a mode of development that values incomplete solutions as sources of inspiration and knowledge. Moreover, to engage animal participants, it is fundamental to create physical interactive objects that will ultimately be deployed by the stakeholders (designers, researchers, animals, carers) as cognitive tools. A physical device can express abstract ideas to an animal—moreover, tangible objects may be more usable for some species than graphical interfaces such as touchscreens, if only because animals might be expected to learn the relevance and purpose of a physical object faster than an abstract representation, since they use this skill as part of their normal behavior [27]. If an animal can make sense of a novel technology-enhanced feature (designed by a human) in their enclosure, through interacting with it and controlling an output, we might then say there is a mutual (multispecies) understanding of the design. The profound experiential knowledge gained from physical interaction with an object is something shared between designer and user, despite their reliance on different modes of perception. If this awareness is positioned within an ACI context, it is possible that craft may bring to the ACI designer a heightened sensibility of an animal user's potential experience. As the design becomes more refined over time and craft skills mature, the point can be made that concurrently, there is an intensification of the insight of the designer into the perspective of the animal.

For ACI designers, it is often the case that early prototypes are developed for a small cohort of users—individual case studies are common before large-scale deployment of solutions. This means that quantitative feedback may be difficult to obtain. While a data-driven scientific approach would require a statistically viable number of animals to test a novel device under same conditions, to authenticate results, it is possible to justify the exploration of one animal's preferences. The knowledge obtained in a single case study can inspire and inform subsequent projects, as well as being an exemplar of the sustainability "3 Rs" approach (Replace, Refine, Reduce) to conducting experiments with animals [28]. In support of this, the Research through Design methodology proposes that particularity can be an advantage [29]. The design of a single, bespoke solution for a specific and unique context can offer valuable outputs, through garnering rich qualitative data relating to the behavioral responses of the animal, generating unexpected knowledge and inspiring future directions for research. Keeping the scope small like this enables the design team to spend time crafting different versions of their design, thereby simultaneously learning more about their user.

It should be noted that what apparently works in a clean laboratory might not work out so well in an elephant shed or a chicken barn. It can be very hard to anticipate all the possible variabilities, and the only way to resolve this issue is to test the prototype in the space for which it is designed, with the target users themselves.

Therefore, another important consideration for designers is the context in which the animal users are housed. For practical reasons, it is crucial to take account of the particularity of the environment as well as the animals themselves. It is therefore recommended to conduct a short (architectural type) survey of an enclosure prior to starting any production work, and of course, negotiate suitable locations and fixings with staff. Additionally, social contexts for managed animals rarely replicate wild conditions and depend on the set-up in each institution, which involves understanding the social dynamics of the situation from each animal's perspective, as well as considering practical aspects such as animal management. For example, a system could be designed so that it requires animals to cooperate, but only the animal stewards would have the knowledge to say if this was beneficial, and how to deploy such a system with animals who have different backgrounds, characteristics, and temperaments.

No solution could ever be "one size fits all" [30], illustrated by work with apes [31] and elephants [32]. In other words, we should not expect enrichment to necessarily be identical for different members of the same species. This appreciation of individual preference can lead to interesting areas for future research. For example, devices could be explicitly designed for investigating social dynamics, particularly in regard to the fluidity of hierarchies, the sharing of resources and the perceived values of different skillsets amongst the animals.

17.5 Reflections

ACI researchers work in a field that is largely uncharted. Although connections between animals and technology have been made for many years, the careful design of novel interactive artifacts that support animals' behavior, whether trained (e.g., tools for working animals) or natural (e.g., enrichment for zoo-housed animals), is a relatively recent topic for investigation.

The zoo jam workshops mentioned earlier [1] include team co-crafting sessions to enable collaborative multi-disciplinary design practice to take place, generating a range of novel technology-enhanced enrichment concepts. The resulting hand-crafted physical prototypes facilitate the design teams to appreciate their devices from the animals' perspective and are ideal for demonstrating scale and functionality to others, because they are easy for an audience to comprehend. The zoo jam format seems to help build trust and respect between technologists and species specialists, such that networks are established, and colleagues can learn from each other through the skills and knowledge shared during the event. Thus, the experience has the potential to be both educational and transformative for participants and stakeholders, as well as the wider academic community.

Moving forward, technology might help us find new ways to capture an animal's *intention* so that we can better serve their purpose. In parallel to humans designing systems that an animal is expected to learn how to use, whereby we incrementally change their behavior, humans may remain receptive to the idea of animals trying to shape OUR behavior (as dogs have learned to do), in order to support their own wellbeing and enable all denizens of earth to flourish. Personally, I'm looking forward to finding out more about Valli's infrasonic musical repertoire and exploring ways that her carers can use non-human acoustic signals to communicate with her. . . But that's another story!

References

1. Zoo Jam. https://www.zoojam.org.
2. Hirskyj-Douglas, I. 2023. Systematic iterative design of interactive devices for animals: guidance and reflections. *Interaction Studies. Social Behaviour and Communication in Biological and Artificial Systems* 24: 225–256. https://doi.org/10.1075/is.22045.hir.
3. French, F. 2022. Expanding aesthetics. *Frontiers in Veterinary Science* 9. Frontiers. https://doi.org/10.3389/fvets.2022.855087.
4. Mills, D., M. B. Dube, and H. Zulch. 2012. Stress and pheromonatherapy in the veterinary clinic. In *Stress and Pheromonatherapy in Small Animal Clinical Behaviour*, 242–256. John Wiley & Sons, Ltd. https://doi.org/10.1002/9781118702642.ch12.
5. *Environmental Enrichment for Captive Animals*. 2003. John Wiley & Sons, Ltd. https://doi.org/10.1002/9780470751046.ch1.
6. Coe, J. 2017. Embedding environmental enrichment into zoo animal facility design. In *Proceedings of the Zoo Design 2017 Conference*, Wroclaw, Poland, pp. 5–7. 2017.

7. Environmental enrichment. 2024. https://www.ufaw.org.uk/why-ufaws-work-is-important/environmental-enrichment-3. Accessed May 4.
8. Ritvo, S. E., and R. S. Allison. 2014. Challenges related to nonhuman animal-computer interaction: usability and "liking." In *Proceedings of the 2014 Workshops on Advances in Computer Entertainment Conference*, 1–7. ACE '14 Workshops. New York, NY, USA: Association for Computing Machinery. https://doi.org/10.1145/2693787.2693795.
9. Team Building with Bite Website.
10. Mancini, C. 2011. Animal-computer interaction: a manifesto. *Interactions* 18: 69–73. https://doi.org/10.1145/1978822.1978836.
11. Norman, D. A. 2008. The way I see it: Signifiers, not affordances. *Interactions* 15: 18–19. https://doi.org/10.1145/1409040.1409044.
12. Vermeulen, J., K. Luyten, E. van den Hoven, and K. Coninx. 2013. Crossing the bridge over Norman's Gulf of Execution: revealing feedforward's true identity. In *Proceedings of the SIGCHI Conference on Human Factors in Computing Systems*, 1931–1940. CHI '13. New York, NY, USA: Association for Computing Machinery. https://doi.org/10.1145/2470654.2466255.
13. Kleinberger, R. 2024. No more angry birds: investigating touchscreen ergonomics to improve tablet-based enrichment for parrots. In *Proceedings of the 2024 CHI Conference on Human Factors in Computing Systems*, pp. 1–16. 2024.
14. Niedderer, K., and K. Townsend. 2014. Designing craft research: joining emotion and knowledge. *The Design Journal* 17: 624–647. https://doi.org/10.2752/175630614X14056185480221.
15. Hållander, F. 2011. Crafted vernacular: a practice based research through craft. *Current Issues in European Cultural Studies* 062: 73.
16. Nimkulrat, N. 2012. Hands-on intellect: integrating craft practice into design research. *International Journal of Design* 6: 1–14.
17. Zheng, C., and M. Nitsche. 2017. Combining practices in craft and design. In *Proceedings of the Eleventh International Conference on Tangible, Embedded, and Embodied Interaction*, 331–340. TEI '17. New York, NY, USA: Association for Computing Machinery. https://doi.org/10.1145/3024969.3024973.
18. Golsteijn, C., E. Van Den Hoven, D. Frohlich, and A. Sellen. 2014. Reflections on craft research for and through design. *Proceedings of the 8th Nordic Conference on Human-Computer Interaction: Fun, Fast, Foundational*. Helsinki Finland: ACM: 421–430. https://doi.org/10.1145/2639189.2639194.

19. Huotilainen, M., M. Rankanen, C. Groth, P. Seitamaa-hakkarainen, and M. Mäkelä. 2018. Why our brains love arts and crafts. *Form Akademisk—forskningstidsskrift for design og designdidaktikk* 11. https://doi.org/10.7577/formakademisk.1908.
20. Nitsche, M., and A. Weisling. 2019. When is it not craft?: materiality and mediation when craft and computing meet. *Proceedings of the Thirteenth International Conference on Tangible, Embedded, and Embodied Interaction*. Tempe Arizona USA: ACM: 683–689. https://doi.org/10.1145/3294109.3295651.
21. Biello, D. 2023. Efforts to resuscitate extinct species may spawn a new era of the hybrid. *Scientific American*. https://www.scientificamerican.com/article/lost-species-revived-from-dna-and-restored-to-nature/. Accessed October 15.
22. Haraway, D. 2023. When species meet. In *The Routledge International Handbook of More-than-Human Studies*. Routledge.
23. French, F., S. Baskin, R. Gupfinger, S. Webber, and A. Zamansky. 2019. ZooJamming: designing beyond human experience. *ICGJ '19: Proceedings of the International Conference on Game Jams, Hackathons and Game Creation Events 2019*. San Francisco, USA. https://doi.org/DOI:10.1145/3316287.3316294.
24. Luck, R. 2018. What is it that makes participation in design participatory design? https://doi.org/DOI:10.1016/J.DESTUD.2018.10.002.
25. Sanders, E. B.-N., and P. J. Stappers. 2014. Probes, toolkits and prototypes: three approaches to making in codesigning. *CoDesign* 10: 5–14. https://doi.org/10.1080/15710882.2014.888183.
26. Lim, Y.-K., E. Stolterman, and J. Tenenberg. 2008. The anatomy of prototypes: Prototypes as filters, prototypes as manifestations of design ideas. *ACM Transactions on Computer-Human Interaction* 15: 1–27. https://doi.org/10.1145/1375761.1375762.
27. Wirman, H. E., and I. K. H. Jørgensen. 2015. Designing for intuitive use for non-human users. In *Proceedings of the 12th International Conference on Advances in Computer Entertainment Technology*, 1–8. ACE '15. New York, NY, USA: Association for Computing Machinery. https://doi.org/10.1145/2832932.2837008.
28. Russell, W. M. S., and R. Burch. 1959. *The Principles of Humane Experimental Technique*. London, UK: Metheun.
29. Gaver, W. 2012. What should we expect from research through design? In *Proceedings of the SIGCHI Conference on Human Factors in*

Computing Systems, 937–946. CHI '12. New York, NY, USA: Association for Computing Machinery. https://doi.org/10.1145/2207676.2208538.

30. Mancini, C., G. Dewsbury, J. Van Der Linden, G. Kortuem, D. Mills, and N. Smith. 2014. Towards smart kennels for supporting canine welfare: 50th Annual Convention of the Society for the Study of Artificial Intelligence and the Simulation of Behaviour, AISB 2014.
31. Kim-McCormack, N. N. E., C. L. Smith, and A. M. Behie. 2016. Is interactive technology a relevant and effective enrichment for captive great apes? *Applied Animal Behaviour Science* 185: 1–8. https://doi.org/10.1016/j.applanim.2016.09.012.
32. French, F., C. Mancini, and H. Sharp. 2020. More than human aesthetics: interactive enrichment for elephants. *Proceedings of the 2020 ACM Designing Interactive Systems Conference*. Eindhoven Netherlands: ACM: 1661–1672. https://doi.org/10.1145/3357236.3395445.

Chapter 18

Why Are Non-Human Animals Neglected in Engineering Ethics Education?

Justin Hess
School of Engineering Education, Purdue University, West Lafayette, Indiana, USA
jhess@purdue.edu

18.1 Introduction

This chapter began with a question: why is it important to write about empathy with/for animals in the context of engineering ethics education? As the title indicates, the chapter came to focus on the question, "Why are non-human animals neglected in engineering ethics education?"

Animals are seldom the explicit focus of engineering ethics teaching and scholarship [1]. In this chapter, I argue that there are three key reasons for the lack of animal-centricity in engineering ethics. I think it is important to clarify these reasons and make them explicit, as by doing so, engineering educators will become better positioned to counteract them. Thus, this chapter is not a theoretical defense of why we should consider, or even prioritize, animals in engineering ethics. Rather, it aims to identify root causes that inhibit foci on animals in extant engineering ethics discourses. Thus, while I

Animals, Ethics, and Engineering: Intersections and Implications
Edited by Rosalyn W. Berne

ISBN 978-981-5129-56-4 (Hardcover), 978-1-003-65231-1 (eBook)
www.jennystanford.com

initially envisioned engaging in such a theoretical defense, my initial question eventually shifted to the question featured in the title of this work.

Here, I briefly summarize the main points in the chapter. First, there is a dominant worldview that guides thinking within a culture, and in engineering cultures, dominant engineering worldviews are largely human-centric. Even if we agree that there is a need to find ways to help engineers consider or prioritize animals in their thinking, there exist historical beliefs, norms, and values in engineering that serve as inhibiting factors to realizing such change efforts. Second, when we examine engineering ethics discourses, we will find that animals have seldom been the focus of engineering ethics research or teaching. Thus, if one references existing engineering ethics artifacts (i.e., codes of ethics, books, scholarly literature), they are unlikely to see an explicit or prominent focus on animals. Third, when scholars operationalize the "social" in engineering ethics, it often is human-centric. Here, I unpack the concept of empathy for two reasons: (1) empathy is critical to ethical reasoning and engineering ethics and (2) by understanding how empathy manifests between humans and non-humans, we will be better positioned to operationalize the "social" engineering ethics.

While pondering the starting question of this chapter, and while generating the points listed above, I considered my journey through engineering education. Here, I open by sharing how my research and teaching journey has informed my thinking regarding animals and engineering ethics. These experiences thus set the stage for this thought piece, and they provide important context for each main point. These reflections are essentially data, and they provide a quasi-auto-ethnographic account of how my perspectives on the relationship between the topics of this book (i.e., animals, engineering, ethics, and empathy) have shifted over time.

18.2 My Research Journey in Engineering Education

I have now spent a dozen years conducting engineering education research. I entered the space of engineering education with a primary interest in sustainability [2], which was largely informed by a sustainability course I participated in as a sophomore in civil engineering, which was led by Dr. Nies at Purdue University. This course motivated me to pursue a concentration in "environmental" civil engineering, with a concerted emphasis on water. While a future in education was not on my mind, I recall my roommate, close friend, and course peer—Mark Krenzke—expressing a desire to teach such a course in his future. Unbeknownst to Mark, he planted the seed of an idea in my brain; maybe a teaching career focused on sustainability could be for me. At the time, I was also pursuing a philosophy minor, where I learned about environmental ethics. I vividly recall writing a paper on Deep Ecology [3], wherein I realized that engineers seem to devalue or even neglect non-humans, even though we quite literally depend on them. In tandem, during my junior year (January 2009), I began a co-op. Alas, 6 months into this position, the construction company I was working for went bankrupt. Due to how my company treated employees amid the 2009 financial collapse, coupled with my growing interest in sustainability and philosophy, I carefully and critically questioned my career trajectory and motivations.

In 2010, I entered engineering education as an undergraduate researcher with Dr. Johannes Strobel. When I started working with Johannes, my personal motivation was to... well, firstly, as I was questioning my career path, it was to learn more about this thing called graduate school, and more specifically, what graduate studies in engineering education might entail. Secondly, the project I had applied for involved a literature review of indigenous practices. One of our goals was to depict how select indigenous practices may be, in effect, engineering practices. It was frankly the only project listed on Purdue's Summer Undergraduate Research Fellowship's website that interested me at the time. As I recall reading the position description, I quickly saw a connection, which I frame here in the form of a question—how might engineers learn from indigenous peoples to inform efforts in sustainability?

As a White male American engineering undergraduate student from a conservative state, indigeneity was... well, let's say that during my undergraduate studies in civil engineering, it was not high on my radar. I soon "discovered" a history of hegemonic development in the West, coupled with a devaluation of all things indigenous [4]. I also "discovered" how indigenous worldviews guided all indigenous engineering practices, which I consider a key learning for myself as a soon-to-be civil engineer graduate. Moreover, I started to question the dominant worldview of engineering and the compatibility of this worldview with the tenants of sustainability, including its environmental and social pillars [5].

My positive undergraduate research experience solidified my interest in graduate school. So, with Johannes's encouragement, in late 2010 I applied to the School of Engineering Education at Purdue University. In March 2011, I received and accepted an offer to join the program. Shortly thereafter, I began my graduate studies as a PhD student under the guidance of Johannes. I must admit, I did not foresee a long-term career in engineering education. Yet, at the time of this writing, I am currently back at Purdue University as an assistant professor in Engineering Education.

As a graduate student, I continued working with Johannes and exploring indigenous engineering practices, or what we later called "ethno-engineering" [4]. I soon began exploring other topics. One of the most formative learning experiences for me involved an exploration of how engineering and non-engineering faculty members perceive empathy and care to manifest in engineering [6]. Like the ethno-engineering project, Johannes prompted my interest in empathy and care. I recall Johannes sharing multiple studies that I could participate in, including one focused on resistance to change and a second on empathy and care. The latter seemed to directly build on our ethno-engineering literature review. Even better, Johannes had already worked with Dr. Carrie Wachter Morris to collect focus group data. As I had the opportunity to enter the project as a data analyst, I was eager to join.

Thus, as a graduate student, I entered this nascent research area on empathy and care in engineering [6]. I again brought a prominent interest in sustainability; I theorized that perhaps engineers could empathize with/for the earth, and this would motivate sustainable practices. Yet, based on our literature review on ethno-engineering, I was increasingly interested in relationships between dominant and minoritized societal groups including but not limited to indigenous peoples. I was also intrigued with how society—writ broadly—valued and devalued distinct worldviews and ways of being. I wanted to change systems to improve the experience of humans across the globe, and I still held onto the premise—motivated by Deep Ecology—that this would involve a thorough consideration of environmental issues. Yet, humans were becoming my research focus, not the environment (nor animals).

As a follow-up investigation on faculty members' perceptions of empathy and care [7], we analyzed survey responses from more than 1500 engineers [8] and conducted interviews with a subset of survey respondents [9]. During interviews, engineers generally discussed empathy and care in human-to-human interpersonal ways. Few engineers prioritized environmental considerations and I do not recall any engineer focusing on empathy/care for animals. Yet, my first interviewee discussed empathy for objects (i.e., filters on a naval ship). I continued to find my research focus following the data, so to speak. The social angle of my research became ever more prominent. I also found myself primarily focusing on empathy, which I viewed, and still view, as an antecedent to care and helping behavior. Nonetheless, I hoped, and I still hope, that promoting empathy would simultaneously help engineers empathize with users across the globe.

As I was developing novel understandings in the spaces of ethno-engineering, empathy, and care, I entered a position as a Teaching Assistant for a graduate-level biomedical engineering ethics course led by Dr. Brightman. The course was associated with a larger NSF project focused on the development of modules for engineering ethics curriculum [10]. We developed four modules focused on pediatric heart valves [11], the Kansas City Skywalk Collapse, diagnosing "osteopenia," and the Deepwater Horizon oil spill [12]. I soon shifted from a teaching role to a research role on this project, exploring the efficacy of the modules [13], including the impact of module use during a graduate-level biomedical engineering ethics course on students' empathic perspective-taking [14, 15].

The role of empathy in promoting ethical reasoning was clear to me from the start of this project, especially the "reflexive principlism" framework that we brought to module development [16]. The "SIRA" pedagogical framework we used even brought an explicit perspective-taking "stage" [13]. Alas, as I reflect on the modules, they seemed to be largely human-centric. Perhaps it is not surprising to the reader that modules focused on the Kansas City

Skywalk or the development pediatric heart valves were human-centric... but alas, in writing this chapter, it is the first time that I have considered the question, "Were any animals lost during the Skywalk's collapse?" I do not share this question in jest, but rather to highlight the human-centric nature of my thinking.

While we often prompted students to consider human perspectives during the modules, this was not exclusively the case. For example, in initial applications of the Deepwater Horizon module [12], we prompted students to consider the perspectives of marine mammals. I can still recall the challenge students experienced in considering mammals' perspectives. I postulate that the dissimilarities between cognitive capacities of humans and non-humans made this challenging. As a result, for better or worse, we shifted to having students consider the perspective of a "representative from the US marine life preservation society." Thus, we revised the prompt to have students take a human (rather than non-human) perspective.

My work on ethno-engineering, empathy and care, ethics module development, and research on the impact of modules overlapped in time. Each of these activities contributed to my perception that there was a critical need to improve the *social* element of sustainability. Moreover, like Obama [17], I felt there was an "empathy gap" in our society which was even wider in engineering [18]. Thus, by the end of my graduate studies, my research focus shifted specifically to empathy and ethics in engineering, and this is where my research and teaching efforts are largely situated today. I have consistently centered humans in my research. I sometimes make animal considerations explicit in my teaching and research, but this is not constant. I do not believe that I am unique in this sense, and I think there are numerous reasons that undergird the human-centric bias in my engineering ethics thinking.

18.3 Reason 1: Worldviews in Engineering

Individual responses to the question, "Why is it important to write about empathy with/for animals in the context of engineering ethics education?," will be grounded in one's beliefs about (1) what engineering *is* and (2) what engineers *should be* (or, more specifically, what they should become; see [19]). Responses to such questions are inevitably grounded in one's worldview [20, 21]. In the case of an engineer responding to such questions, responses will be grounded in the dominant worldview of engineering [4, 5]. In this section, I first unpack the term worldview and then I unpack facets of the dominant Western and engineering worldviews. Throughout, I explicate how such facets lead to human-centered thinking in engineering.

A worldview includes assumptions about reality, including what relationships between oneself and others, including humans and non-humans, are and ought to be. As Kearney [21] describes, a worldview serves as "the basis for acting within the total environment" (p. 120). A worldview thus includes "dynamic interrelated basic cognitive assumptions of a people that determine much of their behavior and decision making" and it serves as a system for "organizing much of their body of symbolic creations—myth, religion, cosmology—and ethnophilosophy in general" (p. 1). Worldviews thus influence ways of thinking about and responding to problems by providing accepted goals, methods, and practices, which may be implicit or explicit.

Worldviews also bring foundational assumptions, which are often implicit. For example, in the context of science education, a prominent "assumption of course is that scientific conceptions are superior" when compared "to other conceptions" [20]. Individuals within a culture generally believe solutions to problems exist within these implicit assumptions of their worldview. Implicit facets of a worldview manifest like the "hidden curriculum" in engineering education [22]. For example, often "[engineering] students quickly learn that the teacher is in control and is the sole knowledge authority and center of attention." This implicit assumption of how an engineering course operates is (likely) never voiced by an instructor; nonetheless, students quickly learn it.

While individuals can be part of multiple cultures, within any culture—including the disciplinary culture of engineering—there is a *dominant* worldview. This dominant worldview guides one's thinking within that culture by sharing what is acceptable, normative, and valued [21]. Individuals who hold a worldview thus present arguments in relation to the dominant facets of that worldview, including its assumptions, norms, and beliefs. The entire premise of this book is, I think, based on the lack of valuing animals in engineering ethics. While it is perhaps unspoken (like the Hidden Curriculum), as a chapter author, I believe it is my duty to question how we can reverse the assumption in engineering cultures that humans are significantly more valuable than other living beings.

In Kearney's [21] model, external systems and cultures serve as "outside sources of change" on a worldview (p. 120). As a concise example, non-engineering cultures influence engineering culture. As a more specific example, I argue that engineering culture was created by (and is grounded in) Western values, beliefs, and norms. This led to a tension in our quest for defining ethno-engineering [4]; not only were we two Westerners, but we were playing with Western tools (or discourses). We even cited Pacey [23] who suggested that "non-Western" engineering practices might be limited to simple construction projects (p. 748). This may not be surprising to the reader, but even as I reflect on this, such a claim seems to, fundamentally, devalue all things indigenous. Even when we think of historical engineering marvels, we probably first think about Egyptian wonders (which are... well, "non-Western"), followed by infrastructure marvels from Rome or Greece [24]. If we do not go so far back in time, we might consider the industrial revolution. Yet it is much more recently that engineering became its own unique profession (at least in the West) and much of this involved interfacing with capitalism and corporate needs [25]. Thus, let us at least say that the *profession* of engineering was largely informed by Western culture.

Like any worldview, the Western worldview is characterized by the collection of values, beliefs, habits, and norms of multiple cultures, including nations. In Hess and Strobel [5], we built on several thinkers [3, 26, 27]. As a slight update to the list we presented in that work, I argue that a dominant Western worldview includes the following beliefs: (1) humans are fundamentally different from nature/environment; (2) humans are superior and thus of greater value than nature; (3) nature (including animals) is, itself, for humans' control; (4) humans are masters of their destiny; (5) the world is vast, and thus provides unlimited opportunities for humans; (6) some resources are infinite, but all finite or limited resources have substitutes; (7) the history of humanity is one of progress and for every problem there is a solution—thus, progress never ceases; (8) all problems are soluble by humans; (9) many problems are soluble by technology and, if not, there are human solutions; and (10) human civilization will survive.

As the reader will note, most of these Western beliefs are human-centric or, at the very least, involve things that humans must do. Since I have framed the engineering worldview as largely a product of the Western worldview, I conjecture that facets of the Western worldview—including assumptions and beliefs about how the world works—trickle-down to engineering worldviews in a similar way to how industrial values inform organizational values [28]. Despite this, there are distinct facets of the engineering worldview that make it unique from other disciplines or professions [29]; indeed, if engineering were not distinct from other disciplines or professions, it would not require its own name [29].

In Hess and Strobel [4], we again build on several thinkers [27, 30–33] and argued that the engineering worldview may be characterized as (1) reductionistic [27]; (2) deterministic, or reframing complex problems as simple problems [33]; (3) mechanistic, or converting non-mechanistic problems into mechanistic ones [33]; (4) reliant on science and technology, perhaps at the expense of "social" considerations [27]; (5) a "refusal to set limits on production and consumption" [27]; (6) "the fragmentation of human labor into separate and autonomous spheres of operation [27]; and (7) progress-oriented (or, perhaps, exploitation-oriented) through the "constant transformation of the natural world into more valuable and more human-made environments" [27]. The engineering worldview is thus largely human-centric. As point seven indicates, nature, which here includes animals, is valued insofar as it can be utilized in resourceful ways.

Here, it bears briefly unpacking the terms "environmental" and "social," which I revisit in more detail in "Reason 3." I conjecture that animals *could* be considered as part of the "social" domain, and I suspect any pet owner would embrace this idea (although the term pet itself implies ownership). But alas, the engineering worldview often devalues societal considerations. This devaluing even manifests in engineering programs, where engineering students tend to exhibit declines in their valuing of societal considerations throughout their undergraduate careers [34]. The good news is that many scholars, programs, and engineering societies are actively working to reverse such trends.

While the above elements characterize "dominant" facets of the engineering worldview, engineering sub-cultures may hold distinct values and worldviews. The engineering discipline which has perhaps most concertedly explored animal ethics is biomedical engineering; alas, even here, animals can sometimes be forgotten or neglected [35]. The point to reiterate is that there are *dominant* facets that serve as the foundation for which any future engineers must practice. To revisit our review of indigenous engineering practices [4], I constantly found myself questioning whether we should craft a definition of ethno-engineering (i.e., were we the right people to do so?). If yes, ought we do it with a blank slate mentality (and, thus, completely grounded in the literature) *or* would it be okay to build on historical definitions of engineering? In truth, simply by using the term "engineering," we had to engage with extant and historical framings of this concept.

Like engineering sub-cultures, individual engineers can bring distinct individual worldviews, which are likely based on the multitude of cultures of individuals. Moreover, engineers wear different hats (for better or worse, see [29]) and, based on the hat the engineer chooses to wear, they might find themselves exhibiting distinct values or making distinct choices. If we look back at the emergence of engineering as a profession, engineering was largely defined in relation to corporations [25, 32]; yet, it is clear that engineers—at least when they wear their engineering hat—bring distinct values to their practice when compared to other professions [28].

While I have depicted a dominant engineering worldview, worldviews (including dominant worldviews) are not static. While one's worldview directs one's actions, these actions then can alter "social and geographical environments," which—in turn and in time—reciprocally influence a worldview ([21], p. 120). Through this process, individuals create or modify culture, specifically (at least in Kearney's model) through the development of "cultural symbols and project system" (p. 120).

Consider this book as an example of an engineering ethics artifact: a collection of individuals (who are, ostensibly, thought leaders in engineering ethics) brought forth their worldview to create artifacts representing stances on animals and ethics in engineering. Through these individuals' actions (i.e., writing chapters), they will tailor the social environment or landscape of engineering ethics. The book itself will exist as a cultural symbol of ideas regarding how engineers should value non-human animals in their ethical practice or how engineering educators should incorporate non-human animal considerations into their curriculums. Alas, the call for this book is due to the *lack* of explicit focus on animals in engineering ethics (see Reason 2). In this sense, the social landscape that came before us has neglected such a concerted focus. To this point, Kearney [21] cites Marx in his attempt to express how the choices we make are transmitted across generations:

> Men [sic.] make their own history, but they do not make it just as they please; they do not make it under circumstances chosen by themselves, but under circumstances directly encountered, given and transmitted from the past. The tradition of all the dead generations weighs like a nightmare on the brain of the living. (Marx, as cited in [21] p. 5).

The good news, at least for advocates of considering animals in engineering ethics, is that individuals create and change culture. While a culture will change slowly over time—and may even resist change—scientific revolutions can lead to rapid changes in culture [36]. Before or after any change (be it incremental or revolutionary), a dominant worldview remains; thus, one's practice is still largely dependent on present norms and beliefs. Yet, I think we can frame engineering ethics as a sub-culture within engineering. As a sub-culture, individuals engaging in engineering ethics scholarship are similarly beholden to the dominant norms, values, and assumptions of the broader engineering discipline, but I think we are also positioned to prompt change. Part of our change strategies should be informed by what came before us. To that end, we must examine historical artifacts developed within this sub-culture. With this goal in mind, I turn to such artifacts next, with a concerted focus on non-human animals.

18.4 Reason 2: Artifacts in Engineering Ethics

Building on the idea of worldview, here I briefly explore the explicit presence of animals in prominent cultural artifacts in engineering ethics, including: (1) codes of ethics; (2) textbooks; and (3) reviews. As I highlight throughout this section, while a focus on animals and ethics in engineering is not absent from these artifacts, animals are seldom the focus of engineering ethics teaching or engineering ethics education research (a similar point to this was made in [1]). Thus, when one examines engineering ethics research, they are unlikely to see a concerted focus on animals *unless* they are looking for it.

Engineering ethics codes are key "artifacts" representing the culture of engineering, and they are oft-cited guides of *professional* engineering practice. Berne [1] provides numerous examples of the devaluation of non-humans in engineering ethics and, like many scholars before her [29, 35, 37, 38], Berne describes the role of engineering codes in guiding ethical engineering thinking. Here she draws attention to the devaluation of animals in such codes. For example, Berne [1] references the NSPE paramountcy clause which states, "The National Society of Professional Engineering decrees that professional engineers should hold paramount the safety, health, and welfare of the *public*" (emphasis added by Berne). I agree with Berne [1] that engineering codes, such as this clause by the NSPE, tend to "explicitly preference human welfare" (p. 5) but I also note that what constitutes "the public" is itself a matter of interpretation [29]; it could be interpreted in such a way so as to include animals. Alas, given the worldview of engineering (see Reason 1), I agree with Berne that are animals are not likely included in this clause.

Second, like codes of ethics, engineering ethics textbooks seldom explicitly focus on animal considerations. Fu et al. [39] conducted a comparative analysis of the first-level and second-level section headers of six textbooks, three from authors in China and three from authors in the United States. No first- or second-level headers used the term "animals." With that said, we can likely infer that second-level headers described as "environmental/ecological ethics" included animals (all six textbooks featured such headings). For example, in a section entitled "Sustainability: The Environment versus Human Development," Harris Jr. et al. [40] wrote:

> Questions about environmental impact include not only questions about environmental damage and pollution, but also, ultimately about the effect on the long-term sustainability of human and nonhuman life. The effect on inanimate nature must also be considered. A major question here is: How can products and processes be improved from the standpoint of their demands on the resources of the earth? (p. 161)

The above text reinforces the above discussion regarding engineering worldview: non-human animals are often depicted as resources. To return to the chapter headings synthesized by Fu and colleagues [39], it is unclear (at least from the headers alone) whether non-environmental headers included animals (e.g., "engineering and stakeholders", "international engineering standards"). Conversely, we can safely infer that such titles included humans. Moreover, despite the omission of explicit headers focused on animals, other titles were human-centric (e.g., "respect people", "public morality", "inform the public", "humanitarianism"). Thus, when compared to animals, we can presume that humans were implicitly present throughout the textbooks, *and* they received explicit attention in certain chapters. Yet, animals did not receive explicit attention in these headings.

Codes and textbooks are not the sole artifacts that represent thinking related to engineering ethics. Another useful reference point is reviews of engineering ethics discourses, including reviews of scholarly literature or compilations of teaching approaches. Yet, a review of such artifacts delivers the same message—animals are missing or minimized. As an example, Harris et al. [37] explored the what, why, and how of engineering ethics. This article is a common reference point for many thinkers in engineering ethics today [41, 42], and while it is admittedly a short paper, animals are not mentioned. Conversely, environmental harm (i.e., avoiding it) is mentioned once. Nearly identical trends manifest in reviews of engineering ethics instructional approaches, be they decades old [43, 44] or more recent [19, 42, 45, 46] even though some of these authors (including myself) have explicitly focused on animals elsewhere [35, 47].

The above artifacts show perspectives from scholars and ostensible leaders in engineering ethics, but my sense is that non-ethics experts in engineering hold similar views. For example, Katz [48] synthesized engineering faculty members' views of "engineering ethics education," and animals were not featured in this study's findings. Likewise, in an ongoing investigation of how 25 engineering faculty and 25 engineering practitioners view connections between ethics and diversity, equity, and inclusion engineering [49], participants never discussed animals (thus, perhaps ironically, diversity seems similar to ethics insofar as it is largely defined in human-centric ways).

As this section highlights, the worldview of engineering, coupled with the artifacts that represent views on engineering ethics, has led to the neglect of animals in historical engineering ethics artifacts and extant engineering discourses. A focus on the "environment" is more prominent in engineering ethics and may include non-human animals. When we look to the "social" element of engineering ethics (e.g., the "public"), it is less clear whether we are considering animals. I think understanding the concept of empathy can help us consider whether animals are (or should be) part of the "social" (or, if not, the "environmental") in engineering ethics.

18.5 Reason 3: Animals and the "Social" in Engineering Ethics

As I have referenced in prior sections, my sense is that when we use the term "social" in engineering ethics, it is generally referencing humans. In her conclusion, Berne [1] asks, "What would it mean for engineering ethics if non-human animals are in fact sentient, as many scientists are beginning to discover?" I think we can modify this question to the following: "If non-human animals are sentient, ought they be depicted as 'social' or 'environmental' considerations in engineering ethics?" Yet, even this question may not yet be precise enough. . . instead, we might ask, "If non-human animals can empathize, why would we not include them as part of the 'social' in engineering ethics?" As this question suggests, I think empathy can help us identify whether non-human animals (or, perhaps, which non-human animals) should be considered part of the "social."

To start this section, I address the question, "What is empathy?" By first addressing this question, I think we will be better positioned to respond to the above questions. As I seek to respond to this question, at least in this section, I am primarily interested in discussing empathy by non-humans towards others (be they human or non-human). Young et al. [50] cited multiple studies and indicated that "empathy towards animals is developed the same way as towards human." Thus, I do not aim to debate whether humans can empathize with non-humans.

To begin, I offer a bi-dimensional definition of empathy which is comprised of two parts: cognitive empathy and affective empathy [51]. Cognitive empathy is one's ability or tendency to understand others, "or the state of understanding, other's internal states (i.e., thoughts and affective states)" ([52], p. 2019). According to these authors, affective empathy refers to "feeling the same affective state as another person" (p. 168). As the reader will note, this latter definition is human-centric, but a definition of this dimension need not be so. For example, the definition offered by Young et al. [50] is more inclusive of non-human animals: "Affective empathy is the ability to physically experience the perceived emotions of another" (p. 329). These two empathy dimensions are comprised of at least eight associated concepts [53, 54]; thus, in the framing that I advance here, empathy dimensions encompass multiple empathy concepts or "types" of empathy.

Cognitive empathy includes multiple empathy concepts, such as perspective-taking, projection, theory of mind, and empathic accuracy [54]. As an example, empathic accuracy depicts one's ability to comprehend another's perspective with some level of accuracy through what Ickes [55] defined as a process of *empathic inference*, where "empathically" accurate perceivers are those who are good at 'reading' other people's thoughts and feelings" (p. 57). Empathy's cognitive dimension includes an empathy concept that scholars have described as "theory of mind" [56] and which theorizes that one can explain others' behaviors by considering their thoughts, beliefs, or desires [57]. One way of describing theory of mind posits "that we use our lay theories about the mind to infer the internal states of others" [54]. Another way of describing theory of mind posits that individuals come to understand others by imagining one's self in another's position, applying one's own decision-making abilities, and then externalizing one's self-conclusion onto the other [58]. Thus, theory of mind can be operationalized in theory-oriented and self-oriented ways. As is the case with many empathy concepts, these ways likely co-manifest.

Premack and Woodruff [59] first used the term theory of mind when studying chimpanzees' responses to "a human actor struggling with a variety of problems." Through their empirical investigation, they concluded that chimpanzees indeed possessed a theory of mind. Since then, extensive research has been conducted regarding the nature of this empathy concept, largely with adolescents [58] and individuals with psychiatric disorders [60, 61]. Like Premack and Woodruff, many scholars have also continued studying theory of mind among non-human animals, with some authors suggesting that some non-human animals possess a theory of mind [62] and others questioning the validity of studies that conclude affirmatively in this regard [63].

My sense is that much of the disagreement regarding whether animals have a theory of mind is due to methodological limitations *or* critiques of inferences made from limited methods. For example, de Waal and Ferrari [64] argued that many conclusions confirming the *lack of theory mind* (as well as other empathic capacities) among non-human animals are often based on a non-human animal's failure "to demonstrate certain capacities." Yet, they argue that such authors make "premature conclusions" based upon the "absence" of such capacities (they cite [65] as an example). As a result of such methodological challenges, disagreements persist regarding whether certain non-human animal species have a theory of mind. To explicitly connect this idea to the main objective of this section, I believe the debate about whether non-human animals have a theory of mind is important to identifying whether we should consider animals as part of the "social" in engineering ethics. If animals possess a theory of mind, limiting their status to the "environment" seems misaligned with extant definitions of "environment."

By further differentiating *cognitive empathy* (an empathy dimension) and *theory of mind* (a type of cognitive empathy), we will develop a better understanding of each. First, theory of mind draws one's attention to other rational entities, with the goal of understanding another's mind. By itself, cognitive empathy does not preclude considering others' thoughts or feelings based on their cognitive capacities. As a result, cognitive empathy is more inclusive than theory of mind. For example, one can consider the perspectives of robots [66], physical spaces [67], or societal groups [68]; yet, unlike non-human animals, it is likely that these entities cannot cognitively empathize with humans. Second, while theory of mind strives to understand how another would *think*, other cognitive empathy concepts often involve considering how another would *feel*. For example, when one engages in perspective-taking, one may need to consider the feelings experienced by another to accurately understand another's perspective.

Above, I have unpacked cognitive empathy. I think affective empathy brings with it fewer concerns regarding whether humans can empathize with non-human animals; however, current literature expresses uncertainty whether the reverse is true [69]. Like cognitive empathy, affective empathy includes multiple concepts [53], such as empathic distress, empathic concern, or emotion contagion [54]. Emotion contagion describes automatic emotional state matching between individuals [70, 71]. Empathic distress draws attention to feelings of distress that manifest when one considers another [72]. Empathic concern is similar, but focuses on one's feelings *for* another [37]. Pérez-Manrique and Gomila [69] operationalized sympathetic concern as "an other-oriented emotional reaction," and I personally view this operationalization as aligned with the definition of "empathic concern" by Davis [53]. Affective empathy can lead to behavior, but distinct affective empathy types may promote such behavior in distinct ways. For example, empathic distress may be the most important driver of helping behavior [72].

Pérez-Manrique and Gomila [69] explored "sympathetic concern" among non-human animals. They differentiated sympathetic concern from other affective empathy types, such as emotional contagion, because it "does not necessarily mimic" the emotions or feelings experienced by a "distressed partner" (p. 250). In their review of literature on studies of non-human animal empathy, they classified an animal species' behavior as sympathetic concern when it included (1) an emotional "reaction" ostensibly related "another's distress, situation or needs" (i.e., *not* mimicry), (2) a "response" to "ameliorate the other's state", and (3) positive "outcomes" in the emotional state of the other.

Through their review, Pérez-Manrique and Gomila [69] found that "studies assessing consolation and empathic targeted helping usually overlook the emotional component of the behavior" thus "undermining" their ability to develop "firm conclusions" regarding animal's capacity to experience sympathetic concern (p. 262). Due to this limitation, they sometimes found evidence that species met two of the three criteria for sympathetic concern (e.g., bonobos, Tonkean macaques, and dogs). They only coded all three criteria for the chimpanzee (p. 263). Thus, they cautiously refrained from inferring sympathetic concern was exhibited by some species given a lack of empirical evidence. Importantly, they also noted some animals' efforts towards "distress alleviation" were with/for "familiar conspecifics" (p. 253); thus, their review leads to questions regarding intra-special empathic biases or limitations.

While Pérez-Manrique and Gomila [69] caution researchers from over-attributing emotional states to animals, especially species who we have more affinity towards, Young et al. [50] conclude that "evidence seems to suggest that non-human animals, much like humans, present not only basic affective empathy but are capable of complex emotional experiences as well." Thus, these authors note that we can infer affective empathy among animals based on their animal responses. To contextualize this idea, if one has or has ever had a pet (or, perhaps a non-human animal friend if we aim to resist resource-like language), they can likely reflect on an experience where that animal seemed to show care or consideration for one's well-being.

To conclude this section, when we focus on select empathy types—such as the theory of mind—we leave with more concerns regarding whether non-humans can empathize with others. As we shift to affective empathy, this becomes less of a concern. To revisit the idea of the "social," which is where this chapter began, I think it behooves the engineering ethics education community to develop criteria for what constitutes the social, with an explicit statement regarding non-human animals. Theory of mind, cognitive empathy, or affective empathy may offer useful language or heuristics to consider in such criteria development. Should animals possess these empathic capacities, it feels unconscionable (at least to me) to relegate them to "environmental" status.

18.6 Conclusion

I started this chapter with a question, "Why is it important to write about empathy with/for animals in the context of engineering ethics education?" To contextualize my response to this question, I summarized my research journey and throughout the remainder of the text, I argued there are three reasons why writing about animals is important in the context of engineering ethics: (1) the human-centric nature of the engineering worldview, (2) the lack of animal considerations in engineering ethics, and (3) the relegation of animals to non-social categories.

To summarize my research journey, during my graduate studies, my research focus shifted to societal rather than environmental considerations. I realized that there is *so much work* needed to improve relationships between humans. While I felt that empathy for the earth was still important by the end of my graduate studies, I felt that empathy may be *most* important in helping engineers value the perspectives of those in society who are often neglected or minoritized. Thus, "empathy for the earth" no longer was the focus of my research. This is not to say that it was missing but rather that my research focus became human-centric. I appreciate the opportunity to contribute to this book, as it provided me with the cognitive space to pause and reflect on my journey as it relates to the inclusion and valuing of animals therein. I hope that by sharing my journey—

including the influence of worldview, engineering ethics artifacts, and empathy on my journey—we will be positioned to develop a greater appreciation of animals in engineering ethics, potentially even elevating animals to the category of the "social" rather than "environmental."

References

1. Berne, R. W. 2023. Non-human animals and a new ethics for engineering, presented at the *ASEE Annual Conference & Exposition*, Baltimore, MD.
2. Brundtland, G. H. 1987. *Our Common Future*. World Commission on Environment and Development.
3. Devall, B., and G. Sessions. 1985. *Deep Ecology: Living as if Nature Mattered: Devall and Sessions on Defending the Earth*. Gibbs Smith.
4. Hess, J. L., and J. Strobel. 2013. Indigenous ways of doing: synthesizing the literature on ethno-engineering. *International Journal of Engineering, Social Justice, and Peace* 2: 55–80. https://doi.org/10.24908/ijesjp.v2i2.4333.
5. Hess, J. L., and J. Strobel. 2013. Sustainability and the engineering worldview, presented at the *IEEE Frontiers in Education Conference*, Oklahoma City.
6. Strobel, J., C. W. Morris, L. Klingler, R. C. Pan, M. Dyehouse, and N. Weber. 2011. Engineering as a caring and empathetic discipline: Conceptualizations and comparisons. *Research in Engineering Education Symposium 2011*, 2011: 622–635.
7. Hess, J., J. Sprowl, R.(Celia) Pan, M. Dyehouse, C. W. Morris, and J. Strobel. 2012. Empathy and caring as conceptualized inside and outside of engineering: extensive literature review and faculty focus group analyses. https://doi.org/10.18260/1-2-21278.
8. Hess, J. L., J. Strobel, R. (Celia) Pan, and C. A. W. Morris. 2017. Insights from industry: a quantitative analysis of engineers' perceptions of empathy and care within their practice. *European Journal of Engineering Education* 42: 1128–1153. https://doi.org/10.1080/03043797.2016.1267717.
9. Hess, J. L., J. Strobel, and R. (Celia) Pan. 2016. Voices from the workplace: practitioners' perspectives on the role of empathy and care within engineering. *Engineering Studies* 8: 212–242. https://doi.org/10.1080/19378629.2016.1241787.

10. Brightman, A. O., J. L. Hess, A. Iliadis, L. Kisselburgh, M. Krane, M. C. Loui, and C. B. Zoltowski. 2016. *PRIME Ethics: Purdue's Reflective & Interactive Modules for Engineering Ethics.* NAE (Ed.). The National Academies Press: 39–40.
11. Merryman, W. D. 2008. Development of a tissue engineered heart valve for pediatrics: a case study in bioengineering ethics. *Science and Engineering Ethics* 14: 93–101. https://doi.org/10.1007/s11948-008-9053-x.
12. Beever, J., and J. L. Hess. 2016. *Deepwater Horizon Oil Spill: An Ethics Case Study in Environmental Engineering*. Paper presented at the *ASEE Annual Conference & Exposition*, New Orleans, LA.
13. Kisselburgh, L. G., J. L. Hess, C. B. Zoltowski, J. Beever, and A. O. Brightman. 2016. Assessing a scaffolded, interactive, and reflective analysis framework for developing Ethical Reasoning in Engineering Students. DOI:10.18260/p.26288.
14. Hess, J. L., J. Beever, C. B. Zoltowski, L. Kisselburgh, and A. O. Brightman. 2019. Enhancing engineering students' ethical reasoning: Situating reflexive principlism within the SIRA framework. *Journal of Engineering Education* 108: 82–102. https://doi.org/10.1002/jee.20249.
15. Hess, J. L., J. Strobel, and A. O. Brightman. 2017. The development of empathic perspective-taking in an engineering ethics course. *Journal of Engineering Education* 106: 534–563. https://doi.org/10.1002/jee.20175.
16. Beever, J., and A. O. Brightman. 2016. Reflexive principlism as an effective approach for developing ethical reasoning in engineering. *Science and Engineering Ethics* 22: 275–291. https://doi.org/10.1007/s11948-015-9633-5.
17. Obama, B. 2015. Remarks by the president at White House Science Fair. https://obamawhitehouse.archives.gov/the-press-office/2015/03/23/remarks-president-white-house-science-fair. March 23.
18. Rasoal, C., H. Danielsson, and T. Jungert. 2012. Empathy among students in engineering programmes. *European Journal of Engineering Education* 37: 427–435. https://doi.org/10.1080/03043797.2012.708720.
19. Hess, J. L., and G. Fore. 2018. A systematic literature review of US engineering ethics interventions. *Science and Engineering Ethics* 24: 551–583. https://doi.org/10.1007/s11948-017-9910-6.
20. Cobern, W. W. 1996. Worldview theory and conceptual change in science education. *Science Education* 80: 579–610. https://doi.org/10.1002/(SICI)1098-237X(199609)80:5<579::AID-SCE5>3.0.CO;2-8.
21. Kearney, M. 1984. *World View*. Novato, Calif: Chandler & Sharp Pub.

22. Villanueva, I., L. Gelles, M. D. Stefano, B. Smith, R. Tull, S. Lord, L. Benson, A. Hunt, D. Riley, and G. Ryan. 2018. What does hidden curriculum in engineering look like and how can it be explored? *Proceedings of the American Society of Engineering Education Annual Conference & Exposition, Minorities in Engineering Division*: 1–16.
23. Pacey, A. 2008. Engineering. In *Encyclopaedia of the History of Science, Technology, and Medicine in Non-Western Cultures*, ed. H. Selin, 748–751. Dordrecht: Springer Netherlands. https://doi.org/10.1007/978-1-4020-4425-0_9009.
24. Kirby, R. S. 1990. *Engineering in History*. Reprint edition. New York: Dover Pubns.
25. Layton Jr., E. T. 1971. *The Revolt of the Engineers: Social Responsibility and the American Engineering Profession*. Baltimore, MD: Johns Hopkins University Press.
26. Ehrenfeld, D. W. 1978. *The Arrogance of Humanism*. New Ed edition. New York: Oxford University Press.
27. Jr, W. W. Q. 1997. *The Only Tradition*. Albany: State University of New York Press.
28. Kim, D., and J. L. Hess. 2023. A multi-layered illustration of exemplary business ethics practices with voices of the engineers in the health products industry. *Journal of Business Ethics* 187: 169–183. https://doi.org/10.1007/s10551-022-05255-2.
29. Davis, M. 1991. Thinking like an engineer: The place of a code of ethics in the practice of a profession. *Philosophy & Public Affairs*: 150–167.
30. de Figueiredo, A. D. 2008. Toward an epistemology of engineering. In *2008 Workshop on Philosophy and Engineering, The Royal Academy of Engineering*, London, November 10–12, 2008
31. Florman, S. C. 1987. *The Civilized Engineer*. 1st ed. New York: St. Martin's Press.
32. Stonyer, H. 2002. Making engineering student—making women: the discursive context of engineering education. *International Journal of Engineering Education* 18: 392–399.
33. Penny, Simon. 1997. The virtualization of art practice: body knowledge and the engineering worldview. *Art Journal* 56: 30–38. https://doi.org/10.2307/777834.
34. Cech, E. A. 2014. Culture of disengagement in engineering education? *Science, Technology, & Human Values* 39: 42–72. https://doi.org/10.1177/0162243913504305.

35. Herkert, J. 2004. *Engineering Research and Animal Subjects.* US Department of Health & Human Services.

36. Kuhn, T. S. 1962. *The Structure of Scientific Revolutions.* University of Chicago Press.

37. Harris Jr., C. E., M. Davis, M. S. Pritchard, and M. J. Rabins. 1996. Engineering ethics: what? why? how? and when? *Journal of Engineering Education* 85: 93–96. https://doi.org/10.1002/j.2168-9830.1996.tb00216.x.

38. Riley, D. M., and Y. Lambrinidou. 2015. Canons against cannons? Social justice and the engineering ethics imaginary. 26.322.1-26.322.19. Presented at *2015 ASEE Annual Conference & Exposition*, Seattle, Washington. 10.18260/p.23661.

39. Lei, Q., D. M. Grzybowski, D. Cheng, and J. Fu. 2018. What should be taught in engineering ethics education under globalization?: Based on the comparative analysis of university textbooks in China and the United States. Presented at *2018 ASEE Annual Conference & Exposition.* DOI:10.18260/1-2–31236.

40. Harris, C. E., M. S. Pritchard, R. W. James, E. E. Englehardt, and M. J. Rabins. 2019. *Engineering Ethics: Concepts and Cases.* 6 ed. Australia; Boston, MA: Cengage.

41. Clancy III, R. F., and Q. Zhu. 2022. Global engineering ethics: what? why? how? and when? *Journal of International Engineering Education* 4. DOI: 10.23860/jiee.2022.04.01.04.

42. Zhu, Q., M. Martin, and R. Schinzinger. 2022. *Ethics in Engineering*, 5th ed., McGraw Hill, New York, NY.

43. Haws, D. R. 2001. Ethics instruction in engineering education: a (mini) meta-analysis. *Journal of Engineering Education* 90: 223–229. https://doi.org/10.1002/j.2168-9830.2001.tb00596.x.

44. Herkert, J. R. 2005. Ways of thinking about and teaching ethical problem solving: Microethics and macroethics in engineering. *Science and Engineering Ethics* 11: 373–385. https://doi.org/10.1007/s11948-005-0006-3.

45. Hess, J. L., A. J. Kerr, A. Lin, and A. Chung. 2023. A systematic review of the 2016 National Academy of Engineering exemplary ethics programs: Revisions to a coding framework. *Science and Engineering Ethics* 29: 36. https://doi.org/10.1007/s11948-023-00456-y.

46. *Infusing Ethics into the Development of Engineers: Exemplary Education Activities and Programs.* 2016. Washington, D.C.: National Academies Press. https://doi.org/10.17226/21889.

47. Hess, J. L., S. Miller, S. Higbee, G. A. Fore, and J. Wallace. 2020. Empathy and ethical becoming in biomedical engineering education: a mixed methods study of an animal tissue harvesting laboratory. *Australasian Journal of Engineering Education* 26: 127–137. https://doi.org/10.1080/22054952.2020.1796045.
48. Katz, A. S. 2019. An inquiry into the nature and causes of the state of U.S. engineering ethics education dissertation. Thesis, Purdue University Graduate School. https://doi.org/10.25394/PGS.8044799.v1.
49. Hess, J. L., A. Katz, I. Anakok, B. K. Jesiek, A. Whitehead, and S. S. Panuganti. 2023. Promoting research quality to study mental models of ethics and diversity, equity, and inclusion (DEI) in engineering. Presented at *2023 ASEE Annual Conference & Exposition*.
50. Young, A., K. A. Khalil, and J. Wharton. 2018. Empathy for animals: a review of the existing literature. *Curator: The Museum Journal* 61: 327–343. https://doi.org/10.1111/cura.12257.
51. Kouprie, M., and F. S. Visser. 2009. A framework for empathy in design: stepping into and out of the user's life. *Journal of Engineering Design* 20: 437–448. https://doi.org/10.1080/09544820902875033.
52. Clark, M. A., M. M. Robertson, and S. Young. 2019. "I feel your pain": A critical review of organizational research on empathy. *Journal of Organizational Behavior* 40: 166–192. https://doi.org/10.1002/job.2348.
53. Hess, J. L., and N. D. Fila. 2016. The development and growth of empathy among engineering students. Paper Presented at the *ASEE Annual Conference & Exposition*, New Orleans, LA.
54. Batson, C. D. 2009. These things called empathy: eight related but distinct phenomena. In *The Social Neuroscience of Empathy*, 3–15. Social Neuroscience. Cambridge, MA, US: Boston Review. https://doi.org/10.7551/mitpress/9780262012973.003.0002.
55. Ickes, W. 2009. Empathic accuracy: its links to clinical, cognitive, developmental, social, and physiological psychology. In *The Social Neuroscience of Empathy*, 57–70. Social Neuroscience. Cambridge, MA, US: Boston Review. https://doi.org/10.7551/mitpress/9780262012973.003.0006.
56. Dziobek, I., K. Rogers, S. Fleck, M. Bahnemann, H. R. Heekeren, O. T. Wolf, and A. Convit. 2008. Dissociation of cognitive and emotional empathy in adults with Asperger syndrome using the multifaceted empathy test (MET). *Journal of Autism and Developmental Disorders* 38: 464–473. https://doi.org/10.1007/s10803-007-0486-x.

57. Iacoboni, M., and M. Dapretto. 2006. The mirror neuron system and the consequences of its dysfunction. *Nature Reviews. Neuroscience* 7: 942–951. https://doi.org/10.1038/nrn2024.
58. Doherty, M. 2012. Theory of Mind. *The Self and Perspective Taking*: 91–107.
59. Premack, D., and G. Woodruff. 1978. Does the chimpanzee have a theory of mind? *Behavioral and Brain Sciences* 1: 515–526. https://doi.org/10.1017/S0140525X00076512.
60. Baron-Cohen, S. 1995. *Mindblindness: An Essay on Autism and Theory of Mind*. The MIT Press. https://doi.org/10.7551/mitpress/4635.001.0001.
61. Poletti, M., I. Enrici, and M. Adenzato. 2012. Cognitive and affective Theory of Mind in neurodegenerative diseases: neuropsychological, neuroanatomical and neurochemical levels. *Neuroscience and Biobehavioral Reviews* 36: 2147–2164. https://doi.org/10.1016/j.neubiorev.2012.07.004.
62. Waal, F. D. 2010. *The Age of Empathy: Nature's Lessons for a Kinder Society*. 1st ed. New York, NY: Crown.
63. Penn, D. C., and D. J. Povinelli. 2007. On the lack of evidence that non-human animals possess anything remotely resembling a 'theory of mind.' *Philosophical Transactions of the Royal Society B: Biological Sciences* 362: 731–744. https://doi.org/10.1098/rstb.2006.2023.
64. de Waal, F. B. M., and P. F. Ferrari. 2010. Towards a bottom-up perspective on animal and human cognition. *Trends in Cognitive Sciences* 14: 201–207. https://doi.org/10.1016/j.tics.2010.03.003.
65. Silk, J. B., S. F. Brosnan, J. Vonk, J. Henrich, D. J. Povinelli, A. S. Richardson, S. P. Lambeth, J. Mascaro, and S. J. Schapiro. 2005. Chimpanzees are indifferent to the welfare of unrelated group members. *Nature* 437: 1357–1359. https://doi.org/10.1038/nature04243.
66. Zhao, X., and B. F. Malle. 2022. Spontaneous perspective taking toward robots: the unique impact of humanlike appearance. *Cognition* 224: 105076. https://doi.org/10.1016/j.cognition.2022.105076.
67. Gunia, A., S. Moraresku, and K. Vlček. 2021. Brain mechanisms of visuospatial perspective-taking in relation to object mental rotation and the theory of mind. *Behavioural Brain Research* 407: 113247. https://doi.org/10.1016/j.bbr.2021.113247.
68. Kohlberg, L. 1976. Moral stages and moralization: The cognitive-developmental approach. In T. Lickona (Ed.), *Moral Development and*

Behavior: Theory, Research, and Social Issues (pp. 170–205). Holt, Rinehart and Winston.

69. Pérez-Manrique, A., and A. Gomila. 2018. The comparative study of empathy: sympathetic concern and empathic perspective-taking in non-human animals. *Biological Reviews* 93: 248–269. https://doi.org/10.1111/brv.12342.
70. Decety, J., and P. L. Jackson. 2004. The functional architecture of human empathy. *Behavioral and Cognitive Neuroscience Reviews* 3: 71–100. https://doi.org/10.1177/1534582304267187.
71. Hatfield, E., R. L. Rapson, and Y.-C. L. Le. 2009. Emotional contagion and empathy. In *The social Neuroscience of Empathy*, 19–30. Social Neuroscience. Cambridge, MA, US: Boston Review. https://doi.org/10.7551/mitpress/9780262012973.003.0003.
72. Hoffman, M. L. 2000. *Empathy and Moral Development: Implications for Caring and Justice*. New York, NY, US: Cambridge University Press. https://doi.org/10.1017/CBO978051180585.

The Great Divide

Major highways that traverse rural areas, connecting people east to west and north to south, sever populations of animals from each other. In all biomes, desert to forest, busy roadways are daunting obstacles to animals travelling in search of food, shelter, water, or the rest of their herd. Species not only lose number due to accidents on the roadways, but hesitation to cross can split groups of animals from each other, and opportunities to reproduce dwindle. Migration is a critical tool for the survival of many animals. Deer, big cats, bears, and smaller animals all must travel during the year to follow warmth, food, and other resources as the seasons change. Construction of freeways over the decades has affected the way animals migrate and jeopardizes endangered and threatened species like the Florida panther, Key deer, bighorn sheep, and Hawaiian goose [1].

Fortunately, the United States has recently increased its efforts to prevent human and non-human casualties on highways. Enormous wildlife bridges, or ecoducts, are bridges customized with brush , rocks, and vegetation to encourage animals to use the overpass instead of risking the road. One might think this is a novel idea, but the first wildlife bridge was built in the 1950s in France [2]. These bridges have been built in many regions and countries, all over the world, and are being used to save endangered species. In Nevada, wildlife overpasses and underpasses have led to a 50% decrease in mule deer collisions with each migration [3].

Ecoducts are an excellent example of how engineering can be used to remediate unforeseen effects of human development on animal populations. Humans are also benefiting from the projects, as a collision with many of these animals can be expensive let alone fatal to the drivers and passengers of vehicles. One day, humans can live in extreme technological advancement in harmony with the natural world if our young engineers apply their minds to life-saving innovation such as wildlife bridges.

Citations

1. Executive Summary - Wildlife-Vehicle Collision Reduction Study: Report To Congress, August 2008 - FHWA-HRT-08-034.2024. https://www.fhwa.dot.gov/publications/research/safety/08034/exec.cfm. Accessed May 9.
2. How wildlife bridges over highways make animals-and people-safer. 2024. https://www.nationalgeographic.com/animals/article/wildlife-overpasses-underpasses-make-animals-people-safer. Accessed May 9.
3. Wildlife Crossings Can Protect Migrating Animals | The Pew Charitable Trusts. 2024. https://www.pewtrusts.org/en/trust/archive/spring-2020/wildlife-crossings-can-protect-migrating-animals. Accessed May 9.

Index